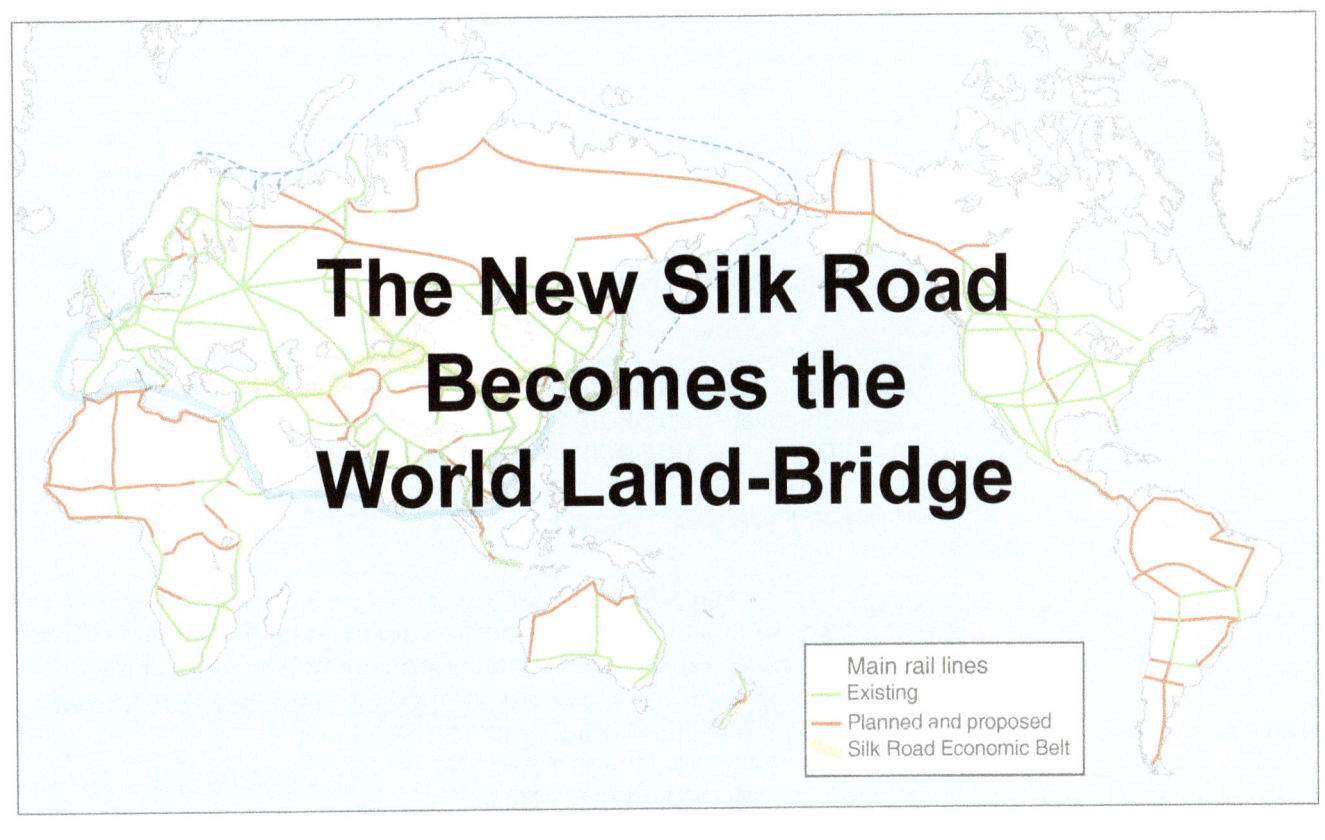

The New Silk Road Becomes the World Land-Bridge

November 2014
Executive Intelligence Review

© EIR News Service, Inc. 2014
Reproduction of all or part of these contents without explicit authorization of the publisher is prohibited.

Third Printing February 2015

Prices:
Hard cover: EIPSP-2014-1-1-0-STD, $300
Soft cover: EIPSP-2014-1-0-0-STD, $250
PDF: EIPSP-2014-1-0-0-PDF, $200

ISBN: 978-0-943235-24-0

Cover Design: Chance McGee
Front Cover: Lower right: China's Yutu rover on the Moon (CNTV); lower left: a maglev train in Germany (Wikimedia Commons); top: camels in the desert—Copyright: <http://www.123rf.com/profile_dimol> dimol / 123RF Stock Photo
Back Cover: The World Land-Bridge Network map (Alan Yue, Asuka Saito/EIRNS 2014.

Credits:

Project Manager: Marcia Merry Baker
Production Manager: Robert Bowen
Editor: Nancy Spannaus
Copyeditor: Ronald Kokinda
Cartography and Design: Asuka Saito, Alan Yue

Contents

Part 1 Introduction
The New Silk Road Leads to the Future of Mankind! .... I: 2
The Coming Fusion Power Economy on the Basis of Helium-3
An Explosion of BRICS Initiatives .... I: 11
World Land-Bridge Network—Key Links and Corridors .... I: 19
  Map .... I: 20
  Descriptions .... I: 21

Part 2 The Metrics of Progress
Energy-Flux Density: Global Measure of Economic Progress ...... II: 2
The Principle of the Development Corridor .... II: 11
1791: A New Method of Financing Infrastructure Discovered ...... II: 14
Expand Nuclear Power for the World's Survival .... II: 36
Solve the World Water Crisis .... II: 52
  Appendix: Initiatives for Nuclear Desalination .... II: 75

Part 3 China: Silk Road to Development and Peace
China Becomes a Model Among Nations: A Science-Driver
  Approach to Lift Up Mankind .... III: 2
China's New Silk Road: Changing the Paradigm Toward
  Global Development .... III: 12
  Appendix: China's Silk Road – Pathway to a New
    Human Civilization .... III: 22

**Part 4 Russia's Mission in North Central Eurasia
and the Arctic**
Russia, Eurasia's Keystone Economy, Looks East .... IV: 2
Tumen River Initiative: A Step for Peace in Northeast Asia .... IV: 54

**Part 5 South and Central Asia: From Arc of Crisis to
Corridors of Development**
India Is Ready to Fulfill Its Legacy of Leadership .... V: 2
Bringing High-Technology Development to South Asia .... V: 10
Central Asia: Ending Geopolitics .... V: 15
  Appendix: The Industrial Development of Afghanistan
    and Central Asia: A Russian Vision .... V: 27

Part 6 Southwest Asia: Crossroads of the Continents
Southwest Asia and the Eurasian Land-Bridge .... VI: 2

The New Silk Road Becomes the World Land-Bridge

Part 7 The Crucial Contributions of East and Southeast Asia
Japan Must Return to Leadership in Nuclear Energy................. VII: 2
The Mekong Development Project: A TVA for
    Southeast Asia.. VII: 6
Thailand's Kra Canal: Keystone for South
    Asian Development... VII: 10
Connecting Indonesia to the Eurasian Mainland....................... VII: 15

Part 8 Australia—Driver for Pacific Development
A Vision To Bring Australia into the Land-Bridge Process........ VIII: 2

Part 9 Europe—Western Pole of the New Silk Road
Germany: The Key to European Integration into the
    New Silk Road... IX: 2
Greece and a Marshall Plan for the Mediterranean .... IX: 9
Italy: Build the Mezzogiorno, and a New Renaissance .... IX: 13
Spain: The World Land-Bridge's Bridge to
    African Development... IX: 21

Part 10 Africa—Test for Global Progress
A Nuclear-Based Infrastructure Platform Is Necessary for
    Africa's Future... X: 2
    Appendices: The Transaqua Project, Africa Pass .... X: 10

Part 11 Bringing the Western Hemisphere On Board
Rediscovering the Americas... XI: 2
North America: Restoring the American System......................... XI: 13
    President Kennedy Would Be Building the World Land-Bridge

Part 12 Fighting for International Development
LaRouche's 40-Year Record — Fighting for International
    Development ....XII: 2
Development Models
    Tennessee Valley Authority: Great Projects Make
        Great Nations... XII: 6
    Deng Xiaoping's China Miracle... XII: 11
    The South Korea Model: How to Transform an
        Impoverished Nation into a Modern Economy............ XII: 15

Part 13 Epilogue
Dump Geopolitics, and Create a Future for Humanity............... XIII: 2

PART 1
Introduction

The New Silk Road Leads to the Future of Mankind!

The Coming Fusion Power Economy on the Basis of Helium-3

by Helga Zepp-LaRouche

August 2014

When the authors of this study decided last year to set out a new concept for peace in the 21st Century, by producing a revised version of the World Land-Bridge program—23 years after the first proposal for the Eurasian Land-Bridge—it was their intention not only to provide a concept for reconstruction of the world economy, but also to present a war-avoidance strategy in the context of an acute strategic crisis at the same time. For in the intervening years, the danger of an intentional—or even an accidental—thermonuclear world war has grown dramatically. The attempt, fed by geopolitical motives, to associate Ukraine with the EU, and thus bring it, de facto, into the NATO sphere of influence, has triggered a series of escalating confrontations, which, in the worst case, could end in the extinction of the human race. But in addition, nearly the entire Near and Middle East is burning; set off by wars built on lies, against so-called rogue states, the seeds of violence were sown that have called to life a million-headed hydra, which not only has leveled the Cradle of Civilization to the ground and created there a Hell on Earth, but also has become an existential threat to the West.

The consequences of this policy of "regime change" have long thrown large parts of Africa into chaos, and overrun the continent with wars of terror and civil wars. But there are also geostrategic conflicts breeding in the Pacific, which have the potential to set loose regional wars and beyond. And since absolutely nothing has been done to remedy the causes for the collapse of Lehman Brothers in 2008, the too-big-to-fail (TBTF) banks are, on average, 30-40% bigger today than they were then, the indebtedness still more massive, and the derivatives bubble grown to nearly $2 quadrillion, so that a new systemic crisis could result at any moment, this time, given the strategic situation we have sketched here, with the danger that chaos will be triggered, making a strategic catastrophe unavoidable.

The entire world thus finds itself in such an alarming condition that one can only wonder how those responsible for the so-called Western community of values could have let things get to this point. Pope Francis, who has characterized the global financial and eco-

Eurasia: Main Routes and Selected Secondary Routes of the Eurasian Land-Bridge

This map represents the schematic of the Eurasian Land-Bridge plan laid out by the Schiller Institute in the early 1990s. Compare this with the map of the status of today's Land-Bridge network, see page I:20.

nomic system as "intolerable," recently put it this way, in an interview with the Spanish newspaper *La Vanguardia*: "In order for the system to continue, wars must be waged, as great empires have always done. But mankind cannot bear a third world war, and so it seizes on regional wars."

Although one could not put it better than the Pope has done, in this case, he underestimates the Satanic energy of the system of globalization, which is ready to defend its privileges with all weapons available. One hundred years after the First World War, we find ourselves in a very similar geopolitical situation, except that this time, there are thermonuclear weapons whose use would wipe out the human race.

There Is an Alternative

Meanwhile, an alternative to the collapsing trans-Atlantic system has been created; the attempts of that system to use supranational institutions such as the IMF, World Bank, WTO, TTP, TTIP, and similar monetarist instruments of globalization, to extend a worldwide imperium, have produced an opposition that might not have been expected to manifest itself as it has. In less than one year, an alliance of nations has been created, which has built a parallel economic order

Russian President Putin and Chinese President Xi, in Shanghai in May 2014, reached a series of bilateral accords, and published a declaration of intent to create a new economic architecture in the Asia-Pacific region.

with giant steps, one which is dedicated exclusively to the building of the real economy, in opposition to the maximization of speculative monetary profit, and which now includes more than half of mankind. This new community of nations represents a power center based on economic growth, and above all, on leading-edge technology, one which belongs to the future, as shown above all by the success of the Chinese lunar exploration program, focused on the idea of bringing large quantities of helium-3 from the Moon back to Earth, for the future economy of thermonuclear fusion power. It points the way to a scientific and technological revolution that will increase, by orders of magnitude, the energy-flux density, both in production processes on Earth, and in fuels for space travel, and thereby introduce a completely new phase in the evolution of the human species.

The first step in the direction of a new economic world order was the announcement by Chinese President Xi Jinping at a conference in Kazakhstan in July of 2013, that China would build a new Silk Road Economic Belt, through Central Asia to Europe, in the tradition of the ancient Silk Road. Then, in October, in a trip to Indonesia and Malaysia, Xi took the initiative to involve all of Southeast Asia in the construction of the Maritime Silk Road.

At the summit meeting between Russian President Vladimir Putin and Chinese President Xi, on May 20, 2014 in Shanghai, and Putin's state visit to China on the occasion of the Fourth Summit of the Conference on Interaction and Confidence-Building Measures in Asia (CICA) in Shanghai on May 21, extensive plans for collaboration of the two great powers were signed, including a 30-year natural gas agreement, and 46 additional bilateral accords. At the end of the summit, the two heads of state published a common declaration of intent, stating that both countries wished to create a new economic architecture in the Asia-Pacific region, oppose interference in the internal affairs of other nations, and intend to coordinate, as much as possible, their responses to important foreign policy questions on which they agree.

They named, among others, one goal of this collaboration as follows: "Increasing the effectiveness of collaboration in high-technology areas, priority projects in the international use of nuclear energy, civil aviation, and a program of cooperation in basic research on space flight, satellite observation of the Earth, satellite navigation, and research into deep space and manned space travel." A further militarization of space should, on the contrary, be prevented, and the unilateral stationing of missile defense installations was judged to be a "destabilizing factor for the world." Other goals include innovative research, improvement of agricultural techniques, and increasing agricultural production. They also expressed the intention to reform the international financial architecture.

The 30-year Russia-China natural gas treaty, with a total value of $400 billion, can be called historic. The two countries' cooperation in the petroleum field is also to be deepened; coal mines in Russia will be jointly developed; additional power plants will be built in Russia to supply electricity to China; and there will be collaboration on many other projects in infrastructure, transportation, water, and nature conservation.

Of still greater importance is President Putin's support for President Xi's strategic initiative to expand the New Silk Road. Their common statement says: "Russia recognizes the enormous significance of the Chinese initiative for the building of the 'Silk Road Economic Belt,' and particularly appreciates the readiness of the Chinese side to take Russian interests into account in its development and realization. Both sides will seek further opportunities to combine the perspective of the 'Silk Road Economic Belt' with the conception of the 'Eurasian Economic Union.' Toward this purpose, they intend to deepen the cooperation of the relevant agencies in the realization of both projects, especially in the development of transportation routes and infrastructure."

The BRICS Summit

Other nations were then drawn into this collaboration at the May 21 Fourth Summit of the CICA in Shanghai. On July 16, the Sixth BRICS Summit was held in Fortaleza, Brazil; on the following day, the Latin American heads of state and government joined the conference, and thus, 48% of humanity was represented at this meeting.

At the BRICS summit itself, and in a series of multilateral and bilateral discussions within and around this summit, the heads of state agreed on the creation of an entirely new economic and financial system, representing a fundamental alternative to the casino economy of the present system of globalization, which is based on maximized profit of the few, and impoverishment of billions of people. Included in the 72 points of the "Declaration of Fortaleza" is the real thunderbolt—the announcement of the creation of a new financial architecture. The new architecture was launched with the formation of a New Development Bank with an initial capitalization of $50 billion, and a Currency Reserve Agreement (CRA) with an initial capacity of $100 billion to help participating nations defend themselves against capital flight and other forms of financial warfare.

China had already previously decided to found an "Asian Infrastructure Investment Bank," the AIIB, to have an initial capitalization of $100 billion, with the invited participation from the start of more than 30 countries. Xinhua quoted Jin Liqun, under whose direction the Chi-

The leaders of the BRICS nations: Putin (Russia); Modi (India); Rousseff (Brazil); Xi (China); and Zuma (South Africa), meeting in Brazil in mid-July, agreed on the creation of a new economic and financial system, representing an alternative to the casino economy of the present system of globalization.

nese Finance Ministry placed the founding of the Bank: "The means of the Asian Development Bank and the World Bank fall far short of satisfying the hunger for more infrastructure.... The Bank will open a new financing channel for developing countries, especially for those with low income.... In October 2013, during a visit to Indonesia, China's President Xi Jinping proposed an Asian Infrastructure Investment Bank to support economic integration."

The general secretary of the Chinese Center for International Economic Exchanges emphasized that the AIIB is to be an open and freely accessible platform, welcoming not only nations in Asia, but also others, such as the United States and the European countries. Up to this point, the nations of the Association of Southeast Asian Nations (ASEAN), at a summit in Myanmar, have declared their intention to participate in the AIIB, including South Korea and Thailand, which resisted U.S. pressure not to do so. In the course of this series of summits, collaboration was decided upon, between the various states, in a large number of projects, above all, the development of nuclear energy, in Russia, China, India, Brazil, Argentina, and South Africa, and also, such groundbreaking projects as a second Panama Canal to be built by China through Nicaragua, and a transcontinental rail connection from Brazil to Peru.

The multiplicity of projects decided on among this community of nations in the areas of infrastructure, energy, industry, agriculture, research, and education, have reached a dimension which puts in the shade the investments made by the U.S.A. and Europe in the same spheres over the past 30 years. The claims that Russia is only a "regional power," and China only a "cheap-production country," as was said at hastily arranged seminars at various think-tanks on the theme of the allegedly minor significance of the BRICS nations, have rather the character of whistling in the dark.

For in reality, there are now two economic and financial systems built on completely different principles. One, the trans-Atlantic system, as an imperial structure, seeks constantly to extend the boundaries of its sphere of power through supranational structures that threaten the sovereignty of other nations. It forces regime change against governments it disapproves of, insists on submission to a "consensus," and in the process, uses methods that do indeed produce an aura of domination for a while, and the feeling of powerlessness among the populations dominated in this way, but it ultimately goes the way of all empires. The moment this aura of power fades, whether because the imperial financial system is bankrupt, or because the people realize the hollowness of the values handed down, then the capability for intimidation disappears.

The newly arising system of the BRICS nations and the countries associated with them, bases itself on entirely different principles. Indian Prime Minister Narendra Modi formulated it most expressively at the plenary session of the summit: "BRICS is unique as an international institution. In the first instance, it unifies a group of nations, not on the basis of their existing prosperity or common identities, but rather their future potentials. The idea of the BRICS itself is thus already aligned with the future."

Modi stressed that the high percentage of young people, in India for example, represents an enormous potential for the future, and proposed forming a BRICS forum for young scientists, and a school of languages "offering language training in all of our languages." Modi made an appeal: "Excellencies, we have an opportunity to define the future—not only for our countries, but for the entire world.... I conceive that as a great challenge."

The Future Lies in Outer Space

Nicholas of Cusa, the founder of modern natural science and a revolutionary scientific method, came to the conclusion, in the 15th Century, that every human being who strives to do so must be capable of reproducing almost the entire evolution of the universe in its essential qualitative levels of development, and that this standpoint makes it possible to determine the necessary next step in scientific progress.

Today, this necessary next discovery, which defines the future for the entire world, is the conquest of the energy source that will bestow energy and raw materials security on mankind for thousands of years into the future—the utilization of thermonuclear fusion power on the basis of helium-3. Therefore, the success of the Chinese Chang'e-3 mission this past December, in achieving a soft landing of the Yutu ("Jade Rabbit") rover on the Moon, was a milestone in achieving this goal. The Chang'e-4 mission will follow immediately this year, in preparation for Chang'e-5 in 2017, which can start the phase in which flight back and forth between the Earth and the Moon, in preparation for the future industrial exploitation of the Moon, can begin. This will bring within reach, the separation of the helium-3 found on the Moon in great quantities, for the nuclear fusion economy on Earth.

In the scientific collaboration among the BRICS nations, but above all, between Russia, China, and India, helium-3 plays a prominent role, because as a fuel for fusion, in contradistinction to deuterium-tritium, it does not produce energetic neutrons, which are very problematic for the reactor materials, but instead produces positively charged protons, which makes possible a revolution in energy generation. Instead of producing energy through the customary method via steam and turbines, in which there is a great energy loss, it will become possible to convert the energy of fusion reactions directly into electricity at much higher efficiencies.

But Russia, too, according to the Russian Federal Space Agency (Roscosmos), plans a mission between 2016 and 2025, which is intended to create the basis for the industrial exploitation of the Moon. In the first phase, this involves the robotic infrastructure for work on the Moon, thus, among other things mobile cranes, dredges, and cable-laying machines. After the landing probe "Luna Globe 1" in 2015, and the orbital module "Luna Globe 2" in 2016, then in 2017 the hard-landing apparatus "Luna Resource," developed together with the Indian Space Research Organization (ISRO), will reach the lunar surface and, among other tasks, convey the Indian Moon vehicle onto the Moon.

The collaboration among China, Russia, and India is paradigmatic for the new era of mankind, in which, instead of plunging ourselves into geopolitical wars, we will concentrate on the common goals of mankind. With the attainment of energy security for at least 10,000 years on the basis of helium-3–fed thermonuclear fusion power, and with the technologies associated with this, such as the fusion torch technique, which enables raw materials security by reducing waste and all types of materials into isotopes that can be recomposed as needed, mankind will reach a completely new economic platform on the basis of a very high energy-flux density. This new economic platform begins a new age of mankind. The utilization of helium-3 sources for the fusion economy will be the game-changer that revolutionizes all relationships in science, economy, and politics on the Earth and in the Solar System.

It is obvious that a continuation of the geopolitical thinking that has already led to two world wars in the 20th Century, into a third, and this time, a thermonuclear world war, will cause the extinction of mankind. Instead of seeing the rise of China as a threat to the West's supposed geopolitical interests, and thus, as the American Joint Chiefs of Staff Chairman Gen. Martin Dempsey has repeatedly warned, groping around in a new "Thucydides Trap," we need a new conception, a new paradigm that considers the development perspective of mankind as a whole.

New Economic Order

The German-American space pioneer Krafft Ehricke described the long arc of evolution as an upward development, in which, at first, life spread from the sea to the continents by means of photosynthesis in the form of the plant world, and then has led gradually to the rise of biological species of high complexity and metabolism with higher energy-flux density. He described how the human species, as the highest expression of this evolution up to now, initially settled on the coasts and the shores of rivers, and then by roads and canals, and finally by railroads and modern infrastructure, made more and more accessible the landlocked regions of the continents.

Visionaries such as German-American space pioneer Krafft Ehricke, shown here, saw in space travel and the colonization of the universe, the natural next phase of the evolution of mankind. Here, Ehricke shows how an orbital hospital might be designed.

This process is still not completed—and exactly this is the goal of the World Land-Bridge presented in this study, to achieve the infrastructural development of the continents of the Earth. Krafft Ehricke saw, in space travel and the colonization of the universe, the natural next phase of the evolution of mankind, and saw in the industrialization of the Moon, in particular, the springboard for excursions of human beings into the Solar System and potentially beyond. Krafft Ehricke was convinced that the evolution of the human species would only effectively reach adulthood with manned space travel; that only the "great challenge of the extraterrestrial imperative," as he called it, will raise mankind to its true purpose and destiny: namely, representing through its power of reason, as the only creative species (known up to now), to act on verifiable universal principles, and not on the illusory world of deceptive sense-perceptions. By doing so, the human species will achieve a considerable advance in bringing its relationships to this planet and to near-space, into harmony with the cosmic order. Perhaps the most important contribution of Lyndon LaRouche consists in that by the further development of the Leibnizian term "physical economy," he has created a theory of scientific economy that corresponds to the real laws of development of the physical universe.

One of LaRouche's basic concepts is that it is indispensable for the continuously sustained existence of the human race that its relative potential population density should increase on the basis of rising energy-flux densities in the production process, because at any arbitrary stage of economic development there is a relative exhaustion of resources. The entire history of human development, particularly the most recent 10,000 years, in which the population potential has risen from a few millions to currently more than 7 billion, demonstrates the correlation of the anti-entropic character of human creativity with the knowable universal principles of the physical universe.

The use of the helium-3 resources on the Moon for the fusion economy on the Earth also recalls in an interesting way the controversy between Plato and Nicholas of Cusa, over whether ideas possess an existence already effectively present in the objective universe, independent of mankind, or whether it is only with human creativity that these ideas are created. Helium-3 supplies on the Moon are, in the first instance, only deposits in the upper layer of the regolith. Only human creativity, in mastering thermonuclear fusion power, makes these isotopes into the fuel which can even exceed the power of nuclear fusion in the Sun!

But mankind has reached a phase-change not only from the scientific standpoint, but also from that of universal history; that is, the end of geopolitics is necessary for the survival of the species. Shortly before the Berlin Wall fell, LaRouche proposed the "Productive Triangle Paris-Berlin-Vienna" infrastructure program, and thereby, the plan to make this triangle the scientific motor and starting point for development corridors for the transformation of the Comecon states (the then-Soviet Union and Eastern Europe).

When the Soviet Union disintegrated in 1991, and thus the Iron

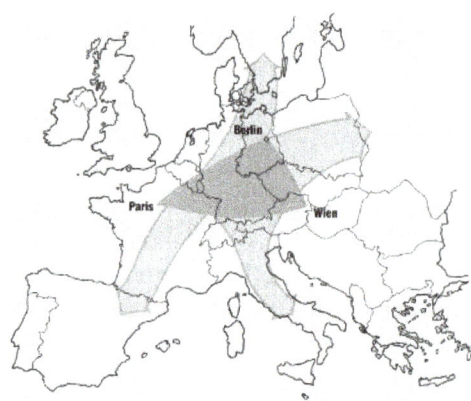

The cover of a Schiller Institute pamphlet from the 1990s, promoting the European "productive triangle," with its spiral arms into southern and eastern Europe.

Curtain disappeared, Schiller Institute teams further elaborated this program into the conception of the Eurasian Land-Bridge. Thus the idea was developed of linking the population and industrial centers of Europe with Asia by so-called development corridors, and thus providing the landlocked regions of Eurasia with the same site-characteristics which the regions with ocean or river access already had.

In the 23 years that have passed since then, this idea has not only been presented in innumerable conferences and seminars in cities across the world, but also further filled out into the idea of the World Land-Bridge. The realization of linking peoples through this World Land-Bridge is now the realistic perspective emerging from the collaboration of the BRICS nations, Latin America, and the ASEAN nations, and in which the U.S.A., Europe, and Africa must urgently participate.

A new strategy for mankind means the ability, from now on, to see the human species as a unity, and to see that unity in the process of mutual development. Thus, with Friedrich Schiller, we see no contradiction whatsoever between the inviolability of national sovereignty, which is guaranteed by the law of nations and by the United Nations Charter, and the sanity of the world citizen who has in view the interests of mankind as a whole. For this unity lies in the higher development of all; the concordance of the macrocosm requires the maximal development of all microcosms to their reciprocal benefit, as Nicholas of Cusa said.

This also signifies a new model of cooperation among the nations of the world. It means that all potential treaty organizations and alliances must be inclusive, that they cannot be for the security and economic interests of some nations, while excluding others. While the support of mutual development is the premise, they must nonetheless respect the different levels of development, history, culture, and social systems, and above all, respect national sovereignty. That is Cusa's idea of unity in multiplicity, and it must be inspired by a tender love for the idea of the community of nations, for the idea of mankind as the creative species.

We must learn to view this mankind from the same perspective as the astronauts, cosmonauts, and taikonauts have seen it, as so wonderfully expressed by one of the Apollo astronauts who walked on the Moon:

"The fact is that evolution is now taking place in space, as much as on Earth. Man has shown that as a species mankind was willing to commit itself to living in environments that were completely different than those in which the species evolved—with a shield of life around ourselves in order to protect the life within. But the willingness to go out there, is there. We've shown that. The curve of human evolution has been bent."[1]

---

1. See the YouTube video, "Apollo 11: For All Mankind." [https://www.youtube.com/watch?v=HxgoV9IMgCg]

An Explosion of BRICS Initiatives

July-September 2014

Immediately before, during, and immediately following the 6th annual conference of the BRICS nations (Brazil, Russia, India, China, and South Africa), July 14-16 in Fortaleza, Brazil, there has been a virtual explosion of announcements of agreements on new major infrastructure projects and new credit measures, which, taken as a whole, constitute a fulcrum for lifting the entire planet onto a new trajectory of development. The following grid, although not complete, gives a snapshot of the new world coming into being. It is divided into new credit arrangements, and physical projects.

I. New Credit Institutions

BRICS New Development Bank and Contingent Reserve Arrangement: The Fortaleza Declaration includes the historic announcement that the BRICS nations agreed to form the New Development Bank (NDB) to fund infrastructure and other development projects in BRICS and other developing economies. The bank is to be headquartered in Shanghai, China, with the first (rotating) six-year presidency held by India. The NDB will have an initial authorized capital of $100 billion, with an initial subscribed capital of $50 billion, equally shared among founding members.

The Fortaleza Declaration calls for "an international financial architecture that is more conducive to overcoming development challenges."

The BRICS Contingent Reserve Arrangement (CRA), with an initial size of $100 billion, is aimed to "help countries forestall short-term liquidity pressures." It is intended, furthermore, to strengthen the global financial safety net and complement existing international arrangements.

China-CELAC Cooperation to Fund Infrastructure Development: The July 17 meeting of heads of state and special representatives of the Community of Latin American and Caribbean States (CELAC) with Chinese President Xi Jinping, discussed deepening their relations on the basis of "equality and mutual benefit, reciprocal cooperation and common development." The Presidents of Brazil and China, current members of CELAC's leadership Quartet (Costa Rica, Cuba, Ecuador, and Antigua and Barbuda), and representatives of all of the rest of South America attended. They agreed to found a China-Latin American-Caribbean Forum, with a mandate to draw up a 2015-19 Chinese-Latin American-Caribbean Cooperation Plan.

President Xi proposed three mechanisms to fund projects: a specific fund to finance infrastructure projects, to start at $10 billion and rise to $20 billion, scheduled to become operational by 2015; a preferential credit line for CELAC, from a Chinese bank, which could be as large as $10 billion; and a Sino-Latin American-Caribbean Cooperation Fund of $5 billion for investment in areas as yet to be defined.

The Declaration's formulation on these funds and projects marked a radical departure from IMF/World Bank conditionalities.

BRICS Energy Association: On July 15, Russian President Vladimir Putin announced in Brasilia plans to establish a BRICS "energy association," which will include a nuclear fuel reserve bank and an energy policy institute.

II. Great Projects

South and Central America

Nicaragua Canal: On July 7, Nicaraguan President Daniel Ortega announced the route of the Great Inter-Oceanic Canal, a great project connecting the Pacific and the Caribbean, which will become a focal point of development for the entire basin. Top Chinese water management, rail, aviation, and port design companies are partners in the project, including institutions that designed the Three Gorges Dam. In the immediately ensuing weeks, survey work began, and the project is fast becoming a national focus.

Russia-Nicaragua Cooperation: On his way to the BRICS summit, President Putin stopped off July 11 in Nicaragua, where he discussed supplying agricultural equipment to that country, the installation of the Russian GLONASS system (space-based global satellite navigation system similar to GPS), and cooperation in other areas.

Russia-Cuba Cooperation: On July 11, President Putin signed 10 agreements with the Cuban government, including one for the modernization of the port of Mariel, the construction of a state-of-the-art airport, the construction of four power units at the Maximo Gomez and East Havana thermal power plants, and exploration for offshore oil deposits.

Russia-Argentina Expanded Cooperation: During his July 12 state visit to Argentina, President Putin signed energy, aerospace, agriculture, communications, and military cooperation agreements with President Cristina Fernández de Kirchner. Of particular importance were the

Russia and Argentina have agreed to cooperate on building more nuclear power plants in Argentina. Here, Argentina's Embalse plant.

An Explosion of BRICS Initiatives

nuclear energy deals, which include an agreement calling for the design, construction, operation, and decommissioning of nuclear power plants and research reactors, as well as "water desalination facilities." Rosatom has submitted a technical and commercial proposal to participate in the construction of Argentina's Atucha III nuclear plant.

Peru-Brazil Transcontinental Railroad: On July 17, China, Brazil, and Peru agreed to initiate feasibility studies on the construction of a transcontinental rail line linking Brazil's Atlantic Coast with Peru's Pacific Coast. Technical teams are to carry out on-site surveys, and each country is to specify the approaches, resources available, and a time frame for the project.

Brazilian President Dilma Rousseff reported that she and President Xi had paid special attention to China's opportunity to bid on the construction of a stretch inside Brazil of "the Brazil-Peru Transoceanic Railroad, which is fundamental to South American integration and an outlet for Brazilian exports to Asia." This Lucas do Rio Verde-Mato Grosso-Campinorte-Goais rail stretch of the transoceanic route is also specifically cited in the joint declaration as one of the projects in which state and private sector investors of both countries should be encouraged to participate.

Bolivia has also asked China for help in developing the Bolivian portion of an alternative transcontinental rail route Brazil-Bolivia-Peru, President Evo Morales reported on Aug. 6.

China-Cuba Cooperation: Twenty-nine agreements for energy, transportation, science, agriculture, telecommunications, and infrastructure development were signed between Cuban and Chinese officials during President Xi's July 23-24 visit to the island. Key among these was the credit line for construction of a multi-purpose terminal at the port of Santiago de Cuba, a framework agreement for participation by the Chinese National Oil Company in development of the Seboruco oil deposit, and a Memorandum of Understanding (MoU) signed by both nations' Industry Ministries for developing Cuba's industrial sector.

Russia-Bolivia Nuclear, Infrastructure Development Cooperation: On July 16, President Putin offered to cooperate with Bolivia for the development of a "comprehensive nuclear energy program" for peaceful purposes. This includes technology transfer and permanent training of Bolivian personnel at various stages of the program. Russia will also help build hydroelectric and thermoelectric plants, and executives from Russian firms will be visiting Bolivia very soon to discuss this. Representatives from Russia's Rosneft oil firm will visit Bolivia to discuss investment in Bolivian oil projects.

China-Bolivia Satellite Cooperation: On July 16, President Xi offered assistance to President Morales in building Bolivia's second satellite. China's Great Wall Industry Corporation built Bolivia's first satellite, Tupac Katari, which was launched from China in December 2013.

Brazil-Russia Trade, Military, Nuclear Cooperation: In a July 14 meeting in Brasilia, Presidents Putin and Rousseff signed an agreement to work toward nearly doubling trade between their two coun-

Bolivian president Evo Morales, shown here at the United Nations, is negotiating with Russia on the development of a "comprehensive nuclear energy program."

tries, to $10 billion a year. The seven bilateral agreements include an anti-air defense system, under which the Brazilian military will participate with the Russian military in the use of Russia's Pantsir-S1 surface-to-air defense system, which the Brazilians are interested in purchasing, and an agreement to expand facilities for Russia's GLONASS satellite navigation system in Brazil.

On July 15, Russian nuclear representative Dzhomart Aliev and Brazilian company Camargo Corrêa signed an MoU to expand bilateral cooperation in nuclear power. A spent-fuel storage facility, the construction of engineering and other technical facilities at Brazil's Angra operating nuclear power plant site, and a "partnership" in the construction of new nuclear plants in Brazil, are included in the MoU.

Brazil-China Infrastructure Development, Scientific and Military Cooperation: In a July 17 meeting, Presidents Xi and Rousseff consolidated a "truly strategic partnership," signing several bilateral agreements, among them projects for the construction and financing of infrastructure inside Brazil; deepening their space cooperation, including joint satellite work with Africa; the sale of Brazilian jets to China; intensified scientific and educational exchanges; and Chinese construction of Brazil's Rio Tapajos hydroelectric project.

They plan to move ahead with their China-Brazil Earth Resources Satellite (CBERS) program, adding a new satellite, and considering more in the future. The two nations have launched a series of four Earth remote sensing satellites, with Brazil building the spacecraft and China providing the launch vehicle.

Argentina-China Infrastructure Development, Nuclear Cooperation: During his July 18-21 visit to Argentina, President Xi signed an agreement with President Fernández for a "Comprehensive Strategic Association" between the two nations. Nineteen agreements were signed in the areas of nuclear energy, infrastructure, communications, transportation, and agriculture; they included $4.7 billion in financing for the Néstor Kirchner-Jorge Cepernic hydroelectric complex in Santa Cruz, a $2.5 billion credit for renovation of the Belgrano Cargas railroad, and an $11 billion currency swap agreement between the two central banks. On Sept. 2, in Beijing, the head of China's National Nuclear Corporation (CNNC) and the head of Argentina's Nucleoeléctrica Argentina, SA signed a $2 billion agreement by which China will provide preferential financing for Argentina's fourth nuclear reactor, the 760-MW Atucha III.

Venezuela-China Economic, Energy, Infrastructure Cooperation: During President Xi's July 22 visit to Caracas, the meeting of the 13th High-Level Mixed Commission in Caracas signed 38 bilateral accords in the context of the "comprehensive strategic alliance" between the two countries.

Agreements involved oil exploration, agriculture, investment in industry, science, and technology to assist the "socio-economic development of both nations." China also signed an agreement to deliver a second VRSS Earth remote-sensing satellite to Venezuela, the first having been delivered in 2012.

China-Mexico Nayarit Port and Rail: On Sept. 5, César Duarte,

Marcelo Camargo/Agencia Brasil

Brazil and China declared the establishment of a "truly strategic partnership," during their meetings in Brazil in mid-July. Here, Xi Jinping and President Dilma Rousseff.

An Explosion of BRICS Initiatives

Governor of Chihuahua, Mexico, announced that the Development Bank of China will provide $1 billion to finance the Nayarit-Chihuahua-New Mexico rail project, with construction to start by the end of 2014. This rail line is part of what is termed Mexico's "Economic Corridor of the North," and features construction of Puerto Nayarit, which, when completed in some three years, will be the largest deepwater port in Ibero-America. The Nayarit government expects that the cornerstone for what they term "the first project of the Mexico-China Alliance," will be personally laid this Winter by China's President Xi and Mexican President Enrique Peña Nieto.

Eurasia

Chinese President Xi Jinping and his wife are greeted by Indian Prime Minister Modi at the start of their Sept. 17 summit in India.

China-India Joint Economic Projects: During the Sept. 17-20 China state visit to India, President Xi and Indian Prime Minister Narendra Modi concluded more than 10 significant economic deals, and pledged to settle long-standing border disputes. Among the projects is collaboration in nuclear science, particularly in developing thorium-fueled nuclear reactors. India plans a 300-MW thorium prototype by 2016, to then scale up. China is working on a pebble-bed solid fuel 100-MW demonstrator reactor by 2024, for full deployment by 2035. Commitments were signed for a faster railroad between Musore City and Chennai via Bengaluru, to allow more of India's pharmaceuticals into China, and other initiatives.

Talks took place on the Bangladesh-China-India-Myanmar trade corridor, which would link the Indian port of Kolkata with Kunming, Yunnan's capital.

Overall, discussion took place on jointly building the Silk Road Economic Belt and the 21st Century Maritime Silk Road, announced in 2013 by President Xi.

Russia-North Korea-South Korea Development Project: On July 18, Russian, North Korean, and South Korean officials opened the port of Rajin, a state-of-the-art port, built by Russia, connecting to the recently completed rail line from Rajin, North Korea to Russia.

Russia and China came to a deal on joint development of floating nuclear plants in late July. Here, an artist's rendition of a Russia floating plant.

Russia-China Nuclear Cooperation: On July 28, the export branch of Russia's Rosatom nuclear company, Rosatom Overseas, signed an MoU with China, to bring the two nations closer to the joint development of floating nuclear power plant (FNPP) technology. Russia is moving toward completion of the first of what will be a fleet of small, marine nuclear energy reactors to be placed on barges for civilian power and desalination. Six of these FNPP units are part of the Russia-India talks. Rosatom Overseas chief executive Dzhomart Aliev said that these small reactors can provide "a reliable power supply, not only to remote settlements but also to large industrial facilities such as oil platforms."

Russia-China Siberian Gas Lines: On Sept. 1, Presidents Putin and Xi attended a ceremony outside Yakutsk, Russia, launching construction on the first section (the China-Russia East Route) of a 4,000-km pipeline, the "Power of Siberia," the mega-project deal that was signed during the Putin-Xi summit on May 21. On Sept. 17, Russia announced that a new China-Russia West Route gas pipeline deal, to provide a 30-year gas supply to China, will be signed in November.

Moscow-Kazan High-Speed Rail Project: Russian Railways on July 31 announced that negotiations are underway with Chinese investment and construction companies about a partnership to construct a high-speed rail line between Moscow and Kazan. Among potential partners in the project are the China Investment Corp., which, in addition to participating in the Moscow-Kazan high-speed rail project, is considering participating in the project to build the Eurasian High-Speed Corridor Russia (Moscow)-China (Beijing) as a whole.

India to Receive High-Speed Trains from Japan: Indian Prime Minister Narendra Modi on Sept. 1 signed an agreement with Prime Minister Shinzo Abe in Tokyo, whereby India will receive Japanese financial, technical, and operational support to introduce Bullet trains. Abe also pledged that Japan would invest $35 billion during the next five years, doubling its investments in both India's private and public sectors. The two countries will accelerate talks on the possible sale of an amphibious aircraft to India's navy.

India-Nepal Hydro-Power Accord: On Sept. 19, India and Nepal signed an agreement for Indian infrastructure builder GMR to construct a 900-MW hydropower project on Nepal's Karnali River. Electricity generation is projected to start in 2021, to the great benefit of both nations. This accord ends years of contention over hydro-development. Nepal has a potential of 40,000 MW of hydropower, of which less than 500 MW has been developed. This breakthrough agreement followed Modi's visit Aug. 3-4 to Nepal, where he pledged commitment to an "HIT" plan for Nepal, comprising highways, information-ways, and transways. For the new dam, Nepal will get a 27% share of equity to begin with, then complete ownership of the project in 25 years after the plant begins generating power. Nepal will get 12% of the power free of cost, with the remainder exported to India, and possibly to Bangladesh.

Kyushu Railway Company
During Japanese Prime Minister Abe's visit to India in early September, he offered financial, technical, and operational assistance to India in building bullet trains. Here, some of Japan's fleet of electric trains.

Russia Supports India and Pakistan Membership in Shanghai Cooperation Organization: The SCO intends to make India and Pakistan full members at its summit next year, Kremlin spokesperson Yuri Ushakov announced on Sept. 12. The summit will be held in the Russian city of Ufa on July 9-10, 2015, along with the 7th BRICS Summit, and Russia will chair both meetings. Russian President Putin said that "the priorities of our chairmanship include enhancing the role of the organization as an efficient mechanism of regional security, launching major multilateral and humanitarian ties, and developing joint approaches to pressing and global issues."

Russia is supporting the inclusion of India and Pakistan into the Shanghai Cooperation Organization, whose logo is shown here.

South and Southeast Asia

China-ASEAN Meeting in Myanmar: On Aug. 10, the ASEAN foreign ministers met in Myanmar, along with representatives of China, India, Russia, the United States, the EU, Japan, South Korea, and

Australia. As reported in the *Daily Times* of Pakistan, China and ASEAN (Brunei, Thailand, Singapore, Cambodia, Laos, Indonesia, Malaysia, the Philippines, Vietnam, and Myanmar) reached an agreement to deepen their strategic partnership, including joint work on China's 21st Century Maritime Silk Road, and projects in the Mekong River development area. China also welcomed all 10 ASEAN nations to join in the Asian Infrastructure Investment Bank as founding members. Thailand has already accepted that invitation.

Maritime Silk Road—China-ASEAN Expo in Nanning: The 11th annual China-ASEAN Expo (CAEXPO) Sept. 16-19 was held in Nanning, the capital of Guangxi Province, on the theme, "Jointly Building the 21st Century Maritime Silk Road," with 4,600 exhibitors. Of these, 1,259 were from the 10 ASEAN countries, for which trade with China is expanding at a rate of 10% a year.

New Southeast Asia Large Dam—Salween River: On Sept. 16, China's Three Gorges Dam Corporation signed a contract with Myanmar's IGE Company to build Southeast Asia's largest dam, on the Thanlwin (Salween) River.

BRICS Young Scientists Forum: Indian Prime Minister Modi proposed such a forum during his July 15 speech in Fortaleza. The BRICS should go beyond "being summit-centric," he proposed; the youth of the BRICS nations should take a lead in expanding people-to-people contact. The Forum would set up schools "to offer language training in each of our languages," exploring the creation of a BRICS University.

China-Singapore Economic Corridor: The first think-tank summit for this project took place Sept. 12. Mayors of cities along the route reached a consensus on the initiative at their meeting, which took place alongside sessions on related aspects of the New Maritime Silk Road, held at the China-ASEAN Expo annual meeting, this year in China's Guangxi Zhuang autonomous region, a gateway to Southeast Asia.

The concept is that the rail, highway, and development corridor starts from Nanning and Kunming in China, going south through the Indochina peninsula, connecting China, Vietnam, Laos, Cambodia, Thailand, Malaysia, and Singapore. Thailand and China have approved the construction of rail lines connecting Bangkok to the north/northeast of Thailand, part of the larger China-Singapore Corridor. China is in negotiations with Laos to construct the Laotian portion of that plan.

Africa

Russia-Egypt Trade Expansion: Following the meeting between Russian President Putin and Egyptian President Abdel Fattah el-Sisi in Sochi, Russia, on Aug. 12, Russian-Egyptian cooperation is expanding. A particular focus of the meeting was trade in food products. Putin also expressed readiness to support Egypt's construction of a nuclear power plant at Dabaa. On Sept. 10, Egypt's Minister of Commerce, Industry, and Small and Medium Enterprises, Mounir Fakhry Abdel Nour, led an Egyptian business delegation to Russia, which included food commodities manufacturers and crop producers.

youtube/euronews

Egyptian president el-Sisi made his first foreign trip to Russia in mid-August, where he made wide-ranging deals on trade and investment. Here, el-Sisi in Moscow before his election.

South African President Jacob Zuma visited Russia in late August, where the two presidents discussed deals on nuclear energy, among other investments.

South Africa Deals with Russia: On Aug. 28, President Jacob Zuma met Russian President Putin in Novo-Ogaryovo, on the outskirts of Moscow, for an agenda topped by trade and investment. Russia offered assistance for a comprehensive nuclear energy industry in South Africa, in light of President Zuma's June announcement that South Africa will greatly expand its nuclear program.

South Africa-China Steel Mill Project: On Sept. 12, South Africa Trade and Industry Minister Dr. Rob Davies confirmed that the Hebei Iron and Steel Group of China will jointly develop steelmaking capacity with the South Africa state-owned Industrial Development Corporation (IDC), in Limpopo Province, near Phalaborwa, where there are large magnetite deposits. Construction is to start in 2015. The initial goal is 3 million tons a year, with 5 million tons, mostly construction grade, by 2019. With this deal, South Africa now resumes having its own, independent steel capacity, after it lost this during a privatization push from 2001 to 2004, when its state-owned Iron and Steel Corporation (ISCOR) was sold off to ArcelorMittal, the British Commonwealth cartel. In the China deal, IDC will have 49% ownership.

Zimbabwe-China Agreements: On Aug. 25, President Xi met Zimbabwe President Robert Mugabe in China. They signed a number of cooperation agreements.

India-South Africa Agriculture: In a statement on Sept. 11, South Africa Agriculture Minister Senzeni Zokwana urged South African farmers to take advantage of the BRICS development bank, including in food processing and farming. His comments came during the first-ever Indo-South Africa Week in India. The week-long seminar began in Mumbai on Sept. 9-10, then moved to Gurgaon on Sept. 11-12. South Africa showcased technologies related to food processing and the agriculture sector. Zokwana met with India Minister of Agriculture Shri Radha Moodan Singh on BRICS financing of farm and food projects.

World Land-Bridge Network—Key Links and Corridors

The New Silk Road Becomes the World Land-Bridge

I: 20 Part 1: Introduction

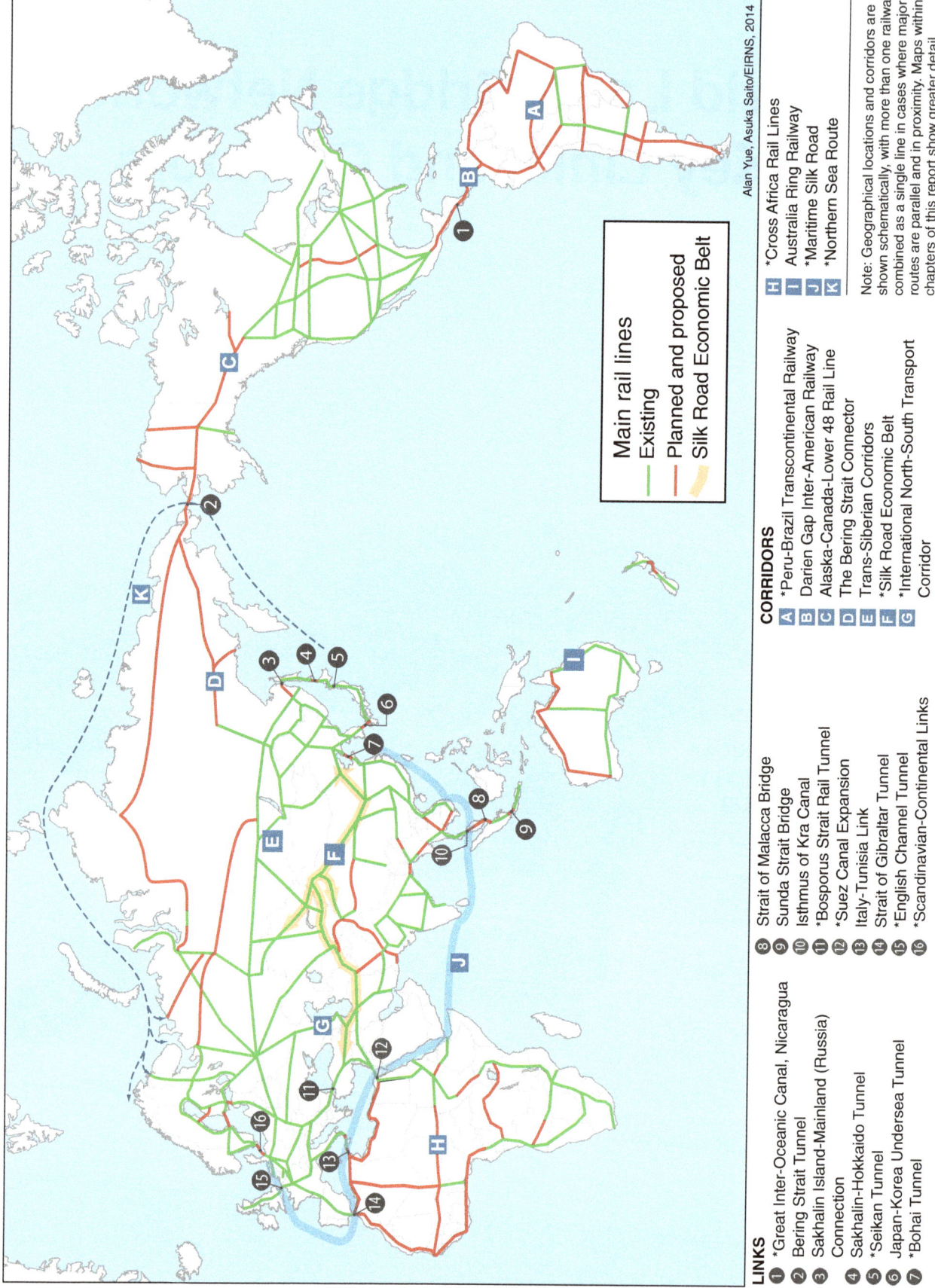

The New Silk Road Becomes the World Land-Bridge

World Land-Bridge Network—Key Links and Corridors

*means committed, underway, or completed

Status descriptions of each of the 27 features identified on the map on the facing page, all of which are important to the emerging worldwide transportation network for development.

LINKS

1 ***Great Inter-Oceanic Canal, Nicaragua**
Construction is scheduled to begin in December 2014 on the cross-Nicaragua Great Inter-Oceanic Canal, connecting the Pacific and Atlantic Oceans (via the Caribbean Sea). It will run 278 km (172.7 mi) from the mouth of the Brito River on the Pacific coast in southwestern Nicaragua, to the mouth of the Punta Gorda River on the Caribbean side, 104.6 km (65 mi) of it passing through Lake Nicaragua. The canal was originally mapped out 118 years ago by U.S. engineers. The new commitment was announced July 7, 2014, by President Daniel Ortega and Wang Jing, head of the Chinese HKND firm. Survey work began in August. More than 50,000 workers will build the complex, including two ports, an international airport, cement and steel factories, and other infrastructure. Completion is projected in five years.

2 **Bering Strait Tunnel**
The gap between Alaska and Siberia can be closed by 85 km (52.8 mi) of tunnels under the Bering Strait, linking the transportation systems of Eurasia and the Americas. Some 3,000 km (1,864.1 mi) of new railway in Eurasia and more than 1,000 km (621.4 mi) in North America will need to be built, under rugged northern conditions on both continents, to complete the connection.

The proposal for a Bering Strait crossing dates from the 1800s; it received a boost in April 2007, when a Moscow conference, convened by Russia's Council for the Study of Productive Forces (SOPS), appealed to leading nations to start feasibility studies. The SOPS design for this link won a Grand Prize at World Expo-2010 in Shanghai. Prof. Wang Mengshu of the Chinese Academy of Engineering said in May 2014, that discussions of the project are under way between China and Russia.

3 **Sakhalin Island-Mainland (Russia) Connection**
The Tatar Strait between Russia's Sakhalin Island and the mainland is 7.3 km (4.5 mi) wide at its narrowest point. This link, combined with a new tunnel from the southern end of Sakhalin

The New Silk Road Becomes the World Land-Bridge

to Japan, will connect Japan to the Eurasian landmass and its railway network, near the terminus of the Baikal-Amur Mainline. A tunnel was partially dug under the Tatar Strait in the early 1950s. Russia is studying tunnel, railway bridge, or dam (sea gate) options. The sea gate design, a gigantic movable dam with transport lines on top of it, could include diversion of the mouth of the Amur River from the Sea of Japan, south of the Strait, to the Sea of Okhotsk north of it, an engineering feat that would help Russia and China with flood control along the Amur.

4 Sakhalin-Hokkaido Tunnel

This tunnel or bridge would link the Russian island of Sakhalin with the Japanese island of Hokkaido, spanning the 45 km (28 mi) LaPerouse Strait. Combined with a Sakhalin-Russian mainland connection, this crossing will provide a rail route from Japan onto the Eurasian land-bridge. The Sakhalin-Hokkaido Tunnel would be shorter than the Seikan Tunnel between Honshu and Hokkaido islands.

5 *Seikan Tunnel

This railway tunnel, which opened in 1988, runs under the Tsugaru Strait, connecting the main Japanese island of Honshu with Hokkaido, to its north. It is currently the longest and deepest in the world, with a total length of 53.85 km (33.46 mi), of which 23.3 km (14.5 mi) is under the seabed, with its track running at a depth of 100 m (328 ft) below the sea floor. First surveyed in 1946, the Seikan Tunnel was seen as an essential project for maintaining a strongly unified nation.

6 Japan-Korea Undersea Tunnel

This proposed tunnel would cross the Korea Strait to connect Japan and South Korea, via the strait islands of Iki and Tsushima. At its shortest, the crossing distance is some 128 km (80 mi). Proposed for a century, the project came under new study when a task force was formed in 2009, headed by Huh Moon-do, former director of the National Unification Board of South Korea. Collaboration on this mutually beneficial project can contribute to easing the political tensions between the two nations.

7 *Bohai Tunnel

The proposed tunnel will run more than 100 km (62.1 mi) under the Bohai Strait, to accommodate a high-speed railway between two Chinese cities, each with a population of about 7 million: Dalian, on the northern end, is a major port in heavily industrial Liaoning Province, while Yantai is a center of industry in Shandong Province. They face each other across the entrance to the Bohai Sea, the westernmost gulf of the Yellow Sea between China and the Korean Peninsula; on the Bohai Sea coast sits the city of Tianjin, population 14 million, which is 130 km (80 mi) southeast of the capital, Beijing. China's State Council announced a commitment to the project in August 2014; work could start during the 13th Five Year Plan, 2016-2020. It will be the world's longest undersea tunnel, taking 10 years to build, according to Prof. Wang Mengshu of the Chinese Academy of Engineering.

8 Strait of Malacca Bridge

The bridge to connect Malaysia and Indonesia will be the longest over-sea bridge in the world. Making use of the mid-strait Rupat Island, the bridge will run 48.7 km (30.3 mi) from Teluk Gong in the Malaysian state of Malake, to the island, and thence to Dumai on Sumatra—a span of 71.2 km (44.2 mi). First proposed in 1995, the project was stalled in the world financial breakdown. In 2006, the Ex-Im Bank of China agreed to provide financing. Chinese and other firms are conferring on the project, among them Danish contractors that built the Øresund Fixed Link between Denmark and Sweden.

9 Sunda Strait Bridge (Indonesia)

It is proposed to close the distance of 27.3 km (17 mi) between the islands of Sumatra and Java—which together account for 80% of the population of Indonesia—by a system of bridges across the Sunda Strait between the Java Sea and the Indian Ocean. Two main spans, 6.5 km (4 mi) and 4 km (2.5 mi) long, will be anchored on Sangiang Island in the strait. Challenges include the area's seismicity. Cutting hours off the current ferry crossing, the Sunda Strait Bridge will boost agricultural supply operations and new industrial development in Indonesia, a major Asian nation of more than 250 million people.

10 Isthmus of Kra Canal

This canal has been proposed for Thailand since at least the 17th century, but consistently opposed by imperial interests. It will loosen the bottleneck of the Malacca Strait (between Sumatra, Indonesia and the Malay Peninsula, Malaysia), now practically the only sea lane between the Pacific and Indian Oceans. The idea was revived at a 1983 Bangkok conference sponsored by the Mitsubishi Global Infrastructure Fund, *EIR*, and Thai government and military officials. It was sidelined during subsequent financial crises, but is now under consideration as part of the BRICS infrastructure revival. The Kra Canal will be 50 to 100 km (31 to 62.1 mi) long, depending on its route, and must cross a 75 m (246 ft)-tall range of highlands.

11 *Bosporus Strait Rail Tunnel

This tunnel is the first all-rail connection between Europe and Asia Minor, crossing the Bosporus at the Istanbul Strait. It was opened October 2013, the 90th anniversary of the Turkish Republic, and hailed as the "Iron Silk Road." Currently, it is a mass transit service, carrying some 3 million passengers daily, but plans call for adding high-speed rail and freight transport, as well as connecting it to the new Istanbul-to-Ankara high-speed rail line. First proposed in the 1860s, the idea awaited construction solutions to deep water and other challenges. The 13.6 km (8.45 mi) tunnel has the deepest immersed tube structure in the world. Positioned 56 m (184 ft) below sea level, a double-tube line of 1.4 km (0.87 mi) was laid on the sea floor; another 2.4 km (1.5 mi) of tunnel was built using "cut and cover" methods; and finally, there are 9.8 km (6.1 mi) of bored tunneling for service and access.

12 *Suez Canal Expansion

Work commenced in Summer 2014 to build a second channel, 72 km (45 mi) long, and a world-class logistics hub, encompassing 47 sq. km (18 sq. mi) surrounding the dual-canal. The new channel will allow two-way traffic, for which the current canal is too narrow in some places. This will shorten the transit time between the Mediterranean and Red Seas from 11 hours to three. The project is a patriotic rallying call in Egypt, and the centerpiece of an aggressive development perspective with implications for all of northern Africa, Southwest Asia, and the Maritime Silk Road.

13 Italy-Tunisia Link

Connecting these two nations and the continents of Europe and Africa across the Mediterranean Sea involves (1) linking mainland Italy and Sicily, and (2) crossing the 155 km (95 mi) distance between Sicily and Tunisia, using four artificial islands, to be made from excavation debris. The first link, the 3.3 km (2 mi) Messina Strait Bridge, would be the longest single-span suspension structure in the world, running from Reggio Calabria on the Italian mainland, to Messina, Sicily. For the 140 km (85 mi) connection between Sicily and Tunisia, a five-tunnel design has been proposed—two for passengers and freight in each direction, and one for service and emergencies. An alternative and complementary plan is the Mediterranean Bridge, using the man-made islands as anchor-points to cover part of the distance; this design was publicized in September 2014 by Enzo Siviero, an Italian professor of construction science who also consults at Tongji University, Shanghai.

14 Strait of Gibraltar

The proposed tunnel between Tarifa, Spain and Tangiers, Morocco, would be 40 km (25 mi) long, with a depth of 300 m (990 ft) below sea level. It would cut the travel time between Barcelona and Casablanca to less than eight hours, and link European with African high-speed rail networks. A detailed feasibility study, commissioned by the Spanish and Moroccan governments, was presented to the EU in 2009, but no action has been taken.

15 *English Channel

The Channel Tunnel opened in 1994, running 50 km (30 mi) between England and France. First proposed in the early 1800s, the tube goes under the Strait of Dover, and is 75 m (250 ft) below the seabed at its lowest point. The "Chunnel" carries the high-speed Eurostar train line, serving London and Paris or Brussels, and car-trains between Folkestone and Calais. As of 2014, the Channel Tunnel still has the world's longest undersea stretch, although the Seikan Tunnel in Japan is longer overall, and deeper.

16 *Scandinavian Peninsula-Continental Links

A set of bridges and tunnels—some operational, some under construction or discussion—link the Scandinavian peninsula with the main part of western continental Europe, via Jutland and the islands of Denmark. First, in 1998, the 18 km (11.2 mi) Great Belt Bridge and railway tunnel was completed between two large Danish islands. The 16 km (9.9 mi), three-part Øresund Fixed Link, from Copenhagen to Malmö, Sweden, was completed in 2000, enabling road and rail transport between Sweden and Norway, on one side, and the rest of continental Europe, on the other. Now under construction is the Fehmarnbelt Tunnel, connecting the Danish islands directly to Germany. This 17.6 km (10.9 mi) tunnel will become the world's longest immersed (as opposed to bored) rail/road tube. The Nordic countries also seek to upgrade road and rail connections with Russia and its transcontinental rail system.

Finnish geologists have designed and proposed east-west tunnel crossings of the Gulf of Bothnia between Finland and Sweden, as well as under the Baltic Sea from Finland southward to Estonia and the Rail Baltica corridor to Warsaw.

CORRIDORS

A *Peru-Brazil Transcontinental Railway

Feasibility studies are now under way, after the July 17, 2014 joint agreement by Peru, Brazil, and China, to construct a rail line across South America, from Campinorte, Goais, Peru to Rio Verde, Mato Grosso, Brazil. This will be the first railroad across the continent, and marks a commitment to launch corridors of development. As early as 1898, an alternative trans-continental route was mapped out, between Brazil and Bolivia, which nation now wants to resume studies for what can be a second corridor.

B Darien Gap Inter-American Railway

Running a rail and road corridor through the Darien Gap, a large tract of swampland and forest on the Isthmus of Panama and straddling the Panama-Colombia border, between the Gulf of Darien in the Caribbean Sea and the Gulf of Panama on the Pacific Ocean side, will finally allow the implementation of a long-envisioned through-route for the Inter-American Railway, running from Alaska to Tierra del Fuego. Full plans were made in the 1890s by the Intercontinental Railway Commission, under the William McKinley Administration. At present, even the Pan American Highway does not go through. The Highway—a series of roads traversing North, Central, and South America—runs about 48,000 km (30,000 mi), with a complete break of some 100 km (60 mi) at the Darien Gap. Marshland rail construction challenges have not been the hold-up. The corridor has been blocked by anti-development opponents under green pretenses.

ARTIST'S CONCEPTUAL VIEW OF THE PROPOSED ALASKA-CANADA RAILROAD NEAR LAKE KLUANE, YUKON

Proposed
ALASKA-CANADA RAIL CORRIDOR
SHOWING UTILITIES AND PIPELINES

Source: Cooper Consulting Co., 2002 J. Craig Thorpe

Labels: ALCAN Highway; Water Pipeline; Fiberoptic Telecommunications Cable Line; Superconducting Electricity Transmission Line; Natural Gas Pipeline

Alaska-Canada–Lower 48 Rail Line

C The route for a rail line from Alaska, through Canada, to the Lower 48 states will be 2,280 km (1,417 mi) long, if it follows the rush survey for a direct route, mapped out in 1942 by the U.S. Army Corps of Engineers. The plan was a defense contingency, and construction never began. In the ensuing more than 70 years, the project was blocked by forces opposed to development in the Americas. Now, it is urgent to build "the missing link." Recent studies have been carried out by the Canadian Arctic Railway Corp. and others. This corridor is vital for the Americas, to pass through the Bering Strait Tunnel, and onward through Eurasia.

The Bering Strait Connector

Russian Railways' Strategy for 2030 calls for a railway from the Baikal-Amur Mainline (BAM) to the Chukotka Autonomous District on the Bering Strait, with a spur to Magadan on the Sea of Okhotsk. The first leg has been built (first freight carried in 2014): the extension of the Amur-Yakutsk Mainline, running northward from the Trans-Siberian Railway (intersecting the BAM) to the bank of the Lena River opposite Yakutsk. The remaining approximately 3,000 km (1,864 mi) railway from the Lena to the Bering Strait will cross rugged, frozen mountain ranges in eastern Siberia. Some Russians envision a second corridor, along the Lena due north from Yakutsk, exiting to the Northern Sea Route at the port of Tiksi.

**LENA RIVER
RAIL-ROAD-UTILITY BRIDGE**

Between Haptagay & Tabaga,
Sakha Republic (Yakutia) Russia

Source: Cooper Consulting Co., 2005 J. Craig Thorpe

The Siberian corridor to the Bering Strait tunnel, with its counterpart on the North American side, will open up vast potential for development in the Far North, radiating across the planet.

E Trans-Siberian Corridors

Russia's Trans-Siberian Railway (TSR), built in 1891-1916, was the first Eurasian transcontinental railroad. It runs 9,289 km (5,771.9 mi) from Moscow to Vladivostok, dipping south of Lake Baikal and following the Russia-China border along the Amur and Ussuri Rivers in the Far East. Three Siberian cities of more than 1 million people each, sit where the TSR crosses major rivers: Omsk on the Irtysh River, Novosibirsk on the Ob, and Krasnoyarsk on the Yenisei. The TSR is double-tracked and fully electrified. Several concepts of a Trans-Eurasian development corridor in Russia center on adding a high-speed railway in the TSR right-of-way.

The Baikal-Amur Mainline (BAM), or "second Trans-Sib," splits from the TSR in east-central Siberia and passes L. Baikal on the northern end, continuing to the Sea of Japan. This single-track line was built in 1974-1991, but only in 2004 did the more than 15 km (9.3 mi) Severomuysky Tunnel become operational, eliminating a steep, 54 km (33.6 mi) bypass and allowing 6 million tons of freight annually to be shifted from the TSR to the BAM. Spurs from the BAM to raw materials deposit areas have been built or are planned.

The planned SevSib ("North Siberian") freight railroad, drawn in red on the map above label E, will run east from the Ural Mountains, joining the BAM between Tayshet and Lake Baikal. Together, the BAM, the SevSib, and the planned Belkomur Corridor, northwest from the Urals to the White Sea, all derive from plans made in 1928 for a Russian Great Northern Railway, a 10,000 km (6,214 mi) diagonal line across the continent from Murmansk on the Kola Peninsula in far northwest Russia, to the Tatar Strait opposite Sakhalin Island on the Pacific Ocean.

F *Silk Road Economic Belt

Chinese President Xi Jinping called for the development of a "Silk Road Economic Belt" (SREB), on Sept. 7, 2013, in a speech in Kazakhstan. The concept is for corridors of rail, agro-industrial activity, water, power, and trade infrastructure, to stretch, as he put it, "from the Pacific to the Baltic"—making use of the old Silk Road. In November 2013, an SREB agreement was signed by representatives of 24 cities in eight countries along the route. Projects are under way on various segments of the SREB, which branches in three main directions westward from China's western Xinjiang Province: a main corridor into Central Asia and on to Iran and Turkey, southern lines to Pakistan's Arabian Sea coast, and northern connections through Kazakhstan to Russia and northern Europe. In October 2014, Chinese and Russian state-owned railway companies discussed one major northern connection, signing a memorandum of understanding on drafting project-designs for a "high-speed Eurasian transportation corridor from Moscow to Beijing." The SREB runs through 18 Asian and European countries, affecting a population of three billions.

G *International North-South Transport Corridor

The International North-South Transport Corridor (INSTC) is a multimodal transportation and economic route from India by sea to the port of Chahbahar in Iran, and thence northward by rail through Central Asia, or northward along (or across, it is planned) the Caspian Sea via Azerbaijan into Russia, and thence to northern Eurasia. The agreement was made in 2000 by Russia, Iran, and India, and work is proceeding.

Additional north-south development corridors, farther east than the INSTC, exist in the form of proposals from Russian circles, and an active policy by China, for railways into Pakistan, from the Trans-Siberian Railway (corridor not shown) and the Silk Road Economic Belt, respectively.

H *Cross-Africa Rail Lines

A cross-Africa network of railways is an urgent planetary task, exemplified by the east-west route, to run from Dakar to Djibouti, and the long-discussed route, to run from the Cape of Good Hope to Cairo. The reality of this prospect was laid out in May 2014 by Chinese Premier Li Keqiang, in a four-nation visit, in which he presented the goal of connecting all African capitals by high-speed rail, so as to boost pan-African communication and development. In 1980, this same perspective—including proposed water projects, nuclear power, and other infrastructure—was issued by the Fusion Energy Foundation of Lyndon LaRouche, in a book, *The Industrialization of Africa*.

I. Australia Ring Railway

Australia will be connected with a high-speed, transcontinental rail network between its major cities and ports, which will open up huge areas of its inland, the "outback," to high-intensity agriculture, mining, and settlement. Currently, there is a standard-rail, east-west transcontinental line in the south, and a standard north-south line through the center of the continent. The plan is to develop a high-speed fast-freight line through the eastern states from Melbourne to Darwin, which continues across the north and down the western side to Perth, connecting to the existing east-west route. This line, and the north-south line, will be upgraded to high-speed rail. Freight will travel from Melbourne to Darwin in just 24 hours, where proposed high-speed shipping services will be able to deliver it to any of the massive ports in Asia within one to four days.

J. *Maritime Silk Road

Chinese President Xi Jinping called for a "Maritime Silk Road" (MSR), on a visit to Indonesia in October 2013. Its purpose is to further mutual development through trade along the sea routes of southern Eurasia, from the Pacific coast to East Africa, the Mediterranean, and eastern Atlantic shores. Xi referred to the same purpose, as it had been furthered historically by the famous voyages of Chinese Admiral Zheng He in the 1400s. The MSR commitment is now active, in terms of projects and plans for port upgrades, canal improvements, and related work. For example, in September 2014, Sri Lanka inaugurated construction of the "Colombo Port City," to participate fully in the MSR. China is providing funding.

K. *Northern Sea Route

The Northern Sea Route (NSR), or Northeast Passage, from Asian ports, through the Bering Strait, and westward along Russia's Arctic coast to its far northwestern ports and beyond, is 5,200 km (3,231 mi) and nine days shorter than shipping to Europe around the southern side of Eurasia. While melting of the sea ice has eased the transit, Russia has also invested heavily in upgrading the NSR, adding new navigation systems, icebreakers, and rescue stations. China backs increased use of the NSR, as do Finnish and Scandinavian railway and shipping interests. New rail links from Finland to Russia's NSR ports of Arkhangelsk and Murmansk have been built. If Russia builds the proposed Belkomur and/or Barentskomur corridors from the Urals to the northwestern ports, traffic in this direction will increase. At sea, potential shipping routes continue westward to Ireland (and its proposed Shannon Super-Port), the British Isles, and, on the so-called Arctic Bridge route, to North America and the famed Northwest Passage. On land, current Russian economic development along the NSR emphasizes onshore and continental shelf oil and gas projects; the world's northernmost railway now runs on the gas-rich Yamal Peninsula. Optimists look ahead to construction of a railway parallel to the NSR, the long-planned Near-Polar Mainline, and habitable cities along Russia's Arctic frontier.

PART 2
The Metrics of Progress

Energy Flux Density: Global Measure of Economic Progress

By Jason Ross

July 2014

Over the period 1948–52, Lyndon LaRouche solidified a fundamental advancement in economic science, a breakthrough that allowed him to become the most accurate economic forecaster of our day.[1] This breakthrough in understanding what Treasury Secretary Alexander Hamilton called "the productive powers of labor" allows him to offer uniquely competent guidance on global economic matters. This section of the report will serve to elucidate several key concepts of Mr. LaRouche's economic method, including, most centrally, that of energy flux density as a measure of economic value.

Starting from Fundamentals: Physical Chemistry as the Origin of Economy

Unlike all other life known to us, human beings are able to discover and apply knowledge of the universe and social functions, to fundamentally transform our relationship to nature and to our fellow man. This occurs uniquely through the process of scientific and artistic creative discovery, and through forms of social organization capable of fostering and implementing those discoveries.

A comprehensive standpoint from which to view such progress is that of *physical chemistry,* from its most humble origins in the use of fire, to the dawn of extractive metallurgy, to the breakthroughs of chemistry proper, to the more modern developments of electromagnetism and nuclear science. Specific periods of development are sometimes known by characteristic chemical knowledge: for example, the Stone Age, Bronze Age (beginning 3200 BCE), and the Iron Age (which began in 1200 BCE in Europe).

Only human beings have economies, because only human beings change their mode of existence from one generation to the next. The source of these changes, creative discoveries of new scientific and cultural principles, is the heart of economic value, and the proper

---

1. LaRouche has written extensively on his discovery, including in his 1984 economics textbook titled *So, You Wish to Learn All About Economics?*

origin of a science of economics.

Against this naturally human development, stands oligarchism.

Against Mankind: Zeus vs Prometheus

Neither history, science, culture, nor economics can be understood as disciplines, without an understanding of the most pertinent conflict between outlooks for human culture: the conflict between Zeus and Prometheus. This story, often wrongly considered only a myth as presented by such as the Greek playwright Aeschylus, tells of the origins of human science and economy, and the opposition, by oligarchy, to such development.

To keep ordinary mortals weak and under his control, the Olympian god Zeus forbade man the use of fire (and, in doing so, forbade *humanity* itself, and all the possible advancements of civilization), reserving it for his use, alone. The god Prometheus, acting as a friend to humanity, took fire from Zeus's heaven and brought it to mankind. For this act, Prometheus was violently punished by Zeus, in a torment designed to last eternally, yet Prometheus did not regret his actions.

The willful use of fire, the first technology, is what sets mankind apart from all animals. Prometheus describes the state of man before giving him fire and knowledge:

The Ukrainian-Russian naturalist Vladimir Vernadsky, shown here, in his 1943 Scientific Thought as a Planetary Phenomenon, *wrote: "The discovery of fire was the first instance in which a living organism took possession, and became the master, of one of the forces of nature. Beyond a doubt, as we now see, this discovery lies at the foundation of the entire subsequent future increase of mankind and of our present power."*

> First of all, though they had eyes to see, they saw to no avail; they had ears, but they did not understand; but, just as shapes in dreams, throughout their length of days, without purpose they wrought all things in confusion. They had neither knowledge of houses built of bricks and turned to face the sun, nor yet of work in wood; but dwelt beneath the ground like swarming ants, in sunless caves. They had no sign either of winter or of flowery spring or of fruitful summer, on which they could depend but managed everything without judgment, until I taught them to discern the risings of the stars and their settings, which are difficult to distinguish.
>
> Yes, and numbers, too, chiefest of sciences, I invented for them, and the combining of letters, creative mother of the Muses' arts, with which to hold all things in memory. I, too, first brought brute beasts beneath the yoke to be subject to the collar and the pack-saddle, so that they might bear in men's stead their heaviest burdens; and to the chariot I harnessed horses and made them obedient to the rein.

Rather than being slaves to ignorance, superstition, and the arbitrary whims of Zeus, human beings could use these gifts of knowledge from Prometheus to guide their own future, increasing their further power over nature through the powers of discovery.

The greatest of the sciences, economics, treats as its subject matter, that unique capability of our species to increase its standard of living and transform its relationship to nature and itself. How can economic progress be measured?

Energy Flux Density: Applying Mankind's Fire

Begin with the first of the gifts of Prometheus, fire, from which he says man "shall learn many arts." The archaeological distinction between humans and apes comes with the first appearance of ancient fire pits, used to control the power of fire for the betterment of the lives of those wielding that then-new power.

From that point on, mankind could no longer be characterized biologically, or as existing in biological evolution—the evolution of the creative powers of the mind became the determining factor, and biology decreased in importance, relative to the power of thought.

Since then, the kernel of economic growth has been expressed in the control over successively higher forms of "fire." First, came increasingly powerful forms of chemical fire: from wood to charcoal, from coal to coke, and on to petroleum and natural gas. The higher types of power not only allowed greater densities of fire-power; they opened up new domains of control and utilization of matter. Metallurgy, materials development, and physical chemistry all developed in dynamic interaction with the development of new forms of fire.

The revolutionary discoveries of the early 20th Century revealed an immense potential, altogether beyond chemical reactions: the fundamental equivalence of matter and energy, as expressed in the domains of fission, fusion, and matter-antimatter reactions. Each in this series of relativistic reactions (reflecting the Einsteinian equivalence between mass and energy) operates at successively higher energy densities, and the entire set is orders of magnitude beyond the entire successive set of chemical reactions. While this distinction is usefully expressed in the immense difference between the *quantity* of energy released in nuclear as compared to chemical reactions (expressing weapons in terms of kilotons or megatons of TNT), the measured quantitative difference is an effect of a *qualitatively* distinct, higher domain of action.

Control over higher energy densities enables the increase in what Lyndon LaRouche has termed the *energy flux density* of the economy, as measured by the density-rate of energy use characteristic of applied technologies, such as the energy concentrated in the beam of a laser used for metal-cutting, compared to a water-mill of the 18th Century. A general value for energy flux density can be measured as the energy use per person and per unit area of the economy as a whole. This increasing power is associated with qualitative changes throughout the entire society—new technologies, new resources, higher levels of living standard, and, essentially new economies (see **Table 1**).

To start with, consider the biological energy usage, the power rate of a human body, roughly 100 watts (corresponding to a 2000-calorie diet). Before the use of fire, all work performed (by human muscle), would be applied at a rate of 100 watts per capita. Compare this

Table I
The Energy Density of Fuels

FUEL SOURCE	ENERGY DENSITY (J/g)
Combustion of Wood	1.8×10^4
Combustion of Coal (Bituminous)	2.7×10^4
Combustion of Petroleum (Diesel)	4.6×10^4
Combustion of H_2/O_2	1.3×10^4 (full mass considered)
Combustion of H_2/O_2	1.2×10^5 (only H_2 mass considered)
Typical Nuclear Fuel	3.7×10^9
Direct Fission Energy of U-235	8.2×10^{10}
Deuterium-Tritium Fusion	3.2×10^{11}
Annihilation of Antimatter	9.0×10^{13}

Fuel energy densities. The change from wood to matter-antimatter reactions is so great that progress must be counted in orders of magnitude, and the greatest single leap is seen in the transition from chemical to nuclear processes.

rate with those seen in the historical development of nations.

For example, at the founding of the United States, the wood fire–based economy of the time provided an estimated 2,400–3,000 watts per capita. Thus, each member of that economy represented a potential application of energy up to 30 times greater than a fire-less society. Clearly, this was not only more energy, but represented a quality of energy that enabled people to create new states of matter and chemistry, states which could never be created by muscle power alone.[2]

By the 1920s, the increasingly coal-powered United States had a per capita power use of 5,000 watts, meaning every individual in the economy expressed nearly twice the power as members of the wood-based economy. This supported the powered machinery, transportation, and early electricity generation that transformed life alongside the development of modern chemistry.

By 1970, the per capita power rate in the United States, which now made extensive use of petroleum, natural gas, and limited applications of nuclear power, had reached 10,000 watts per capita, another doubling over the level 50 years prior.

In each of these transitions, the previous fuel declined in use as a power source, allowing non-combustive uses, as the use of wood for construction and petroleum for plastics and other petro-chemicals, while the array of resources expanded. In today's electromagnetic, and partially nuclear economies, rare earth minerals have become resources, the excellent fusion fuel of helium-3 on the Moon is being eyed by such far-sighted institutions as the Chinese space program, and the future truly fusion-based economy will be able to process mineral deposits far below the quality of ores exploited successfully today.

With these power transitions in mind, it is no surprise that per capita electricity consumption and per capita wealth (as measured by the admittedly quite flawed GDP) are so closely correlated, as seen in **Figure 1**.

Had the advance of nuclear power not been halted, and had fusion power been realized as intended, it would be no stretch to estimate that U.S. power rates would approach 40,000 watts per capita in the first generation of this new century. Such potential boggles the mind, and drives home how unacceptable the current world average of only 2,400 watts per capita (comparable to the newly founded United States more than 200 years ago) truly is.

---

2. Could you cook your meat by beating it with a club, or bake bread by banging it with a rock? Can you produce copper from malachite by using your muscles, without a charcoal-fire?

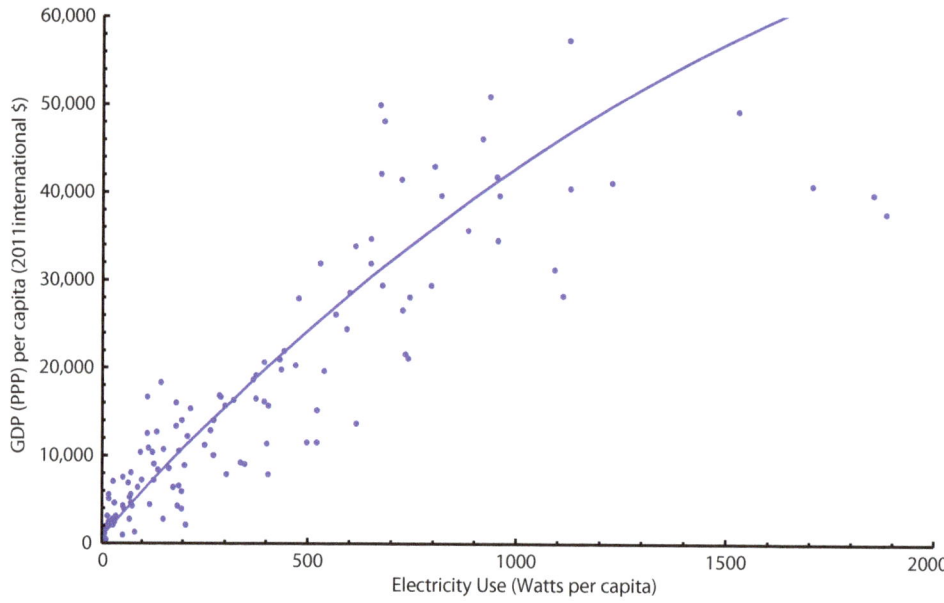

FIGURE 1
Electricity Consumption vs. GDP, Per Capita

EIRNS/Jason Ross, using 2010 data from World DataBank.

The correlation between per-capita electricity consumption and GDP is remarkably clear. Insisting that developing nations use "appropriate technologies" is insisting that they remain eternally poor.

All nations with both indicators were included (N=129). Plot range has been cropped to exclude outliers. Trend line is a best-fit cubic.

Alongside energy flux density, a second key economic metric developed by Mr. LaRouche concerns the demographics of economies powered by increasing levels of energy flux density. This brilliant metric side-steps the principal errors encountered in macroeconomic measurements.

A Global Measure of Economy: Potential Relative Population Density

Most economists seek to determine the overall productivity of a national economy through metrics that add together the monetary value of various components of the economy, resulting in such measures as gross domestic product (GDP). The problems with such an approach are two-fold:

1. Economic activity counting toward the GDP may or may not be conducive or necessary to reaching more developed states of the economy, and may indeed be positively harmful even if not currently illegal (such as drugs, solar panels, prostitution, degrading forms of entertainment, gambling, biofuels, and financial speculation).

2. Rather than looking at economic activity per se (which, at its best, GDP might represent), it is necessary to look at economic activity in the context of development overall. Does our measure include

within it the economic powers which we are capable of reaching? Does it measure progress itself?

Instead of a bottom-up approach, Mr. LaRouche has developed a strikingly simple concept to understand an economy as a whole—*potential relative population density*. The *population density* aspect is the familiar measure of the number of people per square kilometer of land. This must be considered *relative* to the quality of the land, and of human improvements to it. With this in mind, we consider the *potential* level of the relative population density: how many people could a society or economy possibly support in a given area of land? What determines this value?

The potential relative population density (PRPD) is bounded by the scientific principles known to a given culture, and by the capability of that culture to implement such discoveries. The breakthroughs in physical chemistry each transformed the PRPD, by their improvements on the potential productive powers of labor, as have agriculture (including irrigation); windmills (centuries ago); the forging of modern science by the work of Filippo Brunelleschi, Nicolaus of Cusa, and Johannes Kepler; the germ theory of disease; vaccines; steam-powered and internal combustion engines; the Bessemer and later oxygen steel production processes; and such production techniques as standardization and automation—just to name a few.

The combined set of discoveries and cultural framework for their implementation determines the PRPD. Rather than adding up *currently occurring* economic activity (including much undesired activity) the PRPD measure indicates the *potential* economic activity and human life an economy is capable of supporting. The rate of growth of PRPD is the best measure of increasing economic value.

Must We Progress? There Are No Limits to Growth

While no intelligent person would deny the necessity of technologies to make the best aspects of developed, modern life possible, some might argue that technological progress need not continue, that we have reached a sufficient level for our needs, and that perhaps increased economic activity even poses a danger, by more rapidly drawing down limited supplies of raw materials. Such were the ostensible concerns behind the publication of *The Limits to Growth* in 1972.

This silly book, which assumed no fundamental advances in technology (such as nuclear fusion), modeled the world economy, and concluded that in a matter of decades, such factors as galloping pollution, resource scarcity, and food shortages would lead to a maximal human population, and then a rapid decline. The authors, who wanted to prevent economic development for reasons outside those discussed in their book proved, in effect, that *without* technological advances, mankind was doomed, and then used this to argue for preventing technological advance! Instead, they had demonstrated the necessity of such advance—a process which must continue indefinitely.

Opposed to this necessary progression is the current, foolish practice of hydraulic fracturing ("fracking") to recover hydrocarbons. Because more easily accessible supplies of hydrocarbons already have been or are currently being exploited, it becomes necessary to expend more and more (physical) effort to obtain the same resources. While an individual fracking well may offer a monetary return on investment, fracking *as a policy* has negative economic value. Consider the opportunity cost, in the broadest sense: it was possible to have built more fission power plants, and invested the necessary resources into making nuclear fusion a reality, giving a whole new spectrum of potential processes and resources. Instead, we are expending more effort to obtain the same resource.

From the moral imperative to improve the living conditions of the unacceptably large portion of humanity currently in poverty, unable to participate mentally in celebrating and advancing the discoveries that are the common patrimony of all mankind, and from a strict physical standpoint, progress is essential.

An increasing (and increasingly well-educated) population is necessary to tackle the large challenges facing mankind, such as defense against errant asteroids and comets, and long-term management of changing weather conditions. In this regard, humans must pick up where the biosphere has left off.

A Lesson from the Biosphere: Development as Fundamental

The "green" ideology holds that most specifically human behaviors are "unnatural," as though the human species is not part of the natural world. Furthermore, many of the supposedly "natural" virtues extolled by green ideologues—tradition, constancy, eternity, stasis, balance—do *not* describe the biosphere, at least not over evolutionary timeframes. Quite the contrary: the history of our planet, and of its biosphere, is one of evolutionary development that mirrors that of human economic development in surprising respects.

For example, let us apply the concept of energy flux density to the biosphere. We will measure the specific metabolic rates of animals and plants, in units of energy flow per body mass (W/kg). For example, a typical reptile uses 0.3 W/kg, while a typical mammal uses 4 W/kg, an order-of-magnitude difference. As seen in **Figure 2**, the development of new biological "technologies" over time—such as seeds, rather than spores, for plants, and endothermy (warm-bloodedness) for animals—corresponds to higher rates of energy flow per body mass. That is, over evolutionary time, newly developed forms of life require increasingly greater rates of energy flow.

This development process did not occur smoothly. As seen in **Figure 3**, the relative predominance of amphibians, reptiles, birds, and mammals over evolutionary time (as measured in the diversity of lifeforms),[3] shows a marked shift from the relative dominance of am-

---

3. This method avoids the difficulty of trying to estimate the total body masses of the different classes, based on relatively scarce fossil remains.

FIGURE 2

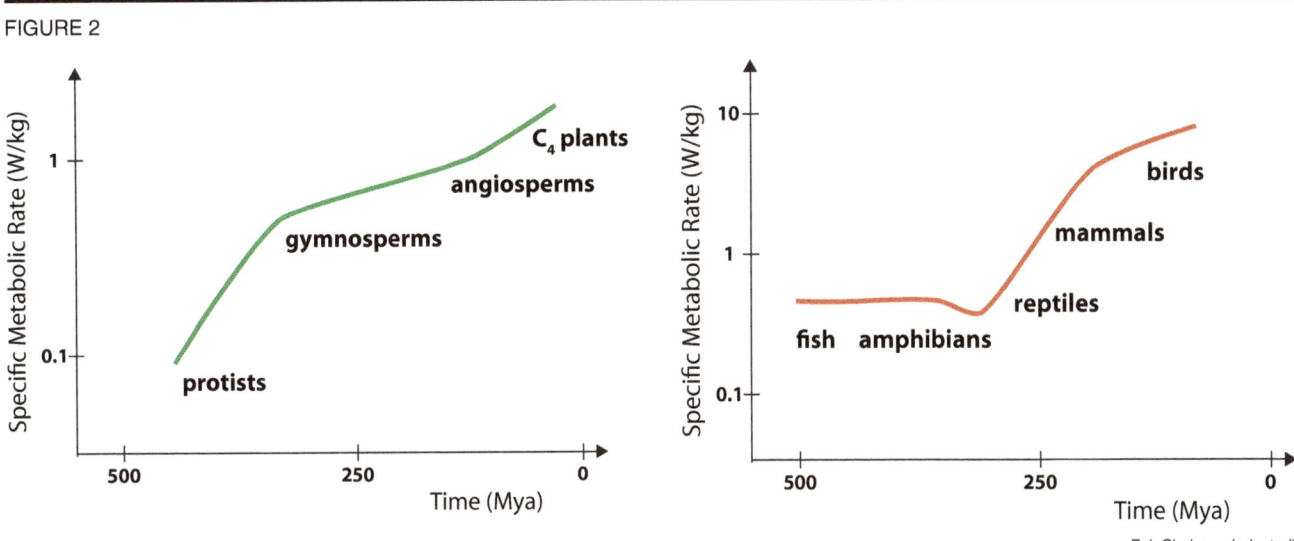

E.J. Chaisson (adapted)

Over time, the rate of energy flow, per g of body mass, for both photosynthesizers and animals, has increased. When this increase is considered in light of specific biological transitions, such as the development of plant seeds, and independence from water and ambient temperature for animals, the transition is understood not as a general increase, but rather one driven by specific improvements in the evolution of life.

FIGURE 3

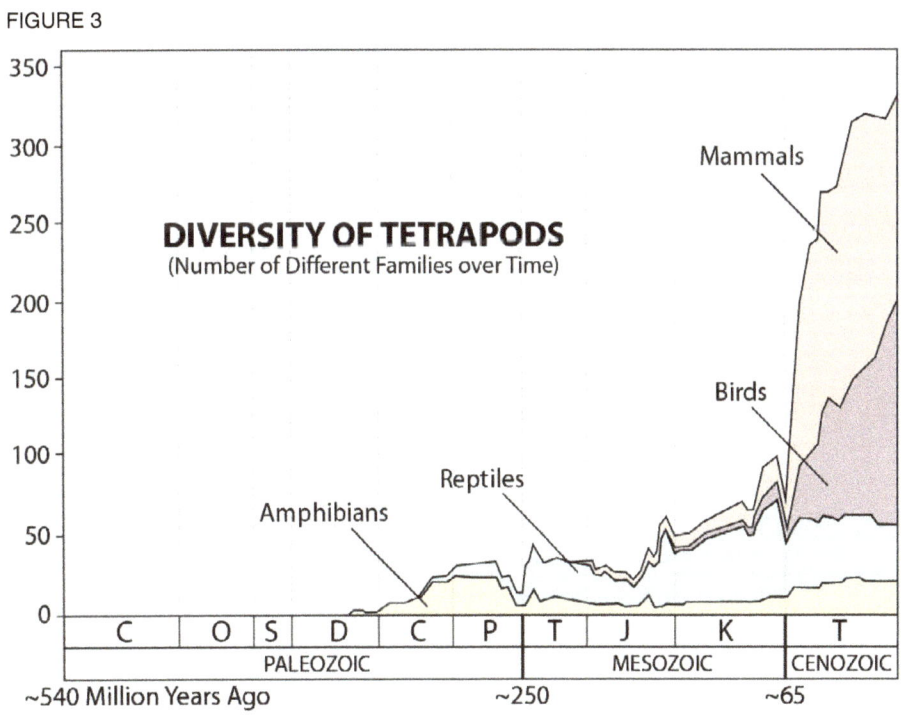

Michael Benton (adapted)

Generalized succession of dominant forms of vertebrates illustrated by the comparative number of known families over geological time.

phibians, then reptiles, and finally birds and mammals. Not only do the developing forms of life themselves have more internal diversity (mammals have more biodiversity than the previously developed class of reptiles), but these changes do not occur gradually, but instead as shifting eras of life, similar to the Stone, Bronze, and Iron Ages of man.

These very cursory[4] examples of changes over evolutionary time reflect human economy in two essential respects: (1) they show increases of energy flux density (and diversity) over time, and (2) these increases are not smooth, typically occurring as almost discontinuous jumps with the introduction of new biological "technologies." This is a remarkable parallel to human economic development, which shows a secular increase in energy use per capita, driven by new technologies which transform that rate quite rapidly.

There can be nothing more "natural" than revolutionary changes in applied technologies, as mankind acts, as one species, in a way that encompasses the biosphere as a whole. All of nature changes—all landscapes, all climates, all forests, and all life. The world is our garden, ours to develop as is best for our human future.

Conclusion

The economic breakthroughs of Lyndon LaRouche allow him (and those who study his work) a greater insight into economic processes and into history. Those who do not wish to consider themselves as historical actors, with a responsibility to cause the continued development of the human species, may not wish to adopt this method, but those who are serious about improving mankind will find much benefit in approaching economics from the standpoint and lessons of this wise man. It has been by adopting his outlook that this report has been assembled.

---

4. For a fuller treatment of the concepts developed in this section, see Benjamin Deniston, "Biospheric Energy Flux Density," in *21st Century Science & Technology*, Spring 2013.

The Principle of the Development Corridor

September 2014

The history of human civilization shows us that economic development and modernization occurred largely along natural and man-made corridors, some of them going back thousands of years to early trading routes, and others formed by major rivers and coastlines. In recent times, man-made canals, roads, and railroads have provided the basis for such corridors, around which cities, industries, and population concentrations have developed. It is a remarkable fact that, as of the mid-1990s, 25% of the entire population of Eurasia, and 70% of its urban population, was concentrated within only three main transport corridors, each approximately 100 km wide, connecting Europe with China. A similar phenomenon is evident in other parts of the world.

The role of the railroad in creating the potential for new development corridors, especially across land-locked interior regions of continents, was demonstrated in an exemplary way with the completion of the Transcontinental Railroad in the United States in 1869. The Russian Trans-Siberian railroad followed in its wake, and simultaneously, there was developed a concentrated grid of railroads across Western Europe, which contributed crucially to its becoming a world center of scientific and technological progress.

The concept of the rail corridor is not, contrary to "common sense," simply to get people or goods from "here to there." This is, indeed, the practice in many parts of the world, where major corporations build rail lines from mines to ports, in order to extract riches for export, often refusing to use any of the wealth created to develop the local populations. In industrialized countries, the same idea applies in developing rail to get people to casinos or resorts. Rather, the corridor should be seen as a center of concentration of industrial, agricultural, and scientific capabilities, and population itself.

Contrary to the prevailing propaganda about "overpopulation" today, it is *lack of sufficient* population and population density in many areas of the world that is limiting development, and keeping people in poverty. This is palpably the case in large parts of Asia, but also Africa, Australia, the Americas, and even Europe itself. The reality is that it takes a certain concentration of population, with certain skills and a

Graphic Representation of a Development Corridor

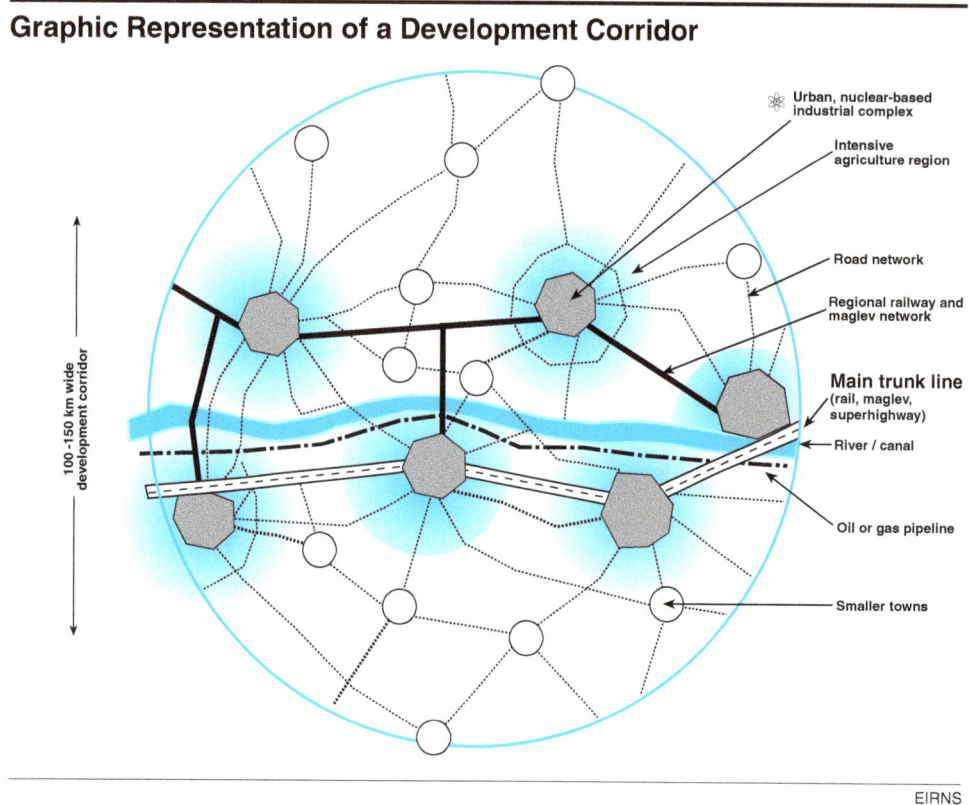

division of labor, to create the basis for raising living standards and the productive powers of labor. Without that concentration, for example, it becomes "too expensive" to develop the basic economic infrastructure—in terms of transportation, energy production and distribution, water supply, communications, education, and health systems—which is needed to advance productive capabilities.

For example, if we compare the estimated average cost per person and per family to provide modern infrastructural services to a unit of population (say 1 million people), living scattered across a large rural area or concentrated in a city, the huge advantage of population concentrations—and of cities themselves—is apparent.

In his economic textbook *So, You Wish to Know All About Economics?* economist Lyndon LaRouche develops the scientific economic concept that underlies this somewhat simple-minded example, with his metric of Potential Relative Population Density (PRPD).

This metric takes us out of the realm of simply "counting noses," and provides a scientific basis for measuring the economy, or productivity, of labor.

LaRouche develops the concept in layers. First, there is the actual population density of an area. But all geographic areas are not equal; they have relative degrees of habitability, depending on the natural condition of the land (say, desert or forest), how the land has been treated by human habitation (improved or depleted), and what kind of technology is available to be applied. These three considerations define the *relative* value of a square kilometer.

Second, there is the usual discrepancy between the number of people that could be supported with the existing level of technology, and the current size of the population. It is the potential number which LaRouche directs his attention to, and which, together with the previous consideration, makes up *potential relative population density*.

It goes beyond that, however, LaRouche emphasizes. The quantity to be measured is not simply the current potential relative population density, but the *rate of increase* of potential relative population density.

A related measure of what is actually the productive power of labor is the *energy flux density* of the power source (discussed in previous chapter).

From this scientific standpoint, at levels of technology either currently available or in sight, the planet could support tens of billions of persons at high standards of living, without overcrowding or any of the horrors environmentalists currently allege threaten us, and while actually enhancing and improving the beauty and usefulness of the Earth.

For purposes of conceptualization, the typical "infrastructure corridor" can be visualized as a continuous strip of land, approximately 100 km wide, centered on a major railroad line. Parallel to the rail line, are high-capacity electric power lines, oil and gas pipelines, water supply systems (including possibly new irrigation canals and aqueducts), fiber-optic communications lines, and so forth. In this way, are created the most essential preconditions for every kind of industrial, mining, agriculture, and urban construction activity within the corridor. From the core can come any number of branches or arteries, supplying surrounding areas.

As new urban centers grow up within the corridors, the eventual result will be a "pearl necklace" effect, of cities and towns, surrounded by regions of intensive agriculture, gardening, forestry, and recreational areas, interspersed with smaller towns and villages.

The economic advantages of such corridor development include vastly increasing the efficiency and economic impact of the infrastructure lines themselves.

A dense fabric of population centers and agricultural and industrial activities located along a rail line actually creates an economic "multiplier," in the sense of increasing the degree of utilization, and thus reducing the per-unit cost. Through this kind of development, a railroad between two points actually takes on the role of a giant "production line," in which value is added to a portion of the goods as they move from one point to the other.

In effect, infrastructure corridors result in a kind of "positive feedback" as a result of the density of population, development, and energy sources. First, the "bundling" of transport, energy, water, communications, and other basic infrastructure along a given route, provides ideal conditions for the growth of a band of intense agriculture, industrial, and population centers along that route. And second, the growth in scale and density of economic activity along the entire length of such a route, greatly increases the efficiency, productivity, and net positive economic effect of infrastructure improvements, as well as every other sort of productive investment. This is in contrast to the contrary approach of oases, or islands, of investment activity.

1791: A New Method of Financing Infrastructure Discovered

By Paul Gallagher and Michael Kirsch

September 2014

I. Preserving a Separated Commercial Banking System: The Glass-Steagall Principle

The issuance of large masses of credits among countries for large-scale and modern new infrastructure platforms requires, first, "Glass-Steagall" bank separation and regulation by the nations involved. Without such legislation urgently soon throughout the trans-Atlantic nations, the major banks are facing another crash. Furthermore, historically, such productivity "driver" projects on a national or global scale have always been financed through national credit. For example, if such credit is issued directly to banks (private or national) that are plugged into securities markets and offshore profit centers, or have large parts of their asset books in high-risk securities and derivatives activities, that is where credit will flow. However, if nationally chartered *commercial banks* have been protected, regulated, and kept out of securities market speculation, those banks will participate in the infrastructure driver projects through vigorous private lending.

The Glass-Steagall principle—strict commercial banking based on mediation of deposits into industrial and commercial, household, and personal loans and leases, supported and regulated by a national bank system—is historically an American development. The United States' first Treasury Secretary, Alexander Hamilton, specified the banks, public and private, which the new government would encourage to form, in his *Report on a National Bank*.

Hamilton defined banks as mediating the investment of otherwise temporarily idle savings into lending to enterprises—and serving the great national purposes of the United States. Whereas the Bank of England was formed fundamentally to lend directly to British governments, Hamilton's Bank of the United States was formed to encourage nascent manufacturing and economic infrastructure. Whereas the merchant banks of Europe were primarily engaged in securities speculations, Hamilton defined U.S. banks as lending to agriculturists, manufacturing enterprises, and households, i.e., commercial banking.

U.S. private commercial banks have fulfilled this role for long

First U.S. Treasury Secretary Alexander Hamilton, shown here in a painting by John Trumbull in 1806, set the standard for American System banking, banking directed to increasing the productive powers of labor.

periods when federally well regulated. The principle of the Glass-Steagall Act was upheld by the U.S. Supreme Court in a landmark 1971 decision (*Camp v. Investment Company Institute*), which held it a proper purpose of Congress to prevent commercial banks from straying from this role, into the lure of securities speculations promising high profits but dangerous to the banks. The U.S. commercial banking sector has in the past proliferated into many thousands of regional and community banks, with—until the past 20 years—no global giants.

The Glass-Steagall Act was enacted June 16, 1933 in the United States after a period in which the largest banks in the country had used their customers' deposits heavily for speculation in securities—including the banks' own securities—and large numbers of banks had thrown the deposit base into stock market speculations such as the infamous Insull electricity utility stock and Morgan railroad stock swindles. After the failure of one-third of all U.S.-based banks by 1933 and government rescue of another third, deposit insurance was introduced together with a strict separation of depository institutions' commercial banking from all other broker-dealer and securities investment activities and companies—"non-banks." The purpose and principle of Glass-Steagall was that investment funds entering the securities markets and their derivatives would have no form of government subsidy or explicit or promised support for their losses, and that commercial banking, which was protected, insured, and allowed government liquidity borrowing, would not be allowed into securities speculation.

The Glass-Steagall Act's regulations basically had four components. First, the requirement that commercial banks, investment banks or broker-dealers/funds and similar entities, and insurance companies (able to underwrite and sell insurance) be entirely separate from one another, and not share directors, ownership, or management. Any commercial bank or bank holding company that has such interconnections must separate completely from them within a reasonable period, usually one year.

Second, the definition of a significant range of securities and derivatives activities as "not sufficiently closely incident to banking as to be proper to it," and therefore not permitted to commercial banks. Third, the provision of Federal deposit insurance exclusively to support commercial banks and their depositors. Fourth, the prohibition against transferring any but AAA securities, within a holding company, onto the books of a Federally insured commercial banking unit, or otherwise causing low-quality securities to be backstopped by government funds.

For more than 60 years after its passage, under Glass-Steagall organization of the commercial banking system, no U.S. bank failure triggered failures or bailouts of other banks.

After the Glass-Steagall Act was progressively eliminated over the course of 1994-99, the effects in U.S. banking were dramatic. The failure of merely a large hedge fund, Long-Term Capital Management, nearly broke the banking system in 1999 because 55 banks had poured leveraged loans into it. The largest banks became impossibly complex, going from typically 1-300 subsidiaries to typically 2,500-4,000 subsidiaries, buying and creating what were overwhelmingly securities and broker-dealer vehicles. The derivatives markets exploded geometrically with the flow from depository giants, from about $70 trillion notional value in 1997 to $700 trillion in 2007 according to the Bank for International Settlements. The largest banks became entirely interconnected with one another, particularly through their mutual derivatives exposures, while their leverage ratios were allowed to rise from typically 16:1 to 30-35:1. Loan/lease assets fell to about half of total assets, while the banks became rapidly larger. The big banks then crashed in 2007-08, saved only by government agency credit extensions to the financial sector—which at one point reached $14 trillion according to the chairman of the U.S. Federal Deposit Insurance Corporation (FDIC) at the time. After being saved, the largest banks' lending fell; the whole banking system's loan/deposit ratio fell to a historically low 70% and the percentage is still today in the low 70s. The largest banks' derivatives exposures are on average 30% larger than they were in 2007. Total bank lending remains below the level of six years earlier. In the EU bank lending is still falling.

Here is the situation today as described by U.S. FDIC Vice Chairman Thomas Hoenig, an advocate of full bank separation on the Glass-Steagall principle, in a May 6, 2014 speech at the Boston Economics Club: "Compared to 2008, the largest financial firms today are in most instances larger, more complicated, and more interconnected. The eight largest banking firms have assets that are the equivalent to 65% of GDP. The average notional value of derivatives for the three largest U.S. banking firms at year-end 2013 exceeded $60 trillion [each], a 30% increase over their level at the start of the crisis.

"The largest banking firms also have tended to increase their complexity. They have used the safety net subsidy to support their expansion across the globe. They have further combined commercial, investment banking, and broker-dealer activities. There have been no fundamental changes in the wholesale funding markets, in the reliance on bank-like money market funds, or in the use of repos, which all are major sources of volatility in times of financial stress.

"While these largest firms highlight that they have added capital to strengthen their balance sheet, they remain excessively leveraged with ratios, on average, of nearly 22 to 1. The remainder of the industry averages below 12 to 1. Thus, the margin for error for the largest, most systemically important financial firms is nearly half of that of other far less systemically important commercial banks and financial firms."

The condition of the largest banks in London and the European Union is much worse than what Mr. Hoenig is describing for the U.S.-based banks. The trans-Atlantic banking system is headed for a general crash despite (and because of) the endless zero-interest-rate money-printing of the central banks.

Against intense opposition from Wall Street and the Obama White House, legislation to restore the Glass-Steagall Act now has bipartisan

support in both Houses of the U.S. Congress: Senate bills S-1282 (prime sponsors Senators Elizabeth Warren, John McCain, Angus King, and Maria Cantwell, with six others) and S-985 (Sen. Thomas Harkin); and House bills HR-129 (prime sponsors Representatives Marcy Kaptur and Walter Jones, with 84 others) and HR-3711 (prime sponsors Reps. John Tierney, Steven Lynch and Walter Jones, with 10 others).

All of the various "alternatives to Glass-Steagall," in which regulators attempt various schemes of "ring-fencing" divisions of banks, have the same fatal disability, and will not produce sound commercial banking. In all the "alternatives," including the much-invoked and disastrously unworkable "bank bail-in" schemes, the large bank holding companies (or whatever agencies try to resolve them into when insolvent) remain responsible for capitalization of all their operating subsidiaries. This capitalization either is taken from the commercial bank division, in violation of the ring-fencing scheme; from a large public taxpayer bail-out in a crisis; or, in the "bail-in" scheme, from both. The "ring-fences" are low ones, and allow holding company senior managements to continue to use deposit bases for securities and derivatives speculation. "Bail-in" simply attempts to expropriate creditors' assets and depositors' money, and besides being chaotic and actually potentially triggering runs on banks, it represents deadly economic austerity.

Only Glass-Steagall separation and regulation of commercial banks provides for depository institutions whose purpose is lending and participation in national banking credit issuance, which are Federally chartered and regulated, and which are barred from any significant exposure either to securities or derivatives markets.

If the Glass-Steagall principle is restored in the U.S. banking system, the Wall Street bank holding companies will have to split off their myriads of investment banks, broker-dealers, and securities investment vehicles, most of which will probably face bankruptcy because they are deep in speculations that require credit backing from Federally insured commercial bank units, and from Federal Reserve money-printing, in order to sustain their high-risk investment models. The commercial banks themselves will have to make loans to business, industry, households, and local government in order to earn profits.

The real economy will lose nothing from those bankruptcies; what will be exposed, is how little real credit there actually is in the economy. A national source of credit will be required to drive the major investments in infrastructure "great projects" and economic productivity. But the separation and protection of the commercial banking system under Glass-Steagall principles will put commercial banks in a position to participate vigorously in national credit creation, including the discounting of their loans to companies and agencies participating in important national and international projects.

II. Generating National Credit for Productivity Projects: Examples from the American Credit System 1652-1945

The U.S. national credit system, called the "American System" in the 19th Century, but essentially abandoned by American leaders since the end of World War II, facilitated infrastructure and industrial development in each historical period of nation building. The period 1865-90 in which the United States emerged as the world's leading industrial nation, followed the Abraham Lincoln Administration's adoption of "Greenback" national credit issuance; investment of that credit into rail, steel, coal, and agricultural infrastructure; and strong tariff protection of national industries—the three fundamental principles of the Hamiltonian, or "American System."

EIR historian Anton Chaitkin has shown ("Leibniz, Gauss Shaped America's Science Successes," *EIR*, Feb. 9, 1996) that in fact *each* surge in industrial growth and scientific/technological revolution in U.S. history, has been associated directly with the implementation of those principles by American governments. Senator and Secretary of State James G. Blaine's two-volume history of the United States in the 19th

Century 125 years ago demonstrated exactly the same conclusion.

The establishment of the fully sovereign nation-state of the United States in 1789-96, and the establishment of the credit system, were one and the same. Alexander Hamilton, U.S. Treasury Secretary during George Washington's Presidency, created a system that closely coordinated the relationship of public credit with the intention of national government to follow through on the actions for which the credit is emitted. The more the nation's government is committed to see through the creation of credit, and properly exercises its complementary necessary powers of finance, trade regulation, and taxation, the better certainty there is on that credit, and the more is credit between parties able to serve as a currency and means of payment.

John Winthrop Jr.'s Plan 1663-1681

The Massachusetts Bay Colony in the 17th Century created the first currency sufficient for payment and trade, in opposition to its British royal governor. The "lack of a medium of exchange" with which to make the needed transactions for building up the economy of Massachusetts Bay and the early Colonies was a constant refrain. From John Winthrop Jr. in 1663 to Benjamin Franklin in 1729, and after, numerous authors wrote of the currency problem. Sufficient currency increases manufacturers, trade, immigration, and foreign returns; keeps interest low; and leads to general improvement. A shortage of currency increases debts, prices, and interests, while property declines in value and trade is stinted.

In 1652, the Massachusetts Bay Colony coined its own money, the "Pine Tree Shilling," to create a sufficient currency with which to trade amongst themselves. This was attacked by the King multiple times in the 1660s-1680s.

The Pine Tree Shilling, shown here, was created by the Massachusetts Bay Colony to promote physical economic development.

In the same period, there were multiple designs by the Winthrop family and others for a means of payment that did not require silver or gold, but would be based on credit. In 1663, John Winthrop Jr. wrote a plan entitled "Some Proposals Concerning the Way of Trade and Banks without Money," and sent it to the Royal Society. Winthrop wrote that his plan would "greatly advance commerce and other public concernment for the benefit of poor and rich," and would answer all the ends which "banks of ready money" in other parts of the world attained. While it would involve land, he wrote that it would not take the land out of use.

The 1686 Bank of Credit

Drawing on these earlier writings and attempts, a more developed plan for a "Bank of Credit" was approved by Wait Winthrop, Adam Winthrop, and other leaders of Boston in 1686. The details of the bank plan were written out by John Blackwell a year later.

"Bank-bills of Credit," signed by several people "with good repute" and emitted on the basis of the mortgages of lands or goods, would be voluntarily accepted by people and business as "ready moneys." The bills would have "at least equal advantages with the current money or coin, of any country." There was no gold or silver reserve in the Bank.

Those who had real wealth and capital could now turn it into a source of credit, instead of requiring money. Weavers could pledge their mills for bills of credit with which to increase their supply of wool; merchants could pledge their land and receive bills to buy additional wares and other commodities from the manufacturer; shopkeepers could mortgage their shop and receiving bills to buy goods from merchants; a mine owner could pledge his mine for bills to obtain additional capital to employ laborers to work the mine. The mine owner could pay interest on the bills in iron, and other tradesman could pay similarly with the value of their goods. The 1687 document presenting the plan concludes:

> By [the Bank], the trade and wealth of this country [will be] established upon its own foundation, and upon a medium or balance arising within itself, viz., the lands and products of this country; and not upon the importation of gold or silver or the scarcity or plenty of them, or of anything else from foreign nations, which may be withheld, prohibited or enhanced, at their pleasures.
>
> Our own native commodities will thus become improved to a sufficiency for our own use (at least) and thereby afford a comfortable subsistence to many ingenious and industrious persons amongst us, who know not at present how to subsist: and this will draw over more inhabitants and planters. It will not be in the power of any, by extortion and oppression, to make a prey of the necessitous.

The Bank of Credit of 1686 was not fully established due to the influence of the King's representative Edmund Andros and the takeover in England by William of Orange.

Cotton Mather 1690-1720

In 1690, Bills of Credit were emitted in Massachusetts as a means to pay soldiers and for supplies. Cotton Mather described in a paper entitled "Some Considerations of the Bills of Credit, Now Passing in New England," that although the colony did not have silver, they could have credit, which would allow the colonists to buy articles just as readily if they were to accept them. Mather wrote that the security of the paper money was "nothing less than the credit of the whole country." The country makes good the credit through its industry, and its inhabitants are "the security of their public bonds."

The taxes later collected were payable in the bills of credit emitted by the colony, and this cycle is essential for issuances of sound currency. "The Credit conveyed by these Bills now Circulates from one hand to another as men's dealings are, until the Publick Taxes call for it." Then the government could put it back into circulation again.

While these bills of credit were first issued for emergency endeavors, such as the Colony's military campaign of 1690, they were later used for general economic purposes—thus similar to the Lincoln Administration's and Congress' "Greenback" issuances of 1861-65.

This is the reconstruction of the forge and iron mill at the Saugus Iron Works in Massachusetts, which in a matter of years in the 1660s became more productive than iron producers in England.

In 1716, an unnamed author, probably associated with Cotton Mather, proposed a Greenback policy in "Some Considerations on Several Sorts of Banks." The author recommended the government should "emit large sums" for "what may be beneficial and of a general good," specifically, infrastructure and industry. The sums were not only to supply the ongoing scarcity of cash, but would "also lay certain and stable foundations of increasing the produce of the country; which is the interest and wisdom of all nations."

The government would pay on credit to a board of trade to lend for "construction of public works and encouragement of industries." They proposed "lending large sums upon good security, without interest for some term of years" to pay for a bridge and cutting a canal for more speedy passage of vessels. A few hundred committed by the government to set up an iron refinery would save the country thousands in a year, the author wrote. The government was expected to take actions to increase the power of the colony. As earlier, the bills would later be accepted to pay taxes to the government.

As in the 1680s, this bank, and similar ones proposed in 1720 and in 1740, were prevented by opposition from the British crown. However, after 1690, bills of credit continued to be issued throughout the colonies. During 1710-1740 the government of Rhode Island successfully "emitted bills of credit to supply the merchants with a medium of exchange, always proportioned to the increase of their commerce," putting their navigation industry ahead of the other colonies.[1] Some of the attempts at currency issued by the governments were successful, while others were failures, depending on how regulated they were, and for what purposes.

Multiple attempts, in 1741, 1751, and 1764, were made by the crown to end the use of all colonial bills of credit. Benjamin Franklin told the parliament in England in 1764 in response that "colonial legislatures [must] be empowered to issue any amount of paper money required for revenue, trade, business, agriculture, to be lent on collateral security, deficiencies guarded against by taxes, and interest on the loans to be used in meeting current expenses."

The 1781 National Bank

Alexander Hamilton addressed the financier of the Continental Congress, Robert Morris, writing, "Tis by introducing order into our finances—by restoring public credit—not by gaining battles that we are finally to gain our object."

Hamilton hypothesized that the solution to the economic crisis of the colonies lay in uniting the influence and interests of individuals in trade, commerce, and industry with the resources and credit of the

---

1. *Records of the Colony of Rhode Island*; Providence, 1860, p. 12.

Hamilton succeeded in establishing the Bank of the United States, which converted the mass of Revolutionary War debt into a base of credit for the prosperity of the nation. Here, a drawing of the First Bank of the United States, which was located in Philadelphia.

government, by the joint subscription to a national bank. The result was the "Bank of North America," chartered in 1781. The alternative to the depreciated continentals it presented and the national unity it signified saved the finances of the country and upheld the credit of the Congress through to the end of the war in 1783.

Hamilton demonstrated a central principle in successful national banking, that no credit currency could be substantial, or durable, which does not unite the resources and growth of the real economy with its establishment and circulation. However, the lack of union of the states and insufficient powers of the Congress did not provide the Bank with proper funding to establish a national economy or uphold federal credit. Without the powers to regulate trade, impose federal taxes, regulate the currency, and coordinate the payments of the debts, there could be no secure funds to establish credit, increase national productivity, or fund the National Bank.

While it was not fully successful, for reasons specified, the bank demonstrated an important principle. What had been defeated throughout the preceding century in 1686, 1720, and 1740 for lack of authority and independence from the crown, was now possible: a sufficient payment system based on the productive capacity of the nation, not determined by artificial restraint.

The System of Public Credit

In 1789, from the standpoint of the gold and silver that had been borrowed for the war, the new republic was bankrupt and had no possible way within the existing system to settle its accounts. By employing the powers of Congress won through the new Constitution, Hamilton implemented the system of public credit he had been developing for a decade.

As his first step, he transformed the seemingly impossible foreign, national, and state debts, and the interest rates on them, into a means to unify the resources of the nation toward one goal through the power of federal revenues. The foreign debt would be refinanced—restructured—and the state debts would be assumed and united with the domestic debt, reissued, and subsequently restructured as new debts of a National Bank. However, in accord with Hamilton's "fundamental maxim, in the system of public credit of the United States," in the same Act of Congress that created these newly transformed debts, the means of extinguishment were built in, initiating the powers of Congress related to economic growth.

Hamilton's recommendations on public credit were passed into law in three Acts of Congress, on August 4, 9, and 12, 1790. New loans for the full amount of the domestic and state debts were taken out, with subscriptions to them made with the old debt certificates. The subscribers received new U.S. debt certificates, with an interest payment on them guaranteed with a permanent appropriation, written into the same August 4 Act that authorized the new loan and state assumption. The funds—Hamilton's "means of extinguishment" of the loans—would come from new protective tariffs and excise taxes passed immediately thereafter. A special fund from the revenues of the new U.S. Post Office was set aside to prevent speculation on the public debt and increase its value.

By ensuring the value of the public debt certificates, they were converted into a real medium of commerce, a vast capital for trade, and basis for a new credit currency, giving life and activity to business. The *funded* debt was now a representation of the new power of government in action, and its value rose from $15 million to $45 million by the end of 1790. Hamilton's actions effectively created a capital resource of $30 million for the economy.

Its value would continue to rise as the strength of the nation's economy increased, and likewise, the increase of the economy was made possible by the creation of the new source of capital, in the form of the funded debt.

The Public Credit Bank of the United States

The step that cemented the credit system was the Bank of the United States, chartered by Congress in 1791 according to Hamilton's next recommendations.

While not circulating as money itself, the capital that Hamilton had created by the funded debt served as the basis for a national currency. Now it could be used to form an enormous (for the time) capital of a national bank, large enough to serve the credit needs of the entire economy. Those who had received new certificates of the public debt could use them to become subscribers to the capital stock. Each share was bought with one part specie, three parts public debt. The government went $2 million further into debt to increase the capital of the bank by one-fifth, which created a sufficiently large circulation. The bank was able to issue notes up to its capital of $10 million, which consisted mostly of the public debt of the United States.

The Bank's main economic functions were the creation of a medium of exchange in which credit could be transferred between parties in commerce, manufactures, agriculture, and industry, and also, directly lending its credit for the same purposes, including economic infrastructure.

The Bank increased the value of the public debt. The act of subscribing to the Bank's capital with public debt securities, increased their value, and the assurance and facilities it provided to the government increased them still further.

Hamilton's Bank was antithetical to the Bank of England, because the National Bank was prohibited from buying and selling public debt, which was the Bank of England's main purpose; also, while the Bank of England's commercial role was secondary, commercial lending was the main function of the Bank of the United States. Its other functions

A drawing of work building the Erie Canal, in the period of the functioning of the Second Bank of the United States.

and benefits included serving as a depository for government revenues, which provided an additional source of credit available at all times until appropriation; creating a unified medium in which taxes could be dependably collected, and enhancing that dependability by loaning to taxpayers in lieu of their possession of money, as in the case of importers; and relieving the nation of the cost of a currency of fluctuating value between states. Hamilton summarized the effects of the system in his final *Report on Public Credit* in 1795:

> Public Credit ... is among the principal engines of useful enterprise and internal improvement. As a substitute for capital, it is little less useful than gold or silver, in agriculture, in commerce, in the manufacturing and mechanic arts.... One man wishes to take up and cultivate a piece of land; he purchases upon credit, and, in time, pays the purchase money out of the produce of the soil improved by his labor. Another sets up in trade; in the credit founded upon a fair character, he seeks, and often finds, the means of becoming, at length, a wealthy merchant. A third commences business as manufacturer or mechanic, with skill, but without money. It is by credit that he is enabled to procure the tools, the materials, and even the subsistence of which he stands in need, until his industry has supplied him with capital; and, even then, he derives, from an established and increased credit, the means of extending his undertakings.

The Bank of the United States credit system put private credit in its proper place, inducing investors to serve the ends of public interest. The Bank's capital was four-fifths subscribed to and owned by private citizens, holders of the restructured United States debt, and it was also privately directed; however, its private status was a means to keep it sound, and was secondary to its nature. By its purpose and functions, its main beneficiary was the nation as a whole. As Hamilton explicitly states in his *Report on the National Bank,* when speaking of the old constitution of the 1781 Bank of North America:

> The interest and accommodation of the public ... are made more subservient to the interest ... of the Stockholders, than they ought to be. It is true, that unless the latter be consulted, there can be no bank ... but it does not follow, that this alone to be consulted, or that it even ought to be paramount. Public utility is more truly the object of public banks, than private profit. And it is the business of Government, to constitute them on such principles, that while the latter will result, in a sufficient degree, to afford competent motives to engage them, the former be not made subservient to it.

The bank notes now issued by the Bank as currency, were made a legal tender and "receivable in all payments to the United States," and could be redeemed for specie if desired, "payable on demand, in gold and silver coin." Because the system was designed to prevent the necessity for this redemption, a circulating currency was created of a magnitude proportional to the active capital of the country, e.g., manufactures, agriculture, and commerce. There was no need to trade in that capital for specie, in order to exchange goods.

Hamilton redefined the meaning of debt within a functional nation-state economy. Rather than a monetary burden to be settled in saved revenues, and contributing to scarcity, the public debt was made an asset, and signified the process of unifying the resources of the national economy.

And as the power of the productive economy grew, particularly in industry, so, reciprocally, did the Bank's value of capital and the general value of the public debt. All of this would be an increased capability of lending for commerce, and increasing the means of payment in the economy available for trade.

The Credit System Elaborated

After the National Bank's charter was allowed to lapse in 1811 under Jeffersonian influences, Hamilton's credit system was revived by Nicholas Biddle, Mathew Carey, and President John Quincy Adams with the Second Bank of the United States, particularly during 1823-36. Under Hamilton and Biddle, who was chairman of the Second Bank, the system was managed to increase the number of transactions occurring on credit rather than liquidating wealth for the present. Transactions were settled by the future resources generated, which gave a credit to the initial borrower. Credits and debts were coordinated according to the cycles of production to defray the time of payment, till each party had sufficient credit to balance their debts. This allowed productive surpluses to be absorbed into future growth and productive investment.

The Bank directly intervened into the economy, not by upholding inflated securities, but by assisting the productive economy or the needed infrastructure projects with capital, in order to maintain the surplus productive capacity.

The protection for manufacturing and support for internal improvements that Hamilton had called for did not come about until the 1820s, under President John Quincy Adams. The major canals, new railroads, and new industries were made possible by federal credit and direct loans and other indirect functions of the Bank. Adams used the nation's stock in the Bank for financing large projects, and under Biddle's direction the Second Bank of the United States loaned and subscribed directly for nearly 50% of all the capital raised to construct the largest canals, which made possible the transport of anthracite coal for the iron industry.

Under the regulation of the credit system under Biddle, the currency bore a proper relation to the real business and exchanges of the country. As more agricultural land was developed, as more manufacturing facilities became established, and as more transportation networks for produce and coal for manufacturing facilities were completed, the amount of bank credit that could safely be put into circulation through loans and discounts increased in proportion.

Lincoln's System of Public Credit

New York banks and the British East India Company opposed the Second Bank of the United States and the internal improvements and domestic manufactures it facilitated. Those interests were politically successful in taking down the U.S. credit system by means of the Jackson and Van Buren administrations. However, Abraham Lincoln, a longtime supporter and advocate of the system of national credit of John Quincy Adams and Alexander Hamilton, revived this system when he began his Presidency.

The first step that Lincoln took was the passage of a strong tariff, the second Morrill Tariff, in the summer of 1861. Hamilton had established that protection of manufacturing is essential for a sound banking and credit system, not only because it generated revenue (customs duties) to fund and support national credit, but also because the specie that was kept on reserve could not be maintained when the country ran a negative trade balance, because imports had to be paid in specie.

Lincoln's next measure, the policy and issue of "greenbacks," then created the medium to revive and accelerate the machine of domestic production and commerce.

At the end of 1861, after buying (at a very high interest rate) an initial round of U.S. Treasury bonds to get the Union mobilized for the Civil War, New York bankers blocked with British and French lenders to stop all revenue streams to the Treasury. These banks suspended payments of gold owed to those who had made deposits in their banks, ceased their purchase or acceptance of government bonds, and blocked foreign loans. The government responded by taking control of the currency, and issued its own U.S. Treasury notes—"greenbacks"— as a circulating medium of payment necessary for commerce and war. The Legal Tender Act, February 25, 1862, read, "To authorize the is-

Under President Lincoln's de facto national banking system, the U.S. built up its infrastructure and industry enormously. One of those major accomplishments was the Transcontinental Railroad, which wasn't finished until 1869. Here, the train carrying one of the principals to the Golden Spike ceremony.

sue of United States Notes and for the redemption or funding thereof, and for funding the floating debt of the United States." Despite widespread doubts in Congress, even in Lincoln's Republican Party, the greenback credit-issue policy was as successful as the Hamiltonian national bank policy on which it was based.

Almost one-half of the circulating currency became greenbacks. The Lincoln Administration increased government spending by 300% by creating $460 million in greenbacks during the Civil War. This legal tender was used, in the first instance, by the Treasury to pay soldiers, contractors, teamsters, manufacturers of weapons and uniforms, farmers, etc. Greenbacks could be used by investors (along with state banks' notes) to purchase bonds sold by the Treasury. From October 1862 to January 1864 the Treasury Department oversaw the selling of more than $500 million in bonds to individual citizens, enough to finance the greenbacks that it issued. And the greenbacks were used to pay the war taxes on imports, industry, and high (more than $800 per year) incomes.

The bonds sold were largely part of the next action taken by the Lincoln Administration, the National Currency and Banking Acts of 1863 and 1864, which, united with the greenbacks measure, and a national funding system, built a system of national banks on the same principle of Hamilton's Bank of the United States. State banks were rechartered as national banks on the basis of the requirement "to purchase United States stocks to hold as securities for their circulating notes."[2] The U.S. bonds purchased by the banks were deposited in the Treasury, and the newly chartered national banks received greenbacks in return, upon which to lend.

Just as the Bank of the United States and its branches had had a large portion of its capital stock in the form of public debt, under Lincoln's Presidency greenbacks and bank notes now circulated on the basis of the public debt, which the nationally regulated private banks purchased and held in the Treasury. The United States bonds, upon which the greenbacks were issued to national banks for lending, were 20-year annuity bonds, paying a dependable interest, but which were not tradable and were callable only by the government prior to their maturity. As with Hamilton, it was the strict regulation of the terms of the public debt by the government that made the credit which circulated upon that debt a reliable medium for growth.

The greenbacks were safely leveraged on the basis of the 20-year bonds, which were held as security, and which themselves were funded by tariffs and taxes. Import duties far exceeded the interest to be paid out on the bonds, in specie. This surplus specie would be a source to redeem any greenbacks or fund other bond issues.

---

2. Wesley Mitchell, *The History of Greenbacks*, 1903, University of Chicago.

Lincoln economist Henry Carey described the similarity of Hamilton and Lincoln's systems, stating, "The U.S. Bank [of Hamilton] did not give us specie, [rather] its notes were current almost on the same fundamental hypothesis, which has given useful circulation to the Legal Tender issues [of Lincoln]."[3]

Following Hamilton's maxim for public credit, Lincoln's Treasury Secretary Salmon Chase funded the public debt and maintained the value of greenbacks through import duties and by implementing the greatest array of internal revenue duties in the nation's history to that point, through an act in 1864 titled, "To provide internal revenue to support the government and to pay interest on the public debt."

Lincoln's issue of Treasury notes as currency had been advocated by Hamilton as an addition to National Bank notes, but on a smaller scale, in 1798. In a letter to Treasury Secretary Oliver Wolcott, Hamilton cited the difficulty in collecting taxes under a "defective circulation" and the unreliability of sources of loans from banks alone. To keep the circulation full and to "facilitate the anticipations which government" will need on occasion, he said he had "come to the conclusion that our Treasury ought to raise up a circulation of its own ... by the issuing of Treasury notes payable, some on demand, others at different periods from very short to pretty considerable—at first having but little time to run."

After the Civil War and Lincoln's death, Lincoln's economic advisor Henry Carey, of the Philadelphia group of leading industrializers, made clear in numerous writings that the greenback issues had launched a great rate of industrial progress in the United States. But Carey warned the Treasury's contraction of greenback circulation from 1866 onwards was the wrong direction for U.S. national credit. Rather, Carey held that with a dozen states reincorporated into the Union and the nation expanding to the west, the greenback issue should have been expanded much beyond the $460 million circulated during the war. Carey described the greenbacks as a "non-exportable" and reliable internal source of credit which was debt-free for its domestic users. With the greenback circulation instead contracted to $330 million by the end of 1867, American businessmen, farmers, and artisans became more dependent on greater amounts of debt, and the United States' general industrial expansion again became dependent for credit on European banking centers and on the use of gold. When the United States "resumed specie currency" in 1879, Americans kept their greenbacks and turned almost none in for gold certificates, proving Carey right that their quantity was much too small to meet the demand for circulating credit. Three decades later, in the debt crisis and panic of 1907, President Theodore Roosevelt considered expanding greenback circulation with a large new issue; he hesitated, however, and let Wall Street bankers take the initiative from him with the 1908 Aldrich Act, allowing private banks to issue "U.S." currency and leading to the Federal Reserve System five years later. The U.S. Treasury has not issued national credit since.

---

3 Henry Carey to Treasury Secretary McCulloch, December 1868.

FDR's recovery plan depended heavily on infrastructure construction, financed by such agencies as the TVA and the Reconstruction Finance Corporation. Here, construction work at the TVA's Douglas Dam in Tennessee, June 1942.

Franklin Roosevelt's RFC

Franklin Roosevelt's makeshift national bank took the form of an expanded Reconstruction Finance Corporation (RFC), which famously loaned $50 billion to every sector of economic activity between 1934 and 1955.

The RFC approximated the Hamiltonian credit system with great success throughout the 1930s, where any corporation, industry, or agriculturalist possessing a productive character was able to obtain credit on reasonable assurance of the loan being repaid, at the discretion of the lender. Growth occurred in a structured way because the process of making good on the credit depended on the productivity increases achieved.

Industry and agriculture were saved from unnecessary bankruptcy, and skilled labor and much needed national enterprises were maintained. Instead of allowing prices to be determined by the random interaction of production cycles or the manipulation of Wall Street, the credit of the RFC offset the economic cycles of the private financial sector.

The RFC operated separately from the authorizations and appropriations of the Federal budget, borrowing from the U.S. Treasury according to limits set by Congress. All loans made through the RFC, as loans, and not appropriations, were repaid, not only with a financial profit to the Treasury, but more importantly, with a productivity increase for the nation as a whole not measurable in dollars, not to mention the profit savings in human and productive capital that would have been lost had the loans not been made.

Under Franklin Roosevelt, the RFC was the embodiment of directed credit and operated almost exactly as the Banks of the United States had under Nicholas Biddle and Alexander Hamilton, increasing the overall indirect and direct long-term credit in the economy, itself directly lending to the economy on non-restrictive terms. The striking differences were that it was not the chief depository institution for United States tax revenues, and thus could not lend them out as a source of credit to banks, industries, and other corporations, as had the Bank of the United States. It also did not receive private subscriptions to its capital stock. The RFC was acting in an environment which included the structure of the Federal Reserve Banks, and therefore was not as efficient as the Bank of the United States, which was acting as the chief institution and the key mover in the banking system.

President Roosevelt's 1934 proposal to create national credit banks for industry, directly *within* the Federal Reserve System, and which would act as depositories for U.S. tax revenues, was blocked in the Congress.

III. International Credit Agreements for Development

The recent critical emergence of two new international development banks for non-austerity-conditioned, infrastructure-specific lending—the BRICS New Development Bank and the Asian Infrastructure Investment Bank (AIIB) initiated by China—open up potentials for credit agreements not seen since the Bretton Woods Conference. The critical great projects or "infrastructure platforms" proposed here require cooperation among several nations, including credit cooperation among the major economic powers providing the bulk of capital goods and industrial products for these projects—but *not supra*national direction. The United States and European economic powers led by Germany easily can, and need to, participate in expanding these banks toward the trillions of dollars-equivalent in new infrastructure credits actually required immediately. But they must give up their "green" hostility to the most productive scientific advances and technologies, in order to do so.

The example of the Bering Strait Tunnel crossing and high-speed rail linkage of Eurasia and North America, now seen as increasingly urgent by China and Russia in particular, or the large-scale water-management breakthrough necessary to stop desertification of western North America, illustrate the general principle. The agreements among the countries involved on joint funds or agencies to carry out these great projects, require agreement on issuing credits over the long term and at low rates of interest. Moreover, these nations remain sovereigns with their own national credit systems, so that the long-term credits are required in several currencies with relatively stable parities over the long term, together with currency-swap arrangements among central banks. A current negative example of this requirement is the serious disruption of trade and development projects in Kazakhstan due to the abrupt drop of the Russian ruble's value in 2014 under increasing sanctions.

Over a period now of more than three decades, economist Lyndon LaRouche and his associates have proposed a return to a New Bretton Woods system of agreements that would return to the credit, currency, and banking arrangements among nations of the post-War period, as exemplified by the credit relationship between the United States with its Marshall Plan and Germany with its reconstruction refinancing institution, the Kreditanstalt für Wiederaufbau (KfW).

The grant and loan aid centered in the Marshall Plan, while brief (1947-51) and small (roughly $125 billion in current-dollar terms), had a relatively powerful impact on post-War European recovery and development because it was firmly embedded in the anti-speculative Bretton Woods system. The aid was in the form of (1) dollar credits, which due to capital controls were *not* re-exported to pay European countries' war and other foreign debts (despite attempts by Great Britain to break these controls and do just that); (2) goods, particularly capital goods, representing capital goods credit and investment within the United States, and which were paid for in

The financing of the "German miracle" of the post-war era followed along the lines of FDR's Reconstruction Finance Corporation, using an institution called the Credit Bank for Reconstruction (KfW), oriented to real physical production.

marks or other European national currencies; and (3) direct dollar aid, used for purchases such as imported construction materials, capital goods, and food. There was no attempt to "integrate the nations back into international capital markets," which would have triggered capital flight and rapid devaluations. The European nations "paid for" the goods and loans by creating equivalent "matching" credit funds in their own currencies, used to generate increasing internal national development credits (the KfW being by far the most successful, high-impact, and long-lasting in this policy). The European Cooperation Agency, which served as a small international development bank under the European Recovery Program (Marshall Plan), was dissolved in 1958, and by that time all the European nations were integrated into the Bretton Woods system; their currencies were convertible at fixed rates. There was no significant use of dollars by these countries except for purchasing U.S. exports and settling trade imbalances; bank accounts in foreign countries' currencies were prohibited under Bretton Woods except for trade purposes.

With imperfections, the principle of international exchange of development credits was there. The KfW played the same internal development-credit role in Germany, relative to credit initially generated from the United States, as Alexander Hamilton's first United States Bank had played for U.S. development, relative to the European banks which heavily invested in Treasury Secretary Hamilton's Bank in 1791. Hamilton's design of the Bank, its sinking funds, and the new tax revenue which supported it, prevented its invested capital from flowing immediately back out to pay the relatively huge debts of the then-bankrupt United States, and directed it instead, into development including of canals, roads, and iron industries. In Henry Carey's phrase, the circulating currency created by Hamilton's bank was "non-exportable," and so was the credit created in Germany by the KfW in the post-World War II period.

Benjamin Franklin and Alexander Hamilton, and later Mathew and Henry Carey, explicitly insisted on *protection* as a feature of national banking, to prevent the newly invested capital of the bank from being rapidly dissipated. For example, without regulations to protect manufacturing and thereby reduce imports, which require payment in real money (then specie, today dollars), the strain on the national bank and its branches for such payment will break the system.

The Bretton Woods system was broken up fundamentally under the impact of the Eurodollar markets, which first appeared in the later 1950s in the form of London (and offshore London) banks

The destruction of the Bretton Woods System in August 1971, by the likes of officials like George Shultz (shown above), went a long way to destroying the basis for financing long-term development globally.

creating accounts for U.S. dollars which paid significantly elevated interest rates, accounts not for trade but for purposes of investment in the international securities markets, sovereign debt markets, and later, for *foreign exchange speculation*. This was allowed by regulators to occur and expand exponentially. By 1979 two-thirds of all U.S. dollars were circulating outside the U.S. economy—"Eurodollars," "petrodollars," etc.—and the resulting inflation had detached the dollar from the gold-reserve basis and broken the Bretton Woods system of fixed currency rates. The resulting "floating exchange rate" regime also seriously negatively impacted the International Bank for Reconstruction and Development (IBRD) (i.e., the World Bank) as a credit mechanism for development, because the capital contributions and the loans of that Bank were overwhelmingly in U.S. dollars and loans had to be repaid in currencies usually devaluing against the dollar.

The United States, China, Russia, and Japan all possess the ability to issue national credit and currency in large amount for development purposes—the United States because of its large, funded, and universally accepted debt that can be converted to development credit by creating a national bank for large projects; the other three nations because they possess large net foreign exchange reserves on which to base national credit issuance through government banks. The new international development banks provide the starting vehicles. China already does this; it has accumulated more than $3.5 trillion in foreign reserves through trade and issued *a multiple of this* in currency emission through state banks since 2007. If a fraction of this emission has fostered real estate and commodity bubbles (aided and abetted by major British and Hong Kong banks and other financial firms), the great majority has created infrastructure, productivity, and growth. If linked to the emissions of other great powers' national banks for specific great projects, China's national development credit will be safer from the speculative obsessions of the world's (particularly London's and Hong Kong's) investment banks and hedge funds. The United States Congress, in any given month, can create a *Third* U.S. National Bank with $1 trillion capital or more, capitalized by holders of United States Treasury debt securities investing them in such a bank in exchange for stock or long-term debentures of the Bank; and issue international project credits through this Bank. Or, the United States can issue a comparable sum of Treasury notes ("Greenback" currency), backed by special long-term and non-callable Treasury bond issues, for the same purpose of international project credit.

The fifth great economic power, India, has created its India Overseas Investment Corp (INOIC) on the lines of a sovereign wealth fund to lend financial muscle for securing access to overseas natural resources. INOIC will not, however, be India's sovereign wealth fund in the conventional sense. It will be patterned on the government's holding arm and registered with the Reserve Bank of India as a non-banking financial institution.

The company will raise funds through rupee bonds of 15-20 years with sovereign guarantee. State-run entities, banks, and financial institutions will subscribe to these papers using their surplus funds. Sovereign guarantee will allow the interest rate to be set marginally higher than government securities. The bonds can also be made part of banks' statutory liquidity ratio (or minimum cash that banks have to keep overnight) to help them subscribe. INOIC will not borrow from the Reserve Bank of India.

India thus can be ready to participate in the BRICS New Development Bank and the new AIIB, as it develops its export capabilities particularly with the other Asian nations.

For an International Development Bank (IDB) to be capable of driving the great projects discussed in this report, some among these great Eurasian economic powers, hopefully in cooperation with the United States, must issue credits in their own currencies to capitalize an IDB, created by treaty, with several trillion dollars equivalent in capital, so that it becomes the ultimate funder and initiator of investments in the great projects.

One or more sovereign wealth funds of other nations may also invest capital in the IDB, but the credit issued to it by the cooperating economic powers must define how it is capitalized—by 20- to 30-year debenture investments of an "annuity" type, paying a dividend but callable only by the Bank itself should it decide to reduce its capital for any reason or to accept other investors. This is the same principle on which national credit banks, able to invest in the IDB, will be created by the United States or other investing nations, insofar as their credit for investment is not created on the basis of trade surpluses and foreign reserves.

In making equally long-term loans for the development of projects in individual nations, the IDB will book a credit with the national development bank of the nation involved, which will use that as the basis to issue credit in its own currency to authorities and enterprises carrying out the work. By design of the national development banks in the borrowing nations, and by capital controls, this currency too must be "non-exportable" except for trade.

The borrowing nations must establish not only capital controls, but more importantly exchange controls, to ensure that no IDB credits are diverted to flight capital or "carry trade" securities investments, and that their use for development projects pre-empts any attempted use for repayment of other sovereign debts of countries receiving credits.

Furthermore, it is necessary to the effectiveness of the IDB's development credit issuance that over-indebted nations with sovereign debts which have been imposed on them illegitimately, in whole or in part, be able to place the illegitimate debt in moratorium, replacing it with much longer term debt if agreements cannot be made to write down, or write off, such debt. Otherwise the borrowing nations' fiscal burden of foreign debt repayment will harm their ability to participate in the IDB's credit issuance for vital great infrastructure projects.

This IDB can be a means of debt reorganization for over-indebted nations or groups of nations requiring IDB credit for great infrastructure development platforms.

Lyndon LaRouche first proposed the framework for a new international credit system in 1975, with his plan for an International Development Bank. That then became a major part of his presidential campaign platform in 1976.

Many nations of the world labor under unpayable, and wholly or partially illegitimate debts resulting from (1) extremely unfavorable terms of trade imposed upon them, or corrupt spending of development loans, or both (the cases of Argentina and Mexico, for example, which dealt with the problem differently), or (2) the rapid loading of debts onto governments in order to bail out private banks' bad debt (the cases of Ireland and Greece, for example). In these cases, the over-indebted nations can, as of a date certain, issue low-interest and long-term sovereign bonds to the IDB *to replace by agreement, their debts owed to major economic powers issuing credit to the IDB as described above; and by agreement, their debts to international lending agencies such as the International Monetary Fund and the European Central Bank. The IDB can use these bonds as the basis for issuing credits to those nations' national development banks, in those nations' currencies.*

Where national and regional authorities receive loans from the IDB *in order to carry out the actual creation of great infrastructure projects and/or scientific and technological developments*, which will generate highly productive economic activity as well as revenues for them, they will repay these IDB credits in the same way—by creating national credit banks, on the model of the KfW in Germany for decades after World War II, both to generate additional internal development credit and to invest in the IDB themselves, using their own national currencies.

Lyndon LaRouche described this process, in his 1982 book-length *Operation Juárez* proposal to the nations of Ibero-America for debt reorganization and development, as being identical in its requirements for debtor nations and for the (then) creditor nation the United States:

1. In no republic must any other issues of credit be permitted, ... excepting (a) Deferred-payment credit between buyers and sellers of goods and services; (b) banking loans against combined lawful currency and bullion on deposit in a lawful manner; (c) loan of issues of credit created in form of issues of national currency—notes of the Treasury of the national government.

2. Loan of government-created credit (currency notes) must be directed to those forms of investment which promote technological progress in realizing the fullest potentials for applying otherwise idled capital-goods, otherwise idled goods-producing capacities, and otherwise idled productive labor, to produce goods or to develop the basic economic infrastructure needed for maintenance and development of production and physical distribution of goods....

3. In each republic, there must be a state-owned national bank, which rejects in its lawfully permitted functions, those private-banking features of central banking associated with the Bank of England and the misguided practices of the U.S.A.'s Federal Reserve System....

4. No lending institution shall exist within the nation except as

they are subject to standards of practice and auditing by the Treasury of the government and auditors of the national bank. No foreign financial institution shall be permitted to do business within the republic unless its international operations meet lawful requirements for standards of reserves and proper banking practices under the laws of the republic, as this shall be periodically determined by proper audit ('transparency' of foreign lending institutions).

5. The Treasury and national bank, as a partnership, have continual authority to administer capital controls and exchange controls, and to assist this function by means of licensing of individual import licenses and export licenses, and to regulate negotiations of loans taken from foreign sources....

8. Sovereign valuation of the foreign exchange value of a nation's currency must be established.... The first approximation of the value of a nation's currency is the purchasing power of that currency within the internal economy of that nation. What are the prices of domestically produced goods and services, relative to the prices of the same quality of goods and services in other nations?

Because trade will increase among the nations participating in the treaty agreements for the building of these great projects, both those issuing credit through the IDB and those receiving loans, the national banks of the participating nations will necessarily create currency swaps large enough for increasing trade payments in each others' currencies. These currency swaps for *increases* in trade, can provide the basis for agreements on stable ranges for exchange rates between and among currencies.

The responsibility and purpose of the International Development Bank is to guarantee that development credits issued by nations go exclusively into the development of the new infrastructure platforms and technological developments most important to increase the productivity of national economies and of the labor forces of the human species.

Expand Nuclear Power for the World's Survival

Much of the world lives in virtual darkness, lacking the electricity essential for modern life; but world leaders are not prioritizing the solution to the problem.

By Ramtanu Maitra

July 2014

More than 1.2 billion people—20% of the world's population—are today without access to electricity, and almost all of them live in developing countries. This includes about 550 million in Africa and more than 400 million in India. It is incumbent on all world leaders to bring this number to zero at the earliest possible date, and thus provide all people with a future to look forward to within a span of 25 years. Can this be achieved with fossil fuels, wind, and solar power? The answer is a resounding "No!"

The only way the world can meet the power requirements of one and all is by fully exploiting the highest energy-flux density power generation achieved through nuclear fission now, and by starting to move to an even higher level by using hydrogen (deuterium), tritium, and, most promising, helium-3, as fuel in generating power through nuclear fusion. As of March 11, 2014, in 31 countries, 435 nuclear power plant units with an installed electric net capacity of about 372 GW were in operation, and an additional 72 plants with an installed capacity of 68 GW in 15 countries were under construction. Altogether, the existing nuclear power plants provide a shade more than 11% of the world's installed generating capacity. Most of the other 89% comes from the burning of fossil fuels.

What becomes evident from those figures is that almost no country—big or small—has made the essential commitment to generate power in the future entirely through nuclear fission. Instead, we see countries such as China and India, among the larger ones that are committed to increasing rates of agro-industrial growth, mining and hauling hundreds of millions of tons of coal on a daily basis to generate power to meet their development requirements.

While coal has played a critical role in man's development and will continue to play an important role for some time to come, as with any technology that is not superseded, the drawbacks are increasingly apparent. It is widely recognized, for example, that coal-fired power generation makes the air less breathable—the technology exists to overcome that problem but at a cost. However, another problem that coal-based power generation systems cause is virtually unsolvable. To begin with, vast amounts of water are needed on a daily basis to clean

The World's Nuclear Plants, in August 2005

International Nuclear Safety Center at ANL

these millions of tons of coal before burning. The polluted water from coal washeries needs to be cleaned before it in turn pollutes waterways and sub-surface groundwater. In addition, handling these vast amounts of coal is burdensome: Millions of tons of coal are shipped from ports or coal mines to the coal washeries. The rule of thumb suggests that an average coal plant burns the contents of approximately 200 coal cars a day, with 100 tons per car. This makes 73,000 cars or 7,300,000 tons per year. The average nuclear plant uses about 0.005 of a rail car of fuel per day—20 tons per year.

The problems that coal-fired power programs cause do not end with the logistical nightmare. Burning vast amounts of coal produces vast amounts of fly ash, which contains acidic chemicals ready to poison the land, clog the waterways, and kill all living things that inhabit the waterways. In the United States alone, coal-fired power plants on average produce 130 million tons of fly ash. All countries that are building up their power generation programs based on coal-fired plants encounter the same logistical nightmare. What that means is that a good part of a nation's railroads remains clogged, hauling in coal from the ports and mines to inland destinations where the power plants are, and then hauling the fly ash out. That situation becomes worse as more such plants are built.

While it should be obvious to policymakers that this policy of basing future growth in energy production on current technology could lead to a long-term disaster, nonetheless these countries have not committed themselves to create the conditions whereby their future electricity generation will come entirely from a clean source, such

The Earth at night: 20% of the world's population has no access to electricity.

as nuclear fission, which uses very little fuel and remains the most reliable and efficient source of power.

The World Power Scene, Briefly

Over the years, the two most populous nations in the world, China and India, have developed indigenous capabilities to manufacture a complete nuclear power plant, with the intent to provide hundreds of millions of their citizens with the electricity that is a vital requirement for living. But while China is making efforts to rapidly enhance its electrical power generation capacity, it is doing so by mining and importing more and more coal, while nuclear power remains a supplementary power source. It is evident that China has not geared up to change that situation in the foreseeable future. According to some analysts, China is expected to add coal-fired capacity of 36 GW in 2014, 42 GW in 2015, 45 GW in 2016, and 47 GW per year starting in 2017. In other words, between 2014 and 2020, China is expected to add about 310 GW of coal-generated electrical power.

By contrast, according to World Nuclear Association reports, while China currently produces about 20 GW, or 2% of its total electricity generation capacity, from nuclear fission, additional nuclear reactors that have been planned, including some of the world's most advanced ones, will help the country to produce a total of 58 GW of electrical power by 2020 using fission.

That means that during the next six years, during which China wants to add 310 GW of electrical capacity from coal-fired plants, nuclear reactors will produce only 38 GW—less than 13% of new coal-based power generation capacity planned. That would bring nuclear power-generated electricity capacity in China's power-generation mix up to 6%. Longer term plans for future capacity show that nuclear-based

Expand Nuclear Power for the World's Survival

power generation is expected to rise to 200 GW by 2030 and 400 GW by 2050. The conclusion is that while China understands the importance of nuclear fission, it has not yet made the necessary commitment to base its future power generation on nuclear, even in the long term.

India's power situation is much worse than China's, although it has well-developed nuclear power generation capabilities, and has been building its own small nuclear reactors for a long time. But the commitment to nuclear power as its primary source of future power generation has remained wholly theoretical. At present, India has installed capacity to generate about 235 GW of electricity, and of that, only 7 GW comes from nuclear, or about 3% of the total. Because India has 400 million people without full access to electricity, it is evident that it needs another 250 GW of power in the short term to provide electricity, education, and productive work to fully exploit the inherent productive potential of its own people. Its short-term nuclear program suggests that it will have about 15 GW of electrical power generated from nuclear reactors by 2020, a negligible amount compared to what the gravity of the situation calls for. By 2030, India's program calls for about 50 GW from nuclear power, which would be much less than 10% of the total power generated.

What Commitment to Nuclear Means

To begin with, the installed electricity-generating capacity of today's world is about 5,200 GW. Five countries (China, the United States, Japan, Russia, and India) account for about 2,900 GW. The rest of the world, which constitutes 55% of the world's population of 7 billion-plus, has a generating capacity of 2,300 GW; much of this is in the European Union, which has a population of 500 million. In other words, much of the world lives in virtual darkness.

However, electricity produced per hour across the world is nowhere near the stated generating capacity. "Capacity" is the maximum electric output a generator can produce under specific conditions, whereas "generation" is the amount of electricity a generator actually produces over a specific period of time. Many generators do not operate at full capacity all the time; their output may vary according to conditions at the power plant, fuel costs, and/or as instructed by the grid operator.

The one major reason that the actual generation of electricity around the world is way below the generating capacity is that only 11% of world's electricity comes from nuclear. Nuclear power plants, on average, have a capacity factor of 92-100%. Only one other power source, hydropower, reaches a capacity factor of 90%. By contrast, coal-fired

China's coal-fired power plants create the country's notorious air pollution, seen here in Beijing. Only the rapid expansion of nuclear power will solve the problem.

power plants, which constitute almost 45% of the world's generating capacity, operate at 50-55%, and natural-gas-burning power plants at about 60%. Solar and wind-based power plants operate at a capacity factor of only 20-30%.

In other words, only nuclear power plants, which can be set up almost anywhere on land, and even at sea, provide power reliably and at the stated generating capacity. By contrast, hydropower can be generated only where the water is flowing, and therefore has severe limitations.

Looking 30 years ahead, it becomes evident that the world's electricity-generating capacity must double to 11,000 GW by 2050. Again, a large amount of this additional power will be required in China and India. It is expected that these two countries, between them, will require an additional 2,500 GW of installed capacity. A similar approach is required for Africa, South America, Central Asia, and parts of South, Southwest, and East Asia. A vast majority of this additional 6,000 GW of power, say 5,000 GW, in the next 30 years, needs to be generated from nuclear plants.

To generate 5,000 GW of nuclear power in the next 30 years means the world will have to manufacture 5,000 nuclear reactors of 1,000 MW capacity. It takes 4-5 years to construct a nuclear plant, thus leaving 25 years in which the world will have to manufacture 5,000 plants with 200-1,000 MW reactors, and associated equipment, ready for installation. As of now, world's capacity to manufacture large reactors (1,000-1,100 MW) and the associated steam turbines, which together form the nuclear power plant (NPP) set, is limited to about 30 annually. India, where pressurized heavy-water reactors (PHWRs) are used for power generation, has the capacity to manufacture a few 600-700 MW installed capacity NPP sets.

That means the world's NPP manufacturers will have to quickly bring up their capacity from 30 to 200 plants per year, on average, to develop an economy based on the highest energy-flux density.

Another issue that has emerged in the manufacture of the new generation of reactors is metallurgy. Generation III+ plants can use existing metal alloys, but Generation IV plants, operating at higher temperatures, will require new materials. At 700°C, degradation problems are much more severe than at today's operating temperatures. Generation IV reactors are being developed by an international task force. Four of these are fast neutron reactors, and all will operate at higher temperatures than today's reactors. Fast neutron reactors have been designated particularly for hydrogen production.

What Rapid Expansion Entails

What both China and India must realize as they are moving towards developing their independent capabilities to build the nuclear reactors they require in the future, is that the optimization of building nuclear power plants lies with each country's heavy engineering capacities. Of the large number of reactors over the next four decades these two countries would require, not even a significant fraction can be met through importation of nuclear power plant sets from abroad.

Moreover, the developed nations, which started developing nuclear power for electricity generation decades ago, are now withdrawing, and as a result, their capability to manufacture nuclear reactors and facilities, and to supply manpower, is now diminishing at a rapid pace.

No country has surplus nuclear power plants sets sitting on the shelf. These are expensive and material consuming items which are made only when they are ordered. As demands for nuclear power plants in developed nations decreased rapidly, and the developing nations lacked the financial strength to buy those nuclear plants in large numbers, the overall engineering capacity in the developed nations diminished significantly.

Moreover, the suppliers of nuclear equipment must be qualified and must practice an extreme level of quality control. Decades ago, when the Westinghouse and GE-designed nuclear reactors were ordered and built, the builders were integrated builders, which means they needed little or nothing from external suppliers. Nowadays, however, a new plant ordered from the West, will require dozens of vendors, each one supplying a part, or parts, of the plant. This makes putting forward a rigid project completion time extremely difficult. Westinghouse itself just looks after design, engineering, and project management.

While everyone recognizes this shortcoming that has developed in the western nations, not much effort is underway in either India or China to rapidly develop their heavy engineering capacity to be able to produce the required 50-75 nuclear reactors every year for the next four decades. Observers have pointed out that the supply challenge is not confined to the heavy forgings for reactor pressure vessels, steam turbines, and generators, but extends to other engineered components as well.

The World Nuclear Association on its website, as of October 2014, pointed out that "for very large generation 3+ reactors, production of the pressure vessel requires, or is best undertaken by, forging presses of about 140-150 MN (14-15,000 tonnes) capacity which accept hot steel ingots of 500-600 tonnes. These are not common, and individual large presses do not have high throughput – about four pressure vessels per year appears to be common at present, fitted in with other work, though the potential is greater than this. Westinghouse was constrained as of 2009 in that the AP1000 pressure vessel closure head and three complex steam generator parts can only be made by JSW. Areva has a little more choice."

That heavy forging or other nuclear power equipment is in short supply should not surprise anyone. For instance, the forging presses that are required to make a 1400MW Pressurized Water Reactor vessel are of about 15,000 tons capacity, and receive steel ingots weighing 500 tons. These forgings do not have much use in any other industry. The reason that the nuclear industry needs such heavy forging is because the reactor vendors prefer large forgings to be integral as single products. While split forgings can be, and are, used, they need welding, and those weldings need checking for the rest of that reactor vessel's life.

It must be noted that the need for nuclear power reactors in China, India, and Russia is bound to grow at a faster pace than in the rest of the world. These three countries, when they increase their NPP sets' manufacturing capacity to the currently desired level, will find it difficult to export a large number of reactors to other countries that will be in need of nuclear reactors.

That means that many other nations in Asia, Africa, and South America have to prepare for rapid development of a nuclear future now. This entails training of manpower using a large number of research reactors, development of heavy engineering capability to forge NPP sets, and other basic infrastructure that will enable them to enhance their power generation. The focus on developing human resources is two-fold: (1) generic capacity-building at the national level in nuclear sciences and technology, to support the government and other stakeholders in making informed decisions on nuclear power; and (2) developing personnel in stakeholder organizations to implement the nuclear power program.

Moreover, the commitment to nuclear power entails developing manpower in all nations, including those that have nuclear power plants, or even just nuclear research reactors. There is already a significant gap between the number of nuclear engineers that are being trained and those that are retiring, which needs to be addressed just to keep the world's existing nuclear reactors running. Therefore, to speed up nuclear generation, countries, one and all, require large-scale training programs to fulfill this need. Developing the right skills base is a priority for the industry to grow to the level that will be demanded.

Why Nuclear?

The world does not have any choice but to go with nuclear fission now and prepare to introduce nuclear fusion at the earliest possible date. Because nuclear power has the highest energy-flux density of all power-generating sources, it generates a vast amount of power using very little fuel. In addition, although the world will run out of other power-generating natural resources, it will never run out of nuclear fuel, because nuclear fuel is renewable: Fast Breeder Reactors (FBRs) produce more fuel than they consume, making nuclear fuel inexhaustible.

Under appropriate operating conditions, neutrons given off by fission reactions can "breed" more fuel from otherwise non-fissile isotopes.

The French Super-Phénix was the world's first large-scale breeder reactor. It was put in service in 1984, and ceased operation as a commercial power plant in 1997. It was the last fast breeder reactor operating in Europe for electricity production—as the result of Green protests.

India's prototype fast breeder reactor at Kalpakkam. The reactors use natural gas as fuel during the current first stage of operation. The third stage reactors will use thorium as fuel.

The most common breeding reaction is that of plutonium-239 (Pu-239) from non-fissionable uranium-238 (U-238). This becomes possible because the non-fissionable U-238 is 140 times more abundant than the fissile uranium-235 (U-235) and can be efficiently converted into Pu-239 by the neutrons from a fission chain reaction. Pu-239 is a fissile material that can be used to generate power.

For instance, the Liquid-Metal Fast Breeder Reactor (LMFBR) is a Pu-239 reactor. In this FBR system, cooling and heat transfer is performed by a liquid metal. The metals that can accomplish this are sodium and lithium, with sodium being the most abundant and most commonly used. Construction of this type of fast breeder requires higher enrichment of U-235 than a light-water reactor, typically 15-30%. The reactor fuel is surrounded by a "blanket" of non-fissile U-238. No moderator is used in the breeder reactor, because fast neutrons are more efficient in transmuting U-238 to Pu-239.

France's Super-Phénix (SPX) was the first large-scale breeder reactor that was built; it was put into service in 1984, and ceased operation as a commercial power plant in 1997. The reactor core consisted of thousands of stainless steel tubes containing a mixture of uranium and plutonium oxides, about 15-20% fissionable Pu-239. Surrounding the core was a region called the breeder blanket, consisting of tubes filled only with uranium oxide. The entire assembly was about 3x5 meters and was supported in a reactor vessel in molten sodium. The energy from the nuclear fission heated the sodium to about 500°C, and it transferred that energy to a second sodium loop, which in turn heated water to produce steam for electricity production. Such a reactor could produce about 20% more fuel than it consumed. Enough excess fuel could be produced over about 20 years to fuel another such reactor. Optimum breeding allowed about 75% of the energy of the natural uranium to be used, compared to only 1% in the standard light-water reactors.

India is now developing a fast breeder reactor that will produce fissile uranium-233 (U-233), which will then be loaded to generate power through fission. Fuelled with uranium-plutonium oxide, these reactors will have a thorium blanket to breed fissile U-233. The plutonium content will be 21% and 27% in two different regions of the core. Initial Indian FBRs will have mixed oxide fuel, but these will be followed by metallic-fuelled ones, to enable a shorter doubling time.

By contrast with nuclear fuel, the most frequently used fossil fuels are not renewable. A 1,000 MW coal-fired power plant needs about 6,600 tons of coal daily—the amount varies slightly according to the quality of coal used. On the other hand, a nuclear power plant requires very little fuel—a tiny fraction of what a coal-burning power plant requires. Used nuclear fuel still contains an immense amount of energy—more than 95% of the potential energy contained in that small amount of material is not even used. Advanced reactors will one day routinely recycle this waste.

In the case of thorium-fueled nuclear power plants, the fuel requirement will be even less. Unlike the pressurized and boiling water reactors that burn about 1% of their fuel before going non-critical and therefore require refueling once every 18-24 months, thorium-fueled power plants can burn more than 90% of the loaded fuel and thus require refueling once every 30 years or so. This means that the overall waste in a reactor's lifespan is a fraction of that in the current generation of uranium-fueled reactors.

Other Benefits

In addition to its low fuel consumption, nuclear power provides mankind with a number of other benefits. Nuclear byproducts are used in some calibration devices, radioactive drugs, bone-mineral analyzers, imaging devices, surgical devices, teletherapy units, and diagnostic devices used in dentistry and podiatry. Some cardiac pacemakers are powered by nuclear batteries. Source material is also used for counterweights in medical devices and for radiation shielding.

Nuclear medicine, developed in the 1950s by physicians using iodine-131 to diagnose and treat thyroid disease, now uses radiation to provide diagnostic information about the functioning of many of a person's organs, or to treat them. In most cases, the information is used by physicians to make a quick, accurate diagnosis of the patient's illness. The thyroid, bones, heart, liver, and many other organs can be easily imaged. In some cases, radiation can be used to treat diseased organs or destroy tumors. More than 10,000 hospitals worldwide use radioisotopes in medicine, and about 90% of the procedures are for diagnosis. The most common radioisotope used in diagnosis is technetium-99, with some 40 million procedures per year (16.7 million in the United States in 2012), accounting for 80% of all nuclear medicine procedures worldwide.

One of the diagnostic techniques used in nuclear medicine is the injection of a radiotracer, or radiopharmaceutical, into the body, which is then traced as it emits gamma rays. Subsequently, the radioactive emissions from the injected radiotracer, that accumulate in the area of the body under examination, are detected by an imaging device, or a gamma camera, that collects the molecular information through pictures. In addition to being injected, a radiotracer can be swallowed, or inhaled as a gas.

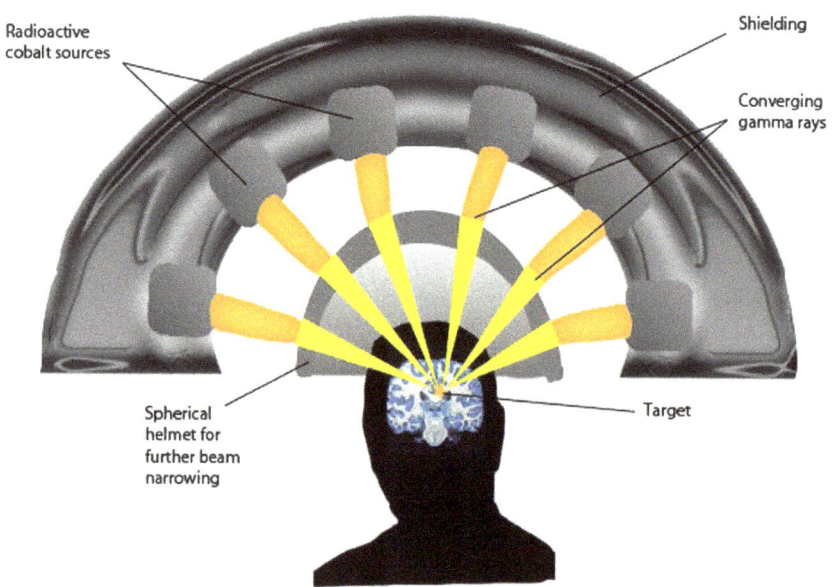

The Gamma Knife concept of stereotaxic radiosurgery. Radioactivity is used for treatment of brain tumors, among many other medical applications.

This is a painless medical process and completely noninvasive. The tracers are short-lived isotopes linked to chemical compounds.

Radiotherapy, on the other hand, uses radiation, such as x-rays, gamma rays, electron beams, or protons, to treat such medical conditions as cancer. It is a localized treatment to kill or damage cancerous cells. The healthy cells, even if they get damaged by radiation, usually repair themselves.

Many radioisotopes are made in nuclear reactors, some in cyclotrons. Generally, neutron-rich ones and those resulting from nuclear fission need to be made in reactors; neutron-depleted ones are made in cyclotrons. There are about 40 activation product and five fission product radioisotopes made in reactors. Tens of millions of nuclear medicine procedures are performed each year, and demand for radioisotopes is increasing rapidly. Sterilization of medical equipment is also an important use of radioisotopes.

Food Preservation, Agro-Industrial Use

Another important use of nuclear radiation is food irradiation. Although food irradiation is decades old, it is still considered a new food safety technology because it has not been widely adopted, even by the countries which have the capability to make its optimum use. Irradiation can eliminate disease-causing germs. It pasteurizes milk and treats food with ionizing radiation (this is high-frequency ultraviolet radiation that possesses enough energy to break chemical bonds), killing the bacteria and parasites that cause health hazards. Currently, the foods that the astronauts eat during their long space journeys are sterilized by irradiation in order to prevent food-borne illness.

There are generally three different food irradiation technologies in use today. These use three different kinds of rays: gamma rays, electron beams, and x-rays. Gamma rays are a radioactive form emitted by the element cobalt (cobalt-60) or the element cesium (cesium-137). Electron beams, a stream of high energy electrons, can penetrate food to a depth of only about an inch. That is why they are used more widely as medical sterilizers.

The third kind, x-ray irradiation, is relatively new and is not widely used. Like cobalt/cesium gamma rays, x-rays can pass through thick foods, and require heavy shielding for safety.

Agriculture itself can be transformed under conditions of a nuclear energy economy. With plentiful power, traditional requirements for farming, e.g., soil, seasonal and diurnal light cycles, or dependence on rainfall or irrigation, no longer apply. Instead, "protected agriculture" can be practiced, in which necessities such as nutrients, water, temperature, and even light itself are completely controlled in an organized growing environment. Inputs are provided to plantlife as needed, whether in a space capsule, the Arctic, or the Sahara Desert. Early work in protected agriculture fell under the rubric "CELSS," for controlled-environment life support systems. However, despite certain applications of hydroponics, such as Japan's produce sector, the technology involved has yet to be widely implemented.

The most direct use of nuclear technology in farming is for seed improvements. Researchers use radiation to modify strains of crops, then breed up desired varieties. Called "mutation breeding," the technique was successfully used, for example, by a team of Kenyan researchers and the International Atomic Energy Agency (IAEA) to produce a new wheat strain, announced in 2013, that is resistant to the damaging UG99 wheat stem rust fungal disease.

Modern industry uses radioisotopes in very many ways to improve productivity and, in some cases, to gain information that cannot be obtained any other way. The continuous analysis and rapid response of nuclear techniques, many involving radioisotopes, mean that reliable flow and analytic data can be constantly available. This results in reduced costs, with enhanced product quality.

Neutrons from a research reactor can interact with atoms in a sample causing the emission of gamma rays, which when analyzed for characteristic energies and intensity will identify the types and quantities of elements present. The two main techniques are Thermal Neutron Capture (TNC) and Neutron Inelastic Scattering (NIS). TNC occurs immediately after a low-energy neutron is absorbed by a nucleus; NIS takes place instantly, when a fast neutron collides with a nucleus. A particular application of this is where a probe containing a neutron source can be lowered into a bore hole where the radiation is scattered by collisions with surrounding soil. Because hydrogen (the major component of water) is by far the best scattering atom, the number of neutrons returning to a detector in the probe is a function of the density of the water in the soil.

With the rapid expansion of coal-fired plants around the world, particularly in two of the most populous nations in the world, China and India, coal ash, a highly toxic waste that needs to be disposed of on land, is becoming a major issue. Large amounts of ash in the coal not only increase the toxic by-product, but also reduce the heat generation capacity of the coal, thus reducing its thermal efficiency. Radiation is now being used to determine the ash content in coal, as a way of determining its quality.

Ash content is correlated with the concentrations of the ash-forming elements, the major ones of which are aluminum and silicon (Al_2O_3 and SiO_2). The industry uses neutron gamma rays to carry out non-destructive, non-contact elemental analysis of solids and liquids. The gamma ray interacts with the coal, the ash part of it in one way, and the bituminous part of the coal in another way. This interaction is measured and used as the matrix for determining the ash content in coal.

Radioisotopes are used as tracers in a wide variety of situations. They can be used within a plant or an animal to trace the movement of certain chemicals. Plants take up phosphorus-containing compounds through their roots, so when an amount of radioactive phosphorus-32 is added to the soil, one can evaluate the rate of uptake of phosphorous by that plant from the soil by analyzing the leaves.

They are also used to measure pesticide levels. A pesticide can be laced with a radioisotope such as chlorine 36 and applied to field test plants. Over a period of time, the plant biologists can determine how much of this pesticide was absorbed by the plant, and how much remained in soil.

Perhaps the most obvious use of nuclear technologies is for peaceful nuclear explosives (PNEs), which technology is critically required to build many infrastructure projects of the World Land-Bridge. Early PNE work was carried out in the Project Plowshare program of the 1950s U.S. Atoms for Peace policy. In 1961, for example, the first underground PNE test, called the "GNOME," was carried out. Subsequently, the R&D was dropped. Today, it is urgent to resume PNE development and application.

Nuclear Desalination, Water Treatment

Another major contribution to mankind from the waste heat generated by nuclear fission is the desalination of sea and brackish water. Freshwater makes up a very small fraction of all water on the planet. While nearly 70% of the world is covered by water, only 2.5% of it is fresh; the rest is ocean-based. Even then, just 1% of our freshwater is easily accessible, with much of it trapped in glaciers and snowfields.

The lack of clean drinking water is a major problem worldwide. The World Health Organization says that more than 1 billion people live in areas where renewable water resources are not available. The problem is especially serious in Africa, followed by Asia and the Pacific, according to a UN report. The lack of clean drinking water around the world forces millions of people to drink unsafe water. This leads to an increase in diseases such as diarrhea, the second leading cause of death in children under five. Unsafe drinking water takes the lives of hundreds of thousands of children every year.

Yet we have the technology to desalinate sea and brackish water, disinfect wastewater through nuclear energy-beam systems, and provide each and every individual with plentiful, potable water and reliable sanitation. But no real effort has been made to do this. Electron beam methods to clean up municipal drinking and wastewater, and industrial waste streams, have been demonstrated to be successful for decades. For example, in Miami, Florida, in the 1980s, the Central District Wastewater Treatment Plant set up a unit to disinfect municipal sludge by Ebeam currents. In Texas, the National Center for Electron Beam Research, the largest of its kind in the world, has programs ready for implementation that harness Ebeam and X-ray technologies for water clean-up.

Nuclear fission-created waste heat has been used sparingly for desalination. Nuclear reactors that help desalinate water will also produce electricity. An exam-

Governmnet of India/Dept. of Atomic Energy

India has been engaged in desalination research since the 1970s. This demonstration plant was set up in 2002, at the Madras Atomic Power Station in Kalpakkam.

The gear box used in the seawater circulating pump at the Hongyanhe nuclear power station in China. The waste heat will provide water to cool the reactors.

ple of a nuclear reactor producing electricity and desalinated water is the BN-350 fast reactor at Aktau in Kazakhstan, which supplied up to 135 MW of electric power while producing 80,000 m^3/day of potable water for some 27 years, about 60% of its power being used for heat and desalination. Japan, Russia, and Canada all have experience with nuclear reactors employed in the desalination of water, and the IAEA strongly promotes this use of nuclear energy.

Early in the 1960s, foreseeing a time when freshwater needs would outstrip supplies, the U.S. Department of the Interior's Office of Saline Water (OSW) authorized funding for five research facilities to develop desalination technologies for the country. The Wrightsville Beach facility on Harbor Island, NC, set up in the early 1960s, was dubbed the "world center for experimental development in saline water conversion," by OSW director C.F. McGowan. It was non-nuclear. The plan did not move forward.

Nuclear desalination is defined to be the production of potable water from seawater, or any saline water, in a facility in which a nuclear reactor is used as the source of energy (electrical and/or thermal) for the desalination process. The facility may be dedicated solely to the production of potable water, or may be used for both the generation of electricity and the production of potable water, in which case only a portion of the total energy output of the reactor is used for water production. In either case, nuclear desalination is taken to mean an integrated facility in which both the reactor and the desalination system are located on a common site, and energy is produced on-site for use in the desalination system. It also involves at least some degree of common or shared facilities, services, staff, operating strategies, outage planning, and possibly control facilities, seawater intake, and outfall structures.

Since most of the economically-weaker countries in need for fresh water do not have extensive infrastructure and adequate nationwide electricity grids, small and medium size nuclear reactors are important for desalination. That means the electricity generated by those reactors is locally consumed, while the heat they generate produces potable water. Smaller reactors are also more appropriate for remote areas that are not close to the existing electricity grid.

Despite the benefits of nuclear desalination, not many countries have developed a full-fledged program for nuclear desalination. (See Appendix) The most wide-ranging experience of desalination has been primarily in two countries—Kazakhstan and Japan. India has recently developed an integrated facility in Kalpakkam. In Japan, 100 reactor-years of desalination have been reported. These desalination facilities are tied to the large pressurized water reactors yielding 14,000m^3/day of potable water. Now that all Japan's nuclear power plants are lying idle, obviously these desalination plants are not producing any potable water.

In Kazakhstan, the fast breeder reactor BN-350, which was decommissioned in 1999, was used for years for desalination. Reports indicate that relevant technical experience has been also accumulated in Russia by its use of floating nuclear power plants on ships.

India has been engaged in desalination research since the 1970s. In 2002, a demonstration plant coupled to twin 170 MW nuclear power reactors (PHWR) was set up at the Madras Atomic Power Station, Kalpakkam, in southeast India. This hybrid Nuclear Desalination Demonstration Project (NDDP) comprises a RO unit with 1,800 m^3/day capacity and a multi-stage flash (MSF) plant unit of 4,500 m^3/day, plus a recently added barge-mounted RO unit. This is the largest nuclear desalination plant based on hybrid MSF-RO technology, using low-pressure steam and seawater from a nuclear power station. The plant incurs a 4 MW loss in power.

A low temperature (LTE) nuclear desalination plant using waste heat from the nuclear research reactor at Trombay, near Mumbai in India, has operated since about 2004, to supply water for the reactor.

Pakistan in 2010 commissioned a 4,800 m^3/day multiple-effect distillation (MED) desalination plant, coupled to the Karachi Nuclear Power Plant (KANUPP), a 125 MWe PHWR near Karachi. It has been operating a 454 m^3/day RO plant for its own use.

China General Nuclear Power (CGN) has commissioned a 10,080 m^3/day seawater desalination plant using waste heat to provide cooling water at its new Hongyanhe project at Dalian, in the northeast Liaoning province. Much relevant experience comes from nuclear plants in Russia, Eastern Europe, and Canada, where district heating for commercial and residential use is a by-product.

The best way to develop large-scale nuclear desalination along the world's coastal areas will be to manufacture large numbers of small modular nuclear reactors of 100-200 MW capacity. These reactors, when put in a cluster, would provide adequate and reliable power to industry and commerce, while supplying the heat to desalinate abundant amounts of seawater.

South Korea has developed a small nuclear reactor design for cogeneration of electricity and potable water. The 330 MWt (thermal) SMART reactor has a long design life and needs refueling only every three years. The main concept has the SMART reactor coupled to four MED units, each with a thermal-vapor compressor (MED-TVC) and producing a total of 40,000 m^3/day, with 90 MWe.

Argentina has designed the CAREM, an integral 100 MWt PWR suitable for cogeneration or desalination alone, and a prototype is being built next to the Atucha nuclear power plant. A larger version is envisaged, which may be built in Saudi Arabia.

China's INET has developed the NHR-200, based on a 5 MW pilot plant.

Russia is in the forefront of developing floating nuclear power plant (FNPP)-based nuclear desalination, although there are reports that China is getting increasingly interested in developing these plants as well. Media reports indicate that Rusatom Overseas, the export branch of the state nuclear reactor monopoly Rosatom, has signed a memorandum of understanding with China on the development of floating NPPs starting in 2019. The plan is to jointly build six plants which will also produce potable water. It is not evident as of now whether these floating plants will be moored along the coast to provide power and potable water to the coastal region, or produce for the consumption of the plant crew only.

However, the Russians are in the process of building a floating nuclear power plant on the ship, *Akademik Lomonosov*. The ship will have two modified KLT-40 reactors that will generate 70MW of electricity, or 300 MW of heat, and will also be used as a desalination plant. This means that the ship could produce 240,000 cubic meters of fresh water every day.

The World Nuclear Association (WNA), in a report on its website in October 2014, pointed out that Russia, using ATETs-80, a twin-reactor cogeneration unit using KLT-40, will be generating 85 MW of electrical power, and at the same time will be producing 120,000 m^3/day of potable water. The WNA also noted that Russia has the smaller ABV-6 reactor which produces 38 MW thermal, and that a pair of them mounted on a 97-meter barge, known as Volnolom floating NPP, will have the capacity to produce 12 MW of electrical power as well as more than 40,000 m^3/day of potable water by reverse osmosis.

Thorium Reactors

The next wave of nuclear reactors that must emerge in large numbers are those fueled by thorium. Thorium has multiple advantages as a nuclear fuel. Thorium ore, or monazite, exists in vast amounts in the dark beach sands of India, Australia, and Brazil. It is also found in large amounts in Norway, the United States, Canada, and South Africa. Thorium-based fuel cycles have been studied for about 30 years, but on a much smaller scale than uranium or uranium/plutonium cycles. Germany, India, Japan, Russia, the United Kingdom, and the United States have conducted research and development, including irradiating thorium fuel in test reactors to high burn-ups. Several reactors have used thorium-based fuel.

India is by far the nation most committed to study and use of thorium fuel; no other country has done as much neutron physics work on thorium. The positive results obtained have motivated Indian nuclear engineers to use thorium-based fuels in their current plans for the more advanced reactors that are now under construction. It is therefore incumbent upon Indian policymakers to make thorium-fueled nuclear reactors their main workhorse and develop the engineering infrastructure to manufacture them in large numbers within a very short period of time.

In addition to thorium's abundance, all of the mined thorium is potentially usable in a reactor, compared with only 0.7% of natural uranium. In other words, thorium has some 40 times the amount of energy per unit mass that could be made available, compared with uranium.

From the technological angle, one reason that thorium is preferred over enriched uranium is that the breeding of U-233 from thorium is more efficient than the breeding of plutonium from U-238. This is because the thorium fuel creates fewer non-fissile isotopes. Fuel-cycle designers can take advantage of this efficiency to decrease the amount of spent fuel per unit of energy generated, which reduces the amount of waste to be disposed of. In addition, the fissionable thorium-232 (Th-232) decays very slowly (its half-life is about three times the age of the Earth).

There are some other benefits as well. For example, thorium oxide, the form of thorium used for nuclear power as fuel, is a highly stable compound—more so than the uranium dioxide that is usually used in today's conventional nuclear fuel. Also, the thermal conductivity of thorium oxide is 10-15% higher than that of uranium dioxide, making it easier for heat to flow out of the fuel rods used inside a reactor. Furthermore, the melting point of thorium oxide is about 500°C higher than that of uranium dioxide, which gives the reactor an additional safety margin, if there is a temporary loss of coolant.

The one challenge in using thorium as a fuel is that it requires neutrons to start its fission process. Thorium is not a fissile fuel like U-235; Th-232 absorbs slow neutrons to produce U-233, which is fissile. In other words, Th-232 is fertile, like U-238. Th-232 absorbs a neutron to become Th-233, which decays to protactinium-233 (Pa-233) and then to fissionable U-233. When the irradiated fuel is unloaded from the reactor, the U-233 can be separated from the thorium, and then used as fuel in another nuclear reactor. U-233 is superior to the conventional nuclear fuels, U-235 and Pu-239, because it has a higher neutron yield per neutron absorbed. This means that once it is activated by neutrons from fissile U-235 or Pu-239, thorium's breeding cycle is more efficient than that using U-238 and plutonium.

Here is a summary of the advantages of using thorium as nuclear fuel:

1. Thorium fuel generates no weaponizable material in its waste profile; the waste consists of the radioisotope U-233, which is virtually impossible to weaponize.
2. Unlike uranium, thorium does not possess any fissile isotopes in its naturally occurring form; consequently, there is no material that can be enriched to weaponizable levels.
3. Thorium fuel can be used to safely incinerate the world's unwanted stockpile of plutonium waste and generate electrical power and heat to desalinate water.
4. Thorium fuel cycle waste has a radio-toxicity period of less than 200 years, which compares favorably with the more than 1 million-year radio-toxicity period estimated to exist for uranium fuel-cycle waste.
5. Thorium fuel has superior fuel economy in various respects; it will generate more energy per unit of mass than uranium fuel by a factor of approximately 40, which means thorium fuel-based power plants do not require re-loading for dozens of years.
6. Thorium fuel-cycle waste can be reprocessed and used as fissile material in a closed fuel cycle, meaning that eventually no new fissile material will be required to power the reactors; however, the reprocessing technology (to separate U-233) does not yet exist.

Solve the World Water Crisis

By Benjamin Deniston

September 2014

Today's world water crises will be solved by recognizing mankind's *obligation* to act as the caretaker of Earth—to be a creative force, continuously improving the conditions across the planet (and beyond). As Lyndon LaRouche has emphasized, this is the scientific conclusion required by the work of the great Russian-Ukrainian scientist Vladimir Vernadsky, who demonstrated that human society expresses a capability absent in all lower forms of animal life, a capability more powerful than the cumulation of the actions of animal and plant life (the biosphere)—the force of scientific and cultural thought (the noösphere). Whether the modern-day environmentalist likes this or not, the scientific reality is that mankind has been born into a responsibility to continuously re-shape and improve the surface of the planet. To deny this is to deny the existence of humanity.

This is the principle at issue in the current global water crisis. Basic progress and development have been thwarted in recent decades, to the point where 4 billion people, more than half the world's population, do not have safe, reliable supplies of water for even drinking and sanitation. Food production is threatened. The industrial base is far below what is required to produce for the future. Sickness and death are occurring, for lack of water.

How can this be tolerated when more than 70% of the Earth's surface is covered in water? To put this in a conceptual perspective, if the entire world population was able to use water at the same per capita levels of the United States currently, there is one hundred thousand times more water on Earth than would be used in a year by 7 billion people at the current U.S.A. per capita rate.[1] In terms of freshwater, there is

---

1. The current U.S. per capita use is four times the global average. The United States Geological Survey (USGS) and U.S. Department of the Interior report, "Estimated Use of Water in the United States in 2005," provides a total freshwater use which translates to a per capita use of about 160 cubic meters per person per year. This is all direct water use for all aspects of society, including public supply (11% of total use), domestic (1%), irrigation/agriculture (31%), livestock/aquaculture (3%), industrial (4%), mining (1%), and cooling of thermoelectric power plants (49%), but it does not include "hidden" water use such as in the production of foods, industrial goods, and other items that require water for their production that are imported.

FIGURE 1
Global Terrestrial Water Cycle

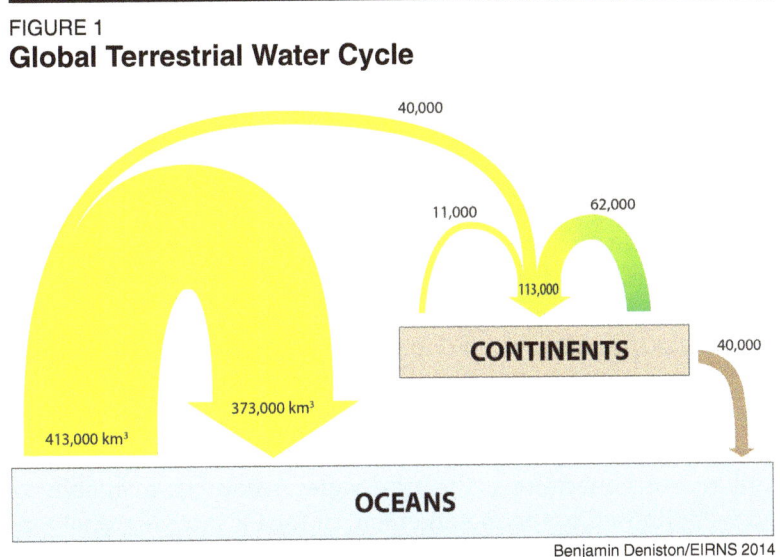

Benjamin Deniston/EIRNS 2014

approximately three thousand times more freshwater on Earth than would be used by the entire world at this hypothetical rate.[2]

But, water supplies cannot be discussed simply in terms of "use." Water is not a "finite resource" that is only used once (such as coal or natural gas). The global water system has cyclical-type characteristics, with water constantly moving from one state to another state (e.g., liquid oceans, frozen ice caps, and atmospheric vapor) and from participating in one system to another system (e.g., oceans, living matter, and human economic processes). *For this reason, any attempt to address the water needs of billions of people, both now and far into the future, must focus on the management—or creation—of cycles, not "use" per se.*

Throughout the history of life on Earth, including human civilization, thus far, the most important water cycle has been that of ocean evaporation, precipitation over land, and surface flow back into the oceans. This will be referred to as the **terrestrial water cycle** (see **Figure 1**). This is what sustains the entirety of life on land (although plants have increasingly augmented and boosted this cycle), and it has improved life in the oceans (by bringing nutrients from land). Today, it is estimated that the rate of this cycle (measured by annual new terrestrial precipitation) is 3.5 times higher than the rate that would be required by our hypothetical case of 7 billion people using water at U.S. per capita levels. However, this falls far short of understanding the water availability of the existing global terrestrial water cycle, because water is used many times in the course of one cycle. For example, the same water could cool a thermoelectric power plant, then irrigate a farm, and then go into a sanitation system, getting "used" three times in the course of one cycle. In many circumstances the reuse rate can be even higher.

Therefore, speaking in terms of cycles, the annual terrestrial precipitation can be used to define the **rate of a cycle**, and the amount of use and reuse can be used to define the productivity of a cycle. These measures can be applied to the global cycle, or divided into continental cycles, or further subdivided into individual river basins, and so on (for example, see the box below, "Increasing the Physical Productivity of the North American Water Cycle"). Examining the global water system from this standpoint, it is clear that the water resources are there; what is lacking is the economic development and energy flux

---

2. Estimates of current global water distribution and water flows are taken from the study, "Estimates of the Global Water Budget and Its Annual Cycle Using Observational and Model Data," by Kevin Trenberth et al., from the National Center for Atmospheric Research in Boulder, Colorado; published in the *Journal of Hydrometeorology*, Volume 8, 2007. This does not include the recent discoveries of large aquifers beneath the oceans, and even larger amounts of water in mineral formations deep within the Earth's crust.

density[3] needed to improve the productivity of existing cycles (with purification, sanitation, and related systems), control or expand existing cycles (e.g., with reservoirs and river diversion systems), and create new cycles (with weather modification and desalination systems).

What Is a Resource?

This becomes a rather basic pedagogy in Mr. LaRouche's science of physical economics. *It is mankind that creates resources.*

The concept of "natural" resources is misleading, if not fraudulent. For mankind, the factor that determines if something is, or is not a "resource" is never simply its "natural" state, but the level of scientific development of a society. Water is simply one excellent example of this principle of humanity.

Until recent generations, the freshwater resources available to mankind were limited to the management of the existing regional terrestrial water cycles, including all the rivers, lakes, groundwater, etc., created and maintained by these cycles. While the use and productivity of the existing cycles could (and can) be improved, the size and availability of this resource had largely remained outside of mankind's control—a situation vulnerable to regional climate changes, such as those associated with changing solar activity (as discussed below).

Now, with new technological developments that can be employed en masse on a global scale with the energy flux density provided by a fission and fusion economy, mankind can, for the first time, *look to managing entire continental cycles and even the creation of new cycles through weather modification technologies and desalination systems.* Before investigating the details of these concepts, reflect upon the broader implications.

Until this point, the entire planetary terrestrial water cycle had been solely under the dominion of the Sun, providing all desalination (ocean evaporation) and water vapor inland transport with solar energy. But now, for the first time in the history of Earth, a new power has emerged. Though relatively small in its beginnings, mankind, in line with Vernadsky's understanding, is beginning to overtake the role of the Sun on Earth, through the manipulation of atmospheric moisture flows (with weather modification), the manufacture of freshwater (with desalination), and the distribution of these new sources of freshwater throughout terrestrial systems.

Lyndon LaRouche has emphasized that the solution to the water crisis requires embracing the scientific realization of Vernadsky—mankind, wielding the power of scientific thought, is a geological force, responsible for the improvement of the global water system as a whole. For an advancing mankind, the needed water resources exist; it is the effective organization of the powers of human society that has been lacking.

This is the scientific basis governing the following section of this report. First, to properly situate the challenges that need be addressed,

---

3. See first article in this section.

the dimensions of the global water crisis are briefly reviewed, with selected examples chosen to illustrate the principled nature of the challenges facing mankind. Then the prospects for the future of a top-down, global approach to the world's water crisis are discussed.

I. Dimensions of the Crisis

The lack of water availability globally can be summarized simply. Of today's total world population, nearly 900 million people do not have safe water to drink and 2.6 billion do not have sanitation systems, for lack of water.[4] When the metric is properly set higher, to include those people without safe and reliable tap water in their homes, the number lacking these arrangements is up to 4 billions. Moreover, for many millions who have had good water—in the Southwest of the United States, for example—their future water security is threatened.

Another dramatic expression of the world water management crisis is seen in the prevalence and increase of waterborne illness. Cholera is a marker disease for lack of basic water management. The number of cases increased 130% worldwide, from 2000 to 2010, according to the World Health Organization (WHO). WHO estimates that every year now, there are 3 to 5 million cholera cases, with 100,000 to 200,000 deaths. This is a conservative guess, given that WHO estimates only 5-10% of cases are officially reported.[5]

A concept-map of the global distribution of the water crisis was featured in the 2012 United Nations World Water Development Report, identifying two aspects to the water crisis: "economic" and "physical" (**Figure 2**).[6] "Economic" refers to locations where the basic infrastructure has not been developed to make use of available water. "Physical" refers to locations where the needs of society have outpaced existing local water supplies.

While Figure 2 serves as a snapshot of the global characteristics and geography of the water crisis, as of this writing the analysis is more than five years old, and conditions in certain regions have gotten worse.

In the following section, on the dimensions of the crisis, four aspects of the global water crisis are examined, starting with a brief focus on insane policies that are unnecessarily accelerating the water crisis and must be ended immediately—hydraulic fracturing and biofuels. Then two aspects of the water crisis are examined—the depletion of ground water stores, followed by the deficiency in or lack of management of surface water supplies. Lastly, we examine the role of changing solar activity, with emphasis on what is known about previous major shifts in regional climate and water patterns associated with the type of solar changes that we may be experiencing in the coming decades.

---

4. United Nations "World Water Development Report 2014—Water and Energy," page 7.

5. World Health Organization Fact Sheet No. 107; reviewed February 2014.

6. World Water Assessment Programme (WWAP), 2012. The United Nations World Water Development Report 4: Managing Water under Uncertainty and Risk. Paris, UNESCO.

FIGURE 2
Global Physical and Economic Water Scarcity

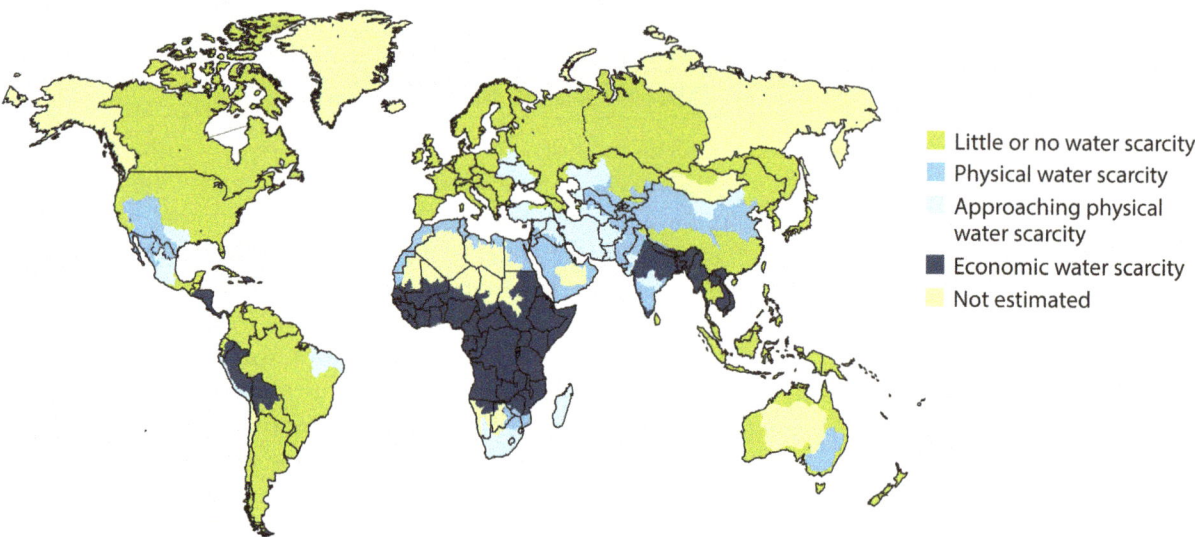

Source: World Water Development Report 4, World Water Assessment Programme (WWAP), March 2012

The already existing depths of the water crisis, seen against the future possibility for solar-driven changes in regional terrestrial water cycles, defines the top-down overview of a single planetary crisis to be addressed in the latter half of this article.

A. Policies Accelerating the Crisis

The combined impact of many factors, such as depletion, physical and economic constraints, and intentional obstruction of infrastructure and technologies, has brought the world to its present point of water availability crisis. Yet, despite this, certain nations are accelerating additional practices making an already bad situation, catastrophic. The most glaring are hydraulic fracturing and biofuels. These policies not only waste water, but they waste water for energy sources that are effectively a net drain on their economies.

Hydraulic Fracturing

The fast-expanding use of water for hydraulic fracturing (fracking) for oil and gas extraction is a direct threat to water availability in certain areas. While the specific water requirements vary per well, depending on the type of shale deposit (e.g., how many times the hole is fracked or what quality of water is used), the practice is clearly detrimental.

In the United States—the world leader in fracking—and Canada, nearly half (47%) of oil and gas wells opened by fracking *are located in areas of high water stress*, including California, North Dakota, and Wyoming in the High Plains and Texas and New Mexico. Drought-stricken Texas leads all states in the number of such wells, with more than 9,000 opened in extremely water-short areas, and another 9,000 in dry-prone locations. Only about 5% of all water used for fracking in these areas has been recycled; that is, 95% is "consumed" and gone. The volume of water consumed in these wells overall in the United States and Canada, over a 2.5 year period, amounted to 367 million m^3 (97 billion gallons). That is an annual rate equivalent to the municipal water use of a city of 1 million people.

While this is already a waste of water occurring in the context of an existing crisis, there is a push to greatly expand this insane policy.

Biofuels

The production of ethanol, biodiesel, and gasohol is underway at levels diverting huge volumes of water for biomass agriculture and processing—a direct loss to food production, as well as a waste of water. World ethanol production for 2014 is expected to reach a record 90 billion liters (23 billion gallons). In the United States, the world leader in ethanol, fully 40% of the annual corn harvest is now going for biofuels.

The water required, ranges from 7 liters of water for every 1 liter of corn-derived ethanol, up to 2,000 liters, depending on whether the corn is irrigated. Thus 90 billion liters of ethanol worldwide consumes at least 637 billion liters of water, equivalent to the annual municipal water use of a city of 4.5 million people.[7]

Backward Policies

This issue is not only that these two specific energy policies, fracking for natural gas and biofuels, use a lot of water. If they also provided an energy source that could upgrade the entire economy, then they could be part of an overall upshifting of the economic system—*but this is not the case.*

The future of mankind's energy needs lies in the domain of nuclear reactions, with fission and especially fusion power. Fracking for natural gas as an energy supply is a step backward, expressing a physical economic phenomenon known as diminishing rates of return—the amount of physical effort and capital supplied is increased in order to acquire the same amount of energy, and thus the energy return per amount of physical economic input is declining. This is characteristic of a typical attrition process, when a resource base is being depleted, and the physical effort of society is increased just to maintain previous levels of production. While technological advance has certainly offset the increased physical cost by increasing the productive powers of labor, much higher order energy sources are available with nuclear reactions, rendering fracking for natural gas as a source of power, a net loss to society.

To a certain degree, the biofuels program is even more insane. In the United States, for example, the production of biofuels from corn is so energy intensive that the energy provided by the combustion of the biofuels is only 1.3 times the energy put into the production of the fuel.[8] When compared to other major energy sources in the United

---

7. See "Measuring Corn Ethanol's Thirst for Water," April 14, 2009, in the M.I.T. Technology Review.

8. United States Department of Agriculture, "The Energy Balance of Corn Ethanol: An Update," by Hosein Shapouri, et al., July 2002.

States, which provide 10 to 100 times more energy than the energy input required, ethanol corn is the lowest. The pitiful energy payback, combined with the water requirements and the diversion of food needed for consumption, shows support of biofuels through government subsidies to be a criminally insane policy.

Again, the future of energy and power lies in the control of atomic reactions. In addition to providing power, these higher energy flux density systems will enable mankind to solve the world water crisis.

Having briefly touched on policies accelerating the water crisis, we now examine the challenges posed by groundwater depletion and surface water deficiency (and/or lack of management).

B. Groundwater Depletion

The location and condition of major world aquifers has been mapped by many science agencies, in particular, UNESCO (U.N. Educational, Scientific, and Cultural Organization), whose International Hydrological Program in 2008 made available an extensive world database. The drawdown of groundwater resources in many places has reached the crisis stage, necessitating ever deeper pumping, while producing poor quality water. Many regions are suffering land subsidence.[9] **Figure 3** shows this for the High Plains (Ogallala) Aquifer in North America.

The general reason groundwater supplies can be problematic is that many aquifers have relatively slow recharge rates.[10] Returning to the opening concept of the terrestrial water cycle, the ultimate source for all groundwater is precipitation brought over land by the action of solar radiation. This is what built up freshwater aquifers, and is the process that maintains them. For many aquifers their cycle is so slow that it is easily outpaced by human activity. This results in a drawing-down of the cycle, and mankind must either accelerate the cycle (through the creation of new re-

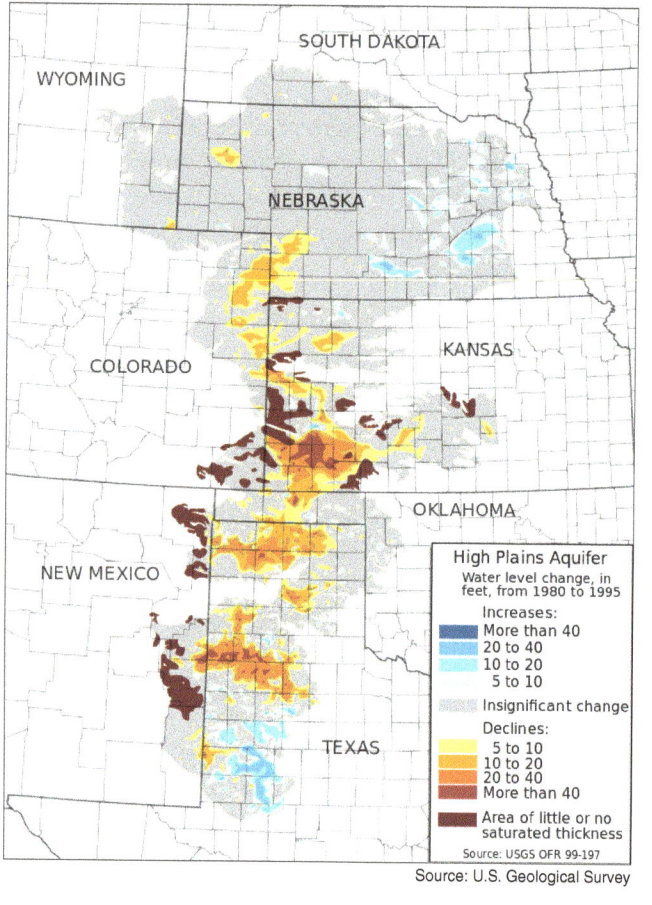

FIGURE 3
Groundwater Level Decline in High Plains Aquifer, North America, 1980-1995

Source: U.S. Geological Survey

---

9. However, there are certain critical locations where accessible, usable groundwater has not been exploited at all, and should be developed to secure immediate water needs as a step toward the development of more advanced systems. For example, in Africa, in northwestern Sudan, where people are desperate for lack of water, there has been no infrastructure installed (pumps and storage and transmission systems) to make use of the Nubian Aquifer. A policy for peace in that region should provide plentiful water for agriculture, food processing, and domestic use for all in this region. This was called for by Dr. Farouk El Baz, space geologist and a specialist in satellite remote sensing and identification of water under the desert. See, "Farouk El-Baz, Ph.D.: Geologist Proposes 1,000 Wells for Darfur; Use Science To Serve Mankind," *EIR*, Sept. 14, 2007.

10. Some aquifers do have fast recharge rates, while others no longer recharge at all, and represent finite stores of "fossil water."

charging systems), or create new cycles to bring in water to augment or replace groundwater use. To illustrate this, examine three examples from the United States.

The Ogallala aquifer, one of the largest in the world, supports about one-quarter of the irrigated land in the entire United States, and provides drinking water for 2 million people. According to a 2007 report by the United States Geological Survey (USGS), the water available in the entire aquifer is nearly 10% less than in 1950, and about 310 km^3 less than its "predevelopment" levels (in some regions, the water level can fall 5 feet in a year).[11] According to a 2002 USGS report to Congress, the annual depletion rate averaged over the 1987 to 1999 interval was about 5 km^3 per year[12] (about equal to California's allocation of the Colorado River), and the depletion rate since then has been increasing.

Another example is California's Central Valley. Covering 60,000 km^2, less than 1% of the United States' total farmland, the Central Valley produces 8% of the nation's agricultural output (by value), making it one of the world's most productive agricultural regions. According to a February 2014 water advisory from the University of California Center for Hydrologic Modeling, the Central Valley aquifer has lost about 75 km^3 between 1962 and 2013; that is, the groundwater is being withdrawn at rate of 1.5 km^3 per year faster than it is being recharged.[13] With the California drought intensifying, groundwater use is accelerating.

A third example from the western United States is the Colorado River basin. The basin covers well more than half a million square kilometers, and the Colorado River itself supplies water for more than 33 million people across seven states, although its flow has significantly diminished over the past decade. In July 2014 a research team led by scientists from NASA and University of California, Irvine, determined that the basin lost 65 km^3 of freshwater between December 2004 and November 2013—nearly twice the volume of Nevada's Lake Mead, the largest reservoir in the United States. This is a rate of loss of more than seven km^3 per year (nearly half the current flow of the Colorado River), and the study determined that 75% of this loss is from groundwater.[14]

Looking at the global picture, the going estimate is that worldwide aquifers have been drawn down some 20% from their former levels in recent, modern history. The annual worldwide groundwater drawdown is increasing by 1-2% each year (having tripled over the

---

11. "Changes in Water Levels and Storage in the High Plains Aquifer, Predevelopment to 2005," V.L. McGuire, http://pubs.usgs.gov/fs/2007/3029/

12. "Report to Congress—Concepts for National Assessment of Water Availability and Use U.S. Geological Survey Circular 1223," 2002.

13. "Water Storage Changes in California's Sacramento and San Joaquin River Basins From GRACE: Preliminary Updated Results for 2003-2013," UC Center for Hydrologic Modeling, University of California, Irvine; UCCHM Water Advisory #1, February 3, 2014.

14. The loss isn't all from pumping, but also from the drought conditions. "AGU: Satellite study reveals parched U.S. West using up underground water," July 24, 2014. NASA, AGU joint release.

past 50 years), with a crude estimate for 2010 being about 1,000 km³ withdrawn.[15]

Again, these aquifers are not finite stores; they are being continuously replenished, but at rates too slow to match the needs of society. As discussed below, one example of a solution for the cited three aquifers and regions of the United States had been proposed a half-century ago—moving to a higher level of control through the management of the continental water cycle that subsumes these particular aquifers—the NAWAPA proposal.

C. Surface Water Deficiency

Despite the fact that an immense amount of freshwater precipitates over land, the distribution of surface water is terribly uneven. The interactions of climate, geography, and weather ensure the "natural" availability of water differs dramatically for different regions of the planet, different regions of the same continent, or even different regions of a nation, state, or province. This is reflected in **Figure 4**, which illustrates the regions of "water stress," where existing water supplies are inadequate to meet the needs of society.

Rather than living with such disparity, mankind can manage and improve these terrestrial water cycles in two categorical ways: by ensuring the surface water is distributed in a useful manner, and by increasing the productivity of the existing cycles. This is largely accomplished through basic infrastructure systems such as dams, reservoirs, canals, pumps, irrigation, water purification, and sanitation.

In most regions this type of development has been held back, and huge potentials remain unutilized. There are two outstanding exceptions—in North America in the first half of the 20th Century, with the Tennessee and the Colorado/California River Basins; and today in China, with the great South-to-North Water Transfer System, regulating and redirecting flow from the Yangtze Basin to the Huang Ho Basin.

In the 1930s, Franklin Delano Roosevelt's Tennessee Valley Authority (TVA) tamed the wild and erratic conditions of the rivers of the Tennessee Valley. Periodic flooding and fluctuations in water availability severely limited the potential for development. The TVA constructed a series of dams and reservoirs to ensure the steady and regular flow of water, during times of excess and during times of scarcity. This greatly improved agriculture, and enabled new navigation and transportation, as well as the development of hydroelectric power. The project transformed a region that suffered from poverty and diseases such as malaria, into a cornerstone of the most advanced scientific research project ever conceived at the time, the Manhattan Project (see section on the TVA).

---

15. Of this, 67% is used for irrigation, 22% for domestic purposes, and 11% for industrial purposes. See, "United Nations World Water Development Report 2014," Chapter 2; "Water and Energy Vol. 1," UNESCO; and "Water Balance of Global Aquifers Revealed by Groundwater Footprint," Gleeson et al., *Nature* magazine, Aug. 8, 2012.

FIGURE 4
Global Water Stress Indicator (WSI) in Major Basins

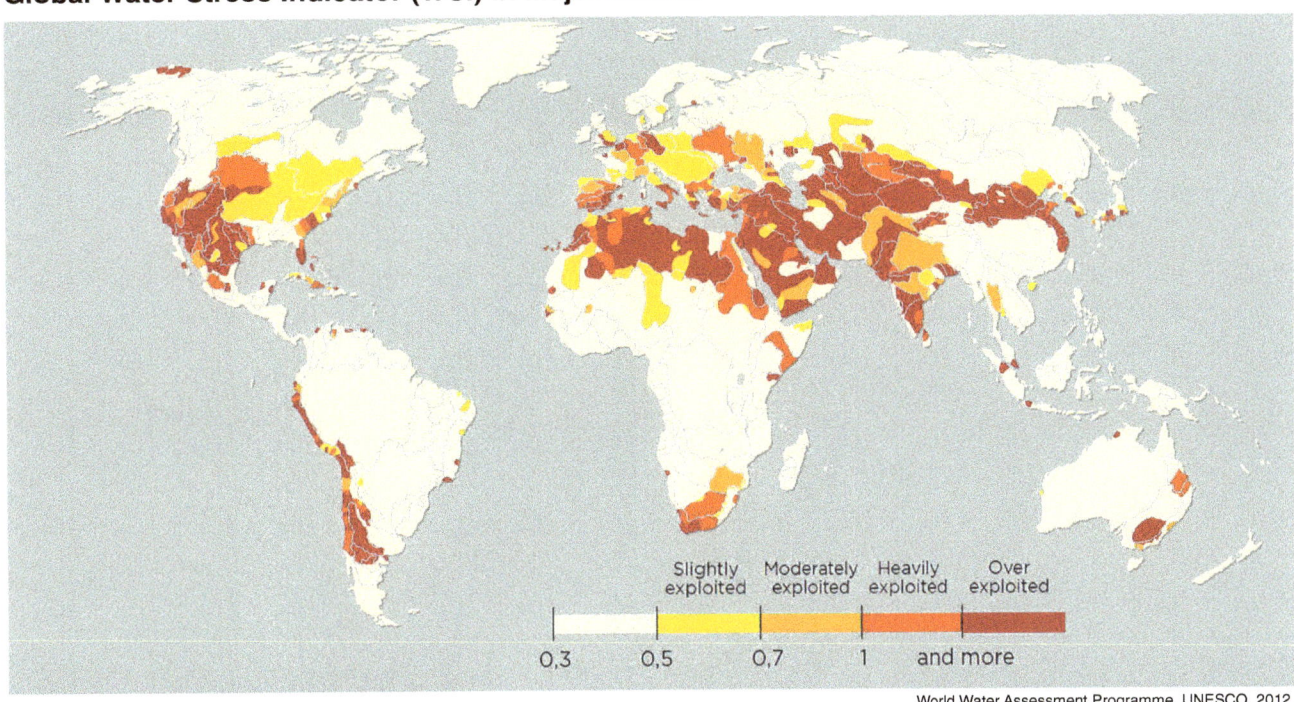

World Water Assessment Programme, UNESCO, 2012.

The dry western region of the United States faced a greater challenge. The larger territory of the Colorado River basin was brought under control through a series of major dams, reservoirs, and irrigation and related systems, led by the famous Hoover Dam and Lake Mead. Currently the agricultural, industrial, and domestic needs of up to 40 million people depend on the management of the Colorado River with dozens of major dams, hundreds of miles of canals, and irrigation water provided for 16,000 km². The development of the West continued with the Central Arizona Project branching off the Colorado River, and the California Water Project regulating the flow of the Sacramento and San Joaquin rivers. This is how California's Central Valley became a breadbasket for the nation, producing nearly one-tenth of the nation's crops on less than 1% of the national farmland.

However, while each of these regional developments in the American southwest have been highly successful, the total water flow of the Colorado, Sacramento, and San Joaquin rivers is relatively small for the size of the land area to be supported by them. Starting in the 1950s and 1960s, it was recognized that the larger issue that needed to be addressed is the great *continental* discrepancy between water excess in the northwest, throughout Canada and Alaska, and the water scarcity in the southwestern United States and northern Mexico. Measured by river flows, this northwestern quarter of the continent has about ten times the water availability of the southwestern quarter.

By the 1960s, designs for the grand North American Water and Power Alliance (NAWAPA) were developed to rectify this great im-

FIGURE 5
NAWAPA, PLHINO, PLHIGON and Photosynthesis

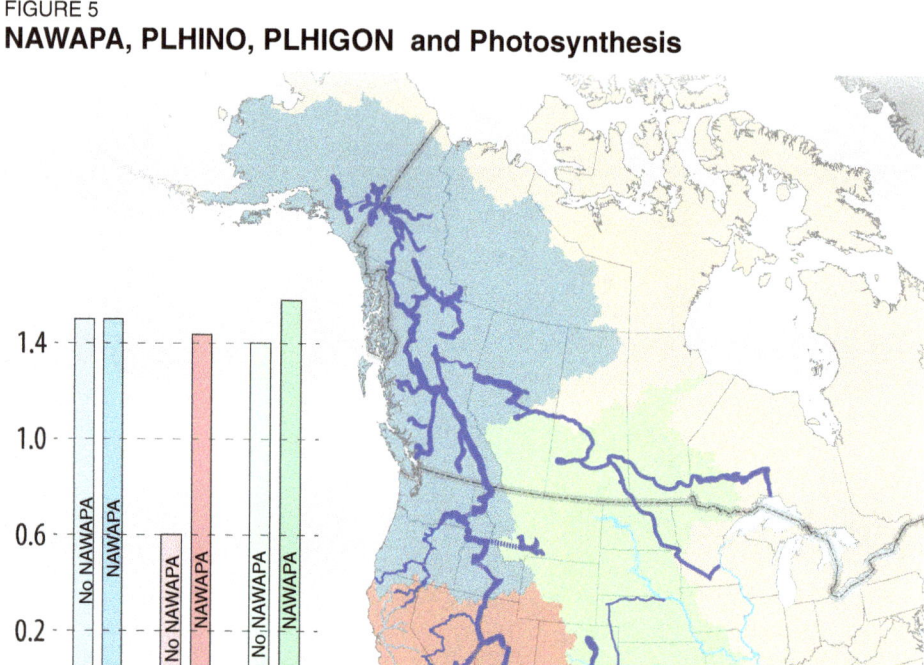

Source: Benjamin Deniston/EIRNS 2014

The waterways shown are the North American Water and Power Alliance system, diverting run-off from Alaska and the Yukon southward through the Southwest; and in Mexico, the diversion northward of water from the Southern and Western Sierra Madre run-off, through the systems of PLHINO (Northeast Hydraulic Plan) and PLHIGON (Northwest Hydraulic Plan). See box on photosynthesis next page.

balance, by proposing a continental water management system that could bring approximately 20% of the freshwater runoff from select rivers in the northwest, down throughout the southwest. From 2010 to 2014 the LaRouche PAC Basement research team re-examined NAWAPA and proposals to further augment and expand the project.[16] (See **Figure 5**) When the potentials for expansion are taken into account, NAWAPA could increase the water available for entire southwestern states by between 50% and 200%, and first order estimates indicate that it could increase the photosynthetic productivity of the water cycles of the western river basins by 30%, and the photosynthetic productivity of the entire North American continental water cycle by 10% (see box, "Increasing the Physical Productivity of the North American Water Cycle").

---

16. See, "Nuclear NAWAPA XXI: Gateway to the Fusion Economy," 21st Century Science and Technology, 2014.

> ### Increasing the Physical Productivity of the North American Water Cycle
>
> The North American continental water cycle can be estimated to be about 3,150 km³ per year (as measured by freshwater river runoff). Of that, 1,466 km³ flows out of the northwest, and only 113 km³ from the southwest. Using measurements and analysis from NASA earth monitoring satellites, the total amount of photosynthetic production can be estimated for these same regions. Comparing these two values allows for a simple, but insightful measure of the productivity of the continental water cycle, and of its respective basins. The figures below are measuring "billions of tonnes of photosynthesis per year" divided by "cubic kilometers of freshwater runoff per year," to measure the productivity as "tonnes of photosynthesis per cubic kilometer of freshwater flow."
>
> **North America:** 7.4 billion tonnes / 3,150 km³ = 2.3 million tonnes per km³
> **Northwest:** 1.5 billion tonnes / 1,466 km³ = 1 million tonnes per km³
> **Southwest:** 0.6 billion tonnes / 113 km³ = 5.5 million tonnes per km³
> **High Plains:** 1.2 billion tonnes / 251 km³ = 4.8 million tonnes per km³
>
> These figures show, in terms of photosynthetic production, the water of the southwest is five and a half times more productive than the water of the northwest. This is a confirmation of what is intuitively clear, there is an excess of freshwater in the northwest, where the cold climate and lack of sunlight limit a more productive use of that water. By these values, a first order estimation of the effect of NAWAPA can be made, by estimating the potential increase in photosynthesis, and the increase in the productivity of the continental water cycle.
>
> **Southwest:** 159 km³ of new freshwater from NAWAPA, at a productivity of 5.5 million tonnes per km³, could increase the annual photosynthesis of the southwest from 0.6 to 1.5 billion tonnes.
> **High Plains:** 37 km³ of new freshwater from NAWAPA, at a productivity of 4.8 million tonnes per km³, could increase the annual photosynthesis of the High Plains from 1.2 to 1.4 billion tonnes.

Because water availability in the northwest is not a limiting factor in photosynthesis, taking this relatively small fraction of water from there would have minimal effects on northwest photosynthetic productivity. Thus the total productivity of the water cycle of the western regions (northwest, southwest, and High Plains) could increase from 1.8 to 2.3 tonnes of photosynthesis per km³—a nearly 30% increase.

The productivity of the entire continental water cycle (including regions not directly affected by NAWAPA) could be increased from 2.3 to 2.6 tonnes of photosynthesis per km³—a 13% increase for the entire continental cycle, done without increasing the net water input, but by better management of the existing cycle.

However, despite significant support, the NAWAPA project was killed by the zero-growth movement gaining power by the late-1960s and 1970s.[17]

The TVA was seen as a model the world over, and variations on it were proposed in many countries. In some, such as the Indus Valley in southern Asia to the North of Scotland and the Murray-Darling River basin in Australia, variations were applied. In other areas, such as Jordan, Africa, South America, and Southeast Asia, plans to apply the TVA model were developed, but blocked.

In dramatic contrast, China's grand inter-basin South-North Water Diversion (SNWD) project now stands as the near solitary, but exemplary, model of large-scale surface water organization. The three-route SNWD complex, shown in **Figure 6**, is now partially complete. The concept is to convey a portion of the abundant water supplies in the monsoonal southern Yangtze

---

17. See the 2011 feature documentary, NAWAPA 1964, http://larouchepac.com/nawapa1964

FIGURE 6
South-North Water Diversion Project

Sources: Chinese Ministry of Water Resources; futuretimeline.net; Will Fox

system, to the arid north. First proposed in the 1950s, designs were debated for decades; then in late 2002, construction began, and since 2009 the project has been accelerated.

The Eastern Route Project (ERP) became operational in December 2013, delivering water to the eastern provinces of Jiangsu, Anhui, and Shandong. By 2015, water in the Middle Route Project (MRP) will flow to Beijing, Tianjin, and environs. In September 2014, testing of water quality began on the MRP, preparatory to activating the full flow. The Western Route, which would capture and divert water from three tributaries of the upper Yangtze River, is still in the planning stages; it involves demanding engineering and construction work.

The SNWD dimensions are significant. The Eastern Route uses upgrades on the 1,500-year-old Grand Canal, a waterway likewise linking the south to north. Today, the ERP transports some 14.8 billion m^3 of water a year. The Middle Route will carry 9-13 billion m^3. This channel required 1,400 km of new construction, with its starting point at the Danjiangkou Reservoir.

The concept of the Western Route is to transfer flows from the headwaters of the Yangtze into the headwaters of the Yellow River, to augment its flow. The hydro-engineering involves major dams and tunnels to move the water across the Qinghai-Tibetan Plateaus and Western Yunnan Plateaus, and to cross the Bayankala Mountains. These watersheds are all within China's borders; initial feasibility studies are in hand.

When complete, the three-route SNWD would transfer 20 to 40 billion m^3 from the Yangtze Basin to the dry north. In addition, there is the idea of diverting northward, some of the flow of the transboundary rivers—the Brahmaputra, Salween, and Mekong, which would vastly increase overall SNWD volume.

Against this background sketch of what has been done, what could be done, and what must be done with respect to surface water management, recent evidence indicates the need to re-examine all aspects of the global water crisis from a higher perspective, starting from the Sun.

D. Solar-Driven Climate Changes

Return again to the basic concept of the terrestrial water cycle. Ultimately all surface and groundwater stores and flows depend on the precipitation of evaporated ocean water, and, as is now being learned in the western regions of the United States, there is no basis to assume that these precipitation patterns are static, unchanging systems. Recent studies of the climate history of this western region indicate that the past thousands of years have seen extreme variations, ranging between so-called mega-droughts to mega-floods, and, against this longer background, the 20th Century had been one of the most stable and wet centuries on record.[18]

It appears the western United States could now be departing from this lucky period of climate stability and relative moisture availability. One example could be the above-cited Colorado River, which averaged a flow of 20 km^3 per year from 1900 to 2000, but a flow of only 15 km^3 per year from 2001 to 2011, and with the accelerating groundwater loss, the river's flow is expected to fall further.

This is just one example of the types of changes in climate and precipitation patterns that regularly occur, challenging existing water management systems. Many factors can be involved in such changes, including cyclical and other changes in the ocean systems and changes in the biosphere, but here we focus on the activity of the Sun. While it is not the only factor involved, changes in solar activity is one of the most ignored and important factors.

---

18. "The West Without Water: What Past Floods, Droughts, and Other Climatic Clues Tell Us About Tomorrow," by B. Lynn Ingram and Frances Malamud-Roam, Berkeley; University of California Press, 2013

Solar Cycles, Grand Minima, and Regional Climates

The Sun goes through a roughly 11-year cycle, as measured by the increasing and decreasing number of sunspots visible on its surface. While sunspot counts are the most long-standing observational measure of solar cycles (with regular records going back to the 17th Century), we now know these are just one expression of much more dynamic, and little understood, periodic changes of the Sun's activity, changes which extend far beyond the Sun's surface, permeating the solar atmosphere which envelops all the planets, including Earth.

While the average length of a solar cycle is 11 years, the actual length of a given cycle can vary, as can its strength. There can be longer periods of a series of strong solar cycles (measured by large numbers of sunspots), periods of a series of weaker solar cycles, or even periods where the sunspots seem to disappear for decades (see **Figure 7**). For example, the first few cycles of the 19th Century were very weak, defining a period known as the Dalton Minimum. Earlier, between 1650 and 1700, there were few or no sunspots at all, as if the solar cycle simply disappeared for more than half a century, a period now known as the Maunder Minimum. This has been called a solar "grand minimum," and was just the most recent of several grand minima over the past 1,000 years.

The period of the Maunder Minimum is famous for another reason; it corresponds to the time of the little ice age throughout Europe. The prospect has been raised, that perhaps these periods of solar grand minima can have significant influences on the Earth's climate systems.

There are now many studies that point in this direction. From a survey of various investigations of past climate and hydrological variations in locations all around the globe, an interesting pattern emerges. During periods of solar grand minima, multiple records from the northern regions show evidence for significant cooling (at least four different sites across Eurasia); records from the tropics show an increase in average precipitation (at least three different sites across Africa and South America); and records from the subtropics show less precipitation and increased drought (at least ten different sites across Asia and the Americas).

Typical of these studies is a 2012 paper by members of the Chi-

FIGURE 7
400 Years of Sunspot Observation

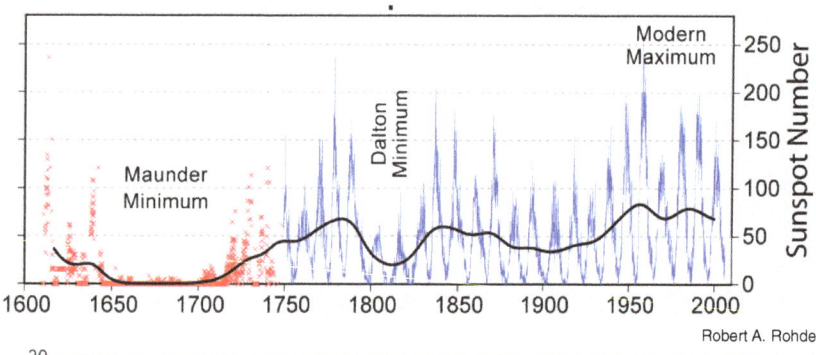

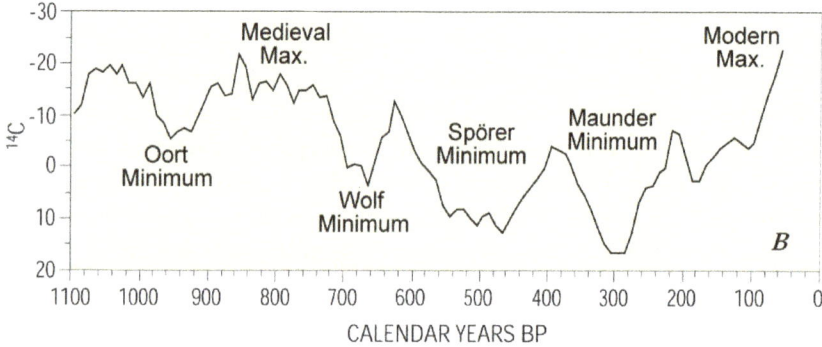

Solar activity over 1,100 years, measured by changes in production of carbon-14 in the atmosphere. More carbon-14 is produced by the increased galactic cosmic radiation the Earth experiences when the solar activity is low.

nese Academy of Sciences, which used tree ring measurements from the Tibetan plateau (where trees are very sensitive to water availability) to show that periods of low solar activity and solar grand minima correspond to periods of drought.[19] Different studies have been done for the South China Sea, Pakistan, Southwest Asia, eastern India, and several sites in the region of the Caribbean, Central America, Florida, and Mexico—all indicating less precipitation during periods of solar grand minima.

While there is still much to understand about the effects of changing solar activity on the Earth's weather, climate, and hydrological systems, there is increasing evidence that the Sun is currently weakening, and could be going into a new period of prolonged lull, perhaps a new grand minimum.

No one knows for certain what the Sun will do, and no one knows for certain what the exact effects of a new solar grand minimum would be. But we do know that dramatic shifts in climate and hydrological patters do occur, and they have been associated with solar variations in the past. However and whenever such shifts occur, mankind must be prepared to handle such changes.

This means that solely relying on existing patterns of precipitation, and the existing levels of surface and ground water flows created by those precipitation patterns, may not be enough. Even large-scale river diversion systems could be vulnerable to such shifts. Ultimately the future of water on this planet requires not only managing historical surface and groundwater flows, but investigations into managing the subsuming atmospheric moisture flows, and large-scale desalination for the creation of completely new, man-made terrestrial water cycles.

This completes the review of the qualitative characteristics of the water challenges facing mankind, laying the basis to examine to concepts needed to solve these issues.

II. The Future of Water: A Global Perspective

The technology exists to develop new water resources. Working from the standpoint of the terrestrial water cycle, we can define two broad categories of action. First, managing and improving use, efficiency, and productivity of existing terrestrial water cycles. Second, recognizing the expected fluctuations in these natural cycles, mankind must be prepared to go to a higher level of control over existing water cycles, and even creating new ones.

In the first category, managing existing cycles, there are two subcategories of action that can be taken. First, improved water management systems can increase the efficiency and reuse of existing water supplies, increasing the productivity of existing cycles. Second, river basins and continental river systems can be developed and improved

---

19. "Tree Ring Based Precipitation Reconstruction in the South Slope of the Middle Qilian Mountains, Northeastern Tibetan Plateau, over the Last Millennium," by Sun and Liu, 2012. *Journal of Geophysical Research: Atmospheres*.

with large-scale river diversions, dams, and irrigation systems. Examples of existing and proposed regional and continental river diversion and management systems were discussed above. These are crucial and must be developed, but, as California and nearby states are now experiencing, they may not always be enough.

This takes us to the need to focus attention on the second category—controlling existing water cycles and even creating new ones, and two distinct subcategories that can be defined therein. First, atmospheric moisture flows and rainfall can be influenced, and potentially controlled, with weather modification technologies. Second, with the development of abundant power from nuclear fission and thermonuclear fusion level economies, new freshwater resources can be produced with ocean desalination systems on a large scale. First, on weather modification.

A. Resources 'In the Sky'

Referring back to the global water cycle estimates, less than 10% of ocean evaporation precipitates over land, meaning there is an immense store of untapped freshwater in the atmosphere. The flux of moisture from the oceans into the atmosphere is incredible, equivalent to about 1,000 Mississippi Rivers, flowing up, from the oceans into the sky at all times.

Can this atmospheric water be induced to fall where it is most needed, or kept from falling where it causes harm? Cloud seeding has shown limited success under certain conditions, and various other schemes have been proposed to induce atmospheric moisture to come down to land. However, here we highlight the potential of a lesser-known approach, based on the electrical properties of the atmosphere and of weather systems—that of atmospheric ionization systems.

This technique uses towers and arrays of wires, through which a precisely tuned current is run, ionizing the surrounding atmosphere. Increasing the ionization of the atmosphere can help to facilitate the formation of clouds and rainfall. Operating on the right scale, these systems could be able to draw more ocean moisture over land, increasing the overall terrestrial water cycle. These techniques have been used in Russia, the United Arab Emirates, Mexico, Israel, and Australia. We examine a few case studies.

Case Study: Mexico

In the 1990s, the then-director of the National University of Mexico's Space Research and Development Program, Dr. Gianfranco Bissiachi, began collaboration with a Russian scientist, Dr. Lev Pokhmelnykh, who had worked on weather modification in Russia since the 1980s. Supported by Heberto Castillo, then-president of Mexico's Senate Committee on Science and Technology, in 1996, Pokhmelnykh and Bissiachi oversaw the development of an initial network of three ionization stations based on Pokhmelnykh's designs. The initial results generated enough interest and support that the system was expanded

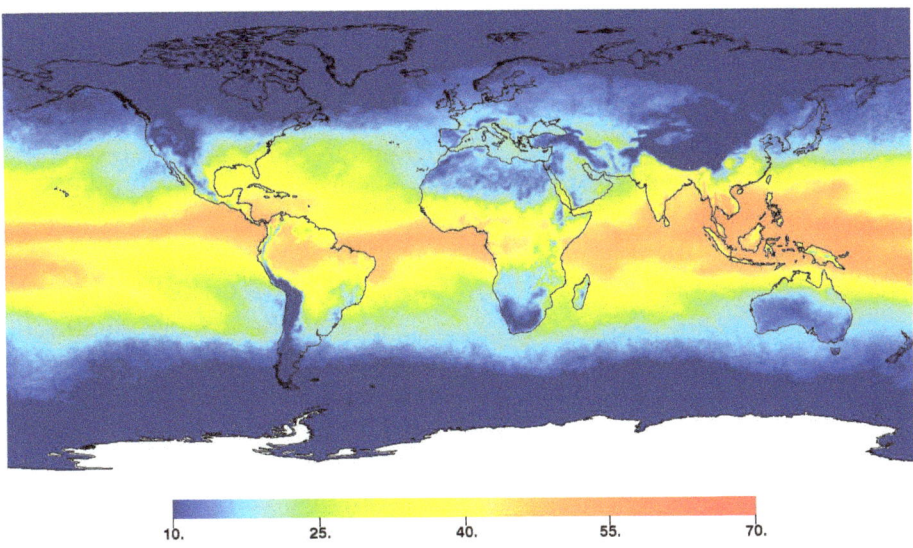

Aqua/Atmospheric Intrared Sounder (AIRS) Total Precipitable Water Vapor (mm) May 2009

from three stations in 1999, to 21 by 2004, and further success led to the expansion to 36 stations by 2006.

In 2003, the Massachusetts science publication *Mass High Tech* ran an article discussing the potential use of ionization systems in the United States, based on the precedent set in Mexico. It describes the success of the first Mexican ionization station as follows:

"That country's first [ionization] station, in the drought-stricken state of Sonora, increased average rainfall from 10.6 inches to 51 inches in the first year, according to Mexican department of agriculture statistics. When a lack of state funds shut down the station the following year, area rainfall measured 11 inches. In the third year, with the station operational again, the area recorded 47 inches of rainfall. [In 2003 the technology was operational] in eight states in the driest regions of Mexico, and some areas [reported] a doubling or tripling of annual rainfall."

In 2004, *IEEE Spectrum* also covered these Mexico operations, citing a doubling of the average historical precipitation in Mexico's central basin, resulting in a 61% increase in bean production in the affected areas. A 2008 paper on the potential use of these ionization systems in Texas analyzed the rainfall levels in the central and southern regions of the Mexican state of Durango. Each year from 1999 to 2003 showed a significant increase in rainfall over the expected levels. The authors of the paper calculated that there was less than a 1 in 400 billion chance that this could have happened by happenstance.[20] Despite these indications of successful results, the Mexico operations have lost the needed financial support.

---

20. See, "Expanding NAWAPA XXI: Weather Modification To Stop Starvation," *EIR*, August 9, 2013.

Case Study: Israel

Lev Pokhmelnykh began developing ionization-based weather modification systems in Israel, using an installation of three stations. Starting in late 2011 they induced increased rainfall in the Golan Heights area, filling seven reservoirs to full capacity, something which has not occurred in the 40 years since the construction of these reservoirs.[21]

Case Study: Australia

In 2007, the weather modification company Australian Rain Corporation was formed, with the intention to develop ionization systems to stimulate rainfall. In 2007 to 2008 the Australian Government's National Water Commission funded some initial trials. From 2008 to 2010, Australian Rain Technologies ran three trial programs:

Paradise Dam, Bundaberg (January-May 2008): Resulted in a 17.6% increase above anticipated rainfall in a 30° downwind arc from the system.

Mt. Lofty Ranges, Adelaide (August-November 2008): Produced an increased rainfall of 15.8% above the anticipated levels over a 120° arc downwind from the system.

Mt. Lofty Ranges, Adelaide (August-December 2009): Generated an increase of 9.4% over an area roughly twice the size of the previous trials.

In 2011, the company submitted a proposal to the Parliament's Standing Committee on Regional Australia, requesting $11 million to construct 14 ionization stations distributed around two catchment areas in southeastern Australia (Gwydir River and Hume-Dartmouth catchment) to increase the rainfall going into the irrigation systems of the Murray-Darling Basin (one of the most significant agricultural areas in Australia, which is facing a major water shortage, largely because of environmentalist-imperial policies).[22]

Case Study: United Arab Emirates

In early 2011, a barrage of media reports covered a leaked report of a weather modification program in the United Arab Emirates. The story broke when the London *Sunday Times* detailed a contract with a Swiss company, Meteo Systems International, to build a series of ionization stations to bring rain to regions of the UAE, including the capital, Abu Dhabi.

The initial coverage claimed evidence for successful operations in 2011, pointing to 52 unanticipated rain showers, and citing interest from numerous scientists involved.

According to the website, the company was started in 2004, ran trials in Switzerland in 2005, and then started trials in the UAE in 2006

---

21. "Inducción Experimental De Lluvias Por Ionización Atmosférica En Las Alturas Del Golán, Israel, En El Período Invernal 2012-2013," by Mario Domínguez and Lev Pokhmelnykh, May 2013.

22. See, "Expanding NAWAPA XXI: Weather Modification To Stop Starvation," *EIR*, August 9, 2013.

and Australia in 2007 before getting funding for an additional trial in Al-Ain, UAE. The website proclaims, "Meteo Systems' WeatherTec™ is an old idea that has been developed and enhanced over years of scientific research and trials."[23]

These four case studies have indicated the potential of these ionization-based weather modification systems. More work needs to be done, and perhaps other methods will be developed, but this opens a critical window into an entire category of action for mankind. Instead of relying on existing precipitation patterns and surface water availability, mankind could potentially take a higher level of control, by affecting atmospheric moisture flows, gaining a greater degree of control over the terrestrial water cycle, and even increasing the rate of the cycle by drawing more ocean moisture over land.

This takes us to the second subcategory of the second general category of action, using higher levels of energy flux density to create completely new terrestrial water cycles through desalination.

B. EFD and Desalination

Everything up to this point has depended on solar evaporation for desalination and transportation of freshwater. But now, for the first time in the entire history of the planet, a new force has emerged.

Mankind can produce freshwater directly from the oceans with desalination systems, opening up the first freshwater production in the biosphere that is not controlled by solar activity. The technology and methods exist, and are improving in their efficiency. What is needed is the mass development of fission and fusion power, in order to be able to expand desalination to the scale needed by mankind. This is a clear expression of the role of higher energy flux density in changing the resources available in an economy.

With the higher quality power sources of fission and fusion, the power available per person in an economy can be greatly increased (the national economic energy flux density), enabling more power to be applied to the development of resources that could not be developed for large-scale use at lower levels of national economic energy flux density. This is the case with the production of water via desalination.

There are already several well-developed industrial methods for the desalination of salt water, processes that have been continually improved over decades, and their merits demonstrated in years of use in large-scale non-nuclear installations and a few small scale nuclear-powered facilities.[24] At present, non-nuclear desalination is providing some freshwater for about 300 million people worldwide, when many millions more are in need, located where no adequate freshwater sources exist.

---

23. Ibid

24. Two main methods involve the use of heat to evaporate water and the use of membranes to filter water. These desalting methods are described in the "Expand Nuclear Power for the World's Survival" section of this report.

The total number of desalination facilities globally, is more than 15,000, almost all of them fossil fuel-powered. The top echelon of large plants in this inventory are concentrated in Southwest Asia—in the Persian Gulf, and recently, in the transJordan in Israel. These big plants account for most of the world's annual capacity of nearly 30 km^3 of freshwater produced by desalination. The thousands of other, smaller, low-volume desalting plants, are mostly located in remote communities, in such places as hotels on resort islands in the Caribbean and Mediterranean and for high-value food processing. While the current world capacity of 30 km^3 per year is impressive—equivalent to twice the Colorado River, or one-fifth the discharge of the Nile River (a sizable increase from the desalination capability of 20 years ago of 5.5 km^3)—today's output is nevertheless far short of what is required to meet the needs of those in the many water-short dryland areas internationally.[25] Recall that in 2011 about 1,000 km^3 of groundwater were depleted, 33 times more than the current global desalination rate.

While the oil- and gas-rich nations of the deserts of southwest Asia (led by Saudi Arabia) have pioneered the development of hydrocarbon-powered desalination, it will be the energy flux density of the nuclear era that will enable the true breakout of desalination on the global scale needed, matching and outpacing the depletion rates of regional water cycles. This can be illustrated with a pedagogical example. As cited above, the Colorado River basin is losing water at a rate of 7 km^3 per year. To provide this much water with the most efficient desalination systems currently available[26] would require a very large amount of power, and an incredible supply of fuel, when using anything but the power of the atom. If coal were used to desalinate as much water as is being lost from the Colorado basin, it would require 6.7 million tonnes of coal per year—enough to fill 67,000 rail cars, equivalent to a train that would stretch the entire length of California, from Mexico to Oregon.

But if the desalination system were powered by a typical uranium fuel cycle for nuclear fission, it would require 100,000 times less fuel by weight, or roughly 50 tonnes per year, which could be transported by a single semi-trailer truck. If the advanced fusion fuel of helium-3 could be developed and used, then only one-third of one tonne of helium-3 would need to be delivered from the Moon to provide the power needed to match the water deficit of an entire

---

25. The geography of priority locations for nuclear mass-output of desalinated water is obvious. It includes the entire Middle East–North Africa region; the southern Indian Subcontinent; Southwest Asia from the Mediterranean through to Pakistan; the water-short areas of the Pacific Rim in northern China, southwestern North America, and along the west coast of South America; parts of the South Atlantic, including northeastern Brazil and Southwestern Africa; and parts of Australia. In addition, there are priority inland regions of Eurasia, including the Aral Sea Basin, dry-lands of Mongolia, and elsewhere.

26. Using reverse osmosis, operating at the expected efficiency of the new desalination plant being developed in Carlsbad, California (10.8 megajoules per m^3 of desalinated water).

river basin (20 million times less fuel than coal). This one-third of one tonne could fit in the back of a regular pickup truck.

This is just one illustration of the five to seven orders of magnitude difference in the energy density of nuclear reactions over any form of chemical reaction. While fusion power is being developed, the most immediate concern will be the development of the nuclear fission systems that can open up this entire new era of water resources for mankind—effectively creating rivers, flowing from the ocean inland. This is beyond water cycle management, and in the domain of water cycle creation, demonstrating the truly unique power of mankind as a creative force on this plant.

See Appendix for a survey of specific nuclear desalination initiatives underway in Eurasia and elsewhere.

C. A Conceptual Synthesis

The current global water crisis is less about where water is and is not, and more about what mankind is, as a uniquely creative force on the planet. Mankind has before him, either the existing capabilities, or the potential to develop the needed capabilities to handle global water systems as a whole (see **Figure 8**).

As discussed above, the hydrological actions available to mankind fall into distinct principled categories.

Category 1 – Managing and improving the productivity and distribution of existing terrestrial water cycles:

  Subcategory A – Improved water management systems can increase the efficiency and reuse of existing water supplies, increasing the productivity of an existing water cycle by ensuring there is a higher amount of productive use per cycle.

  Subcategory B – River basins and continental river systems can be developed and improved with large-scale river diversions, dams, and irrigation systems, to ensure the equitable distribution of water across a given land area.

Category 2 – Modulating, increasing, and creating terrestrial water cycles:

  Subcategory C – Atmospheric ionization technologies are perhaps the beginning phase of a new focus on influencing and controlling atmospheric moisture flows and rainfall, opening the potential to begin to control terrestrial water cycles on a higher level—moving beyond simply dealing with the water that has fallen on land, and into influencing the atmospheric moisture flows that determine the water distribution on land.

  Subcategory D – With the development of nuclear fission and thermonuclear fusion energy flux densities new freshwater resources can be produced with ocean desalination systems on a large scale.

There is no single technology that will solve the global water crisis. All these categories of action must be developed and employed

FIGURE 8
Global Terrestrial Water Cycle Under Mankind's Control

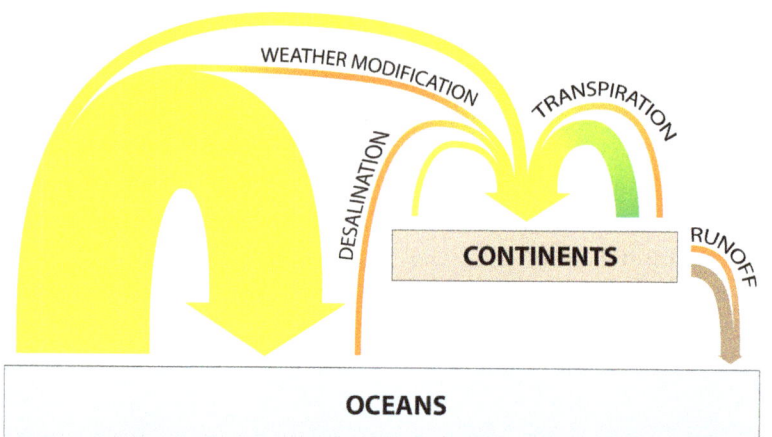

Benjamin Deniston/EIRNS 2014

in the varying degrees required for a particular region. The solution to the global water crisis is for mankind to realize his obligation to develop, scientifically and technologically, as a creative force on this planet.

The great Ukrainian-Russian scientist Vladimir Vernadsky scientifically defined the absolute distinction of the human species from forms of simply animal life. This was expressed, for Vernadsky, by the emergence of the domain of mankind's action, the noösphere, which came to dominate and overpower the biosphere. Today, a new stage of the noösphere is within reach, the expansion of mankind's creative influence throughout the Solar System. Perhaps it is only a small beginning, but the prospect for mankind beginning to control and create our own terrestrial water cycles signifies the emergence of this process.

This is mankind beginning to play a role on planet Earth that was otherwise fully reserved only to the action of the Sun itself. Only in that scientific understanding of the significance of mankind's role on Earth, and beyond, will the global water needs of the human species be addressed far into the future.

Research contributed by Mary Burdman and Marcia Merry Baker

APPENDIX
Initiatives for Nuclear Desalination

By Marcia Merry Baker

September 2014

China, Russia, India, and South Korea all have nuclear desalination commitments and projects at sites in Eurasia, and there are now collaborative projects on other continents, for example, in South America, between Argentina and Russia, China, and South Korea. There are none in the United States, whose western states are in the worst drought in 500 years, and where engineers had active plans in the 1950s and 1960s, under Atoms for Peace and President John F. Kennedy, to build large-scale nuclear desalination facilities for North America and desert locales worldwide. Former president of the American Nuclear Society Edward L. Quinn, in July, called for an urgent revival of nuclear desalination in California.

The International Atomic Energy Agency, founded in 1957, set up the Coordinated Research Project (CRP) in 1998, focused on "Optimization of the Coupling of Nuclear Reactors and Desalination Systems," with participation from nine member nations. The program now has more than 20 countries cooperating, and is a potential framework, on stand-by, for expanded action. At the April 2012 Global Water Summit in Paris, numerous water specialists spoke out on behalf of the prospect of co-location of desalination and nuclear power units.

San Diego County Water Authority 2014

The Carlsbad Desalination Plant, shown under construction in March 2014, on the Pacific coast in southern California, will be the largest in the Western Hemisphere, but is non-nuclear powered and will provide only seven percent of San Diego's water needs.

On Sept. 18, 2014, Russia hosted the first International Expert Council on Desalination, in Moscow, where Russia's Rosatom Overseas nuclear export company stated its readiness to build, or collaborate in building and operating, nuclear desalination plants around the world.

This is a survey of what is in progress, country by country:

China

• China General Nuclear Power (CGN) has the new Hogyanhe nuclear power project in the northeast, at Dalian in Liaoning Province, which will use waste heat to desalinate 10,080 m³/day of seawater to provide its cooling water.

- ACP100 reactors. CNNC New Energy Corporation, a joint venture of CNNC (51%) and China Guodian Corp., began in April 2011 a commitment to R&D for a battery of small modular nuclear reactors for desalination and other applications in industry, as well as power generation. The projects include two small modular integrated ACP100 reactors in Fukian Province (Zhangzhou City), ACP100 reactors at two sites in Jiangxi Province (in Shangrao City and Ganzhou City), and reactors for export.

- Chinese authorities are looking at a seawater desalination facility on the Shandong Peninsula, in the Yantai area, to produce up to 160,000 m^3/day by the multiple-effect distillation (MED) process, using a 200-MWt NHR-200 reactor.

India

- In southeast India in 2002, the Nuclear Desalination Demonstration Project (NDDP) was set up, involving twin 170-MWe nuclear power reactors (PHWR) at the Madras Atomic Power Station in Kalpakkam. The NDDP's desalination comes from a hybrid operation, involving a reverse osmosis (RO) unit (1,800 m^3/day capacity) and a multi-stage flash (MSF) plant unit (4,500 m^3/day), and most recently a barge-mounted RO unit was added. This is the world's largest nuclear desalination plant based on the hybrid MSF-RO technology, according to the World Nuclear Association.

- In Tamil Nadu, the go-ahead was given in April 2013 for an additional desalination operation—using mechanical vapor compression (MVC) to supply 7,200 m^3/day—to be set up at the nuclear power installation at Kudankulam, which already in 2009 had a 10,200 m^3/day MVC plant to supply fresh water to the reactor and to the town.

Nuclear Power Corporation of India

The Kudankulam Nuclear Power Plant on the Indian Ocean, in Tamil Nadu, shown here under construction, now supplies desalinated seawater (MVC—mechanical vapor compression) to the town and the reactor.

Russia

Desalination is integral to several innovative designs for floating nuclear power plant (FNPP) projects under construction in Russia. Small units derived from Russian icebreakers—the KLT-40S reactors—can be either land-based or floating. A pair of such reactors can produce 85 MWe and 120,000 m^3/day of potable water. There are several other configurations. One, known as the Volnolom FNPP, involves a pair of reactors mounted on a 97-meter-long barge with a RO system, to produce 12 MWe plus 40,000 m^3/day of potable water. A larger concept involves a 170-meter-long barge, pontoons, and larger reactors. This plant would have a service life of 60 years and need to be overhauled after 20 years.

Africa

* **Egypt.** On Sept. 6, 2014, President Abdel Fattah el-Sisi, in a national TV address, called for a $12 billion investment over the next five years in power generation. This pushes forward the existing commitment (going back to 1980) for completing the dual-purpose facility underway at Al-Dabaa, on the Mediterranean coast, for power and desalinated seawater. The installation's plans are for four 1,000-MWe-class reactors.
* **Morocco.** China has completed a pre-project study for an Atlantic coast nuclear facility at Tan-Tan. It would use a 10 MWt heating reactor and produce 8,000 m^3/day of potable water by MED. Russia has collaborated on plans for an initial nuclear reactor at Sidi Boulbra, to be completed in 2016-1017; Atomstroyexport is assisting in the feasibility studies.
* **Algeria.** A study was released in 2012 that showed the benefits of locating a dual-purpose nuclear power desalination facility at Mostaganem, on the Western Algerian coast, where rainfall is a very low. Goals were set for power output, in line with the national Indicative Program of Electricity Generation, and for meeting water needs of the region's population for 25 years beyond that projected for 2026.
* **Libya.** In 2007, a memorandum of understanding was signed with France for a mid-sized nuclear reactor on the coast, coupled with seawater desalination. Areva TA was to supply this. This lapsed amid the ensuing bloody regime change forced on the country.

Asia

* In Iran the two new reactors (in addition to the one already in operation) to be built with Russia's Rosatom in its southern Bushehr Province, will each be 1,000 MW and have a desalination unit attached. The first plant already has a desalination unit, which went into operation in the summer of 2014.
* In Jordan, there are active plans under study for nuclear power for desalination and electricity.
* In the Persian Gulf, there are several active plans. Kuwait has

under consideration cogeneration projects, ranging up to producing 140,000 m³/day, from a facility coupled with a 1,000 MWe reactor. The U.A.E. has four nuclear power plants under construction. Qatar has nuclear power desalination under consideration.

• In Southeast Asia, Indonesia has under consideration a large-scale facility in Batan. For Madura Island, a feasibility study conducted with South Korea has been completed, focusing on building a nuclear System-Integrated Modular Advanced Reactor (SMART) with a cogeneration unit using MSF.

South America

• In Argentina, on July 12, 2014, a nuclear power collaboration agreement was signed between Russia's Rosatom and the Argentine government, with Russia's President Vladimir Putin and Argentina's President Cristina de Kirchner stressing that the deal is to include "water desalination facilities." This latest accord is a follow-on to agreements struck in 2010 and since, by Argentina with Russia, China, and South Korea. Russia's Rosatom has submitted a technical and commercial proposal to participate in the construction of the Atucha-III nuclear plant.

Compiled from the World Nuclear Association, and other sources, September 2014.

PART 3
China: Silk Road to Development and Peace

China Becomes a Model Among Nations: A Science-Driver Approach to Lift Up Mankind

By Creighton Jones and Michael Billington

October 2014

Although still a developing nation facing huge problems of poverty and under-development in several regions of the country, the nation of China has become a model among nations. This fact is defined by its determination to base its domestic and international policies on an unswervable commitment to a science-driver economic policy, defined first and foremost by its program of space exploration and development. By so doing, China is adopting the truly human policy of expanding the creative powers of man into and over the Solar system, where his true future lies.

China's leading role in space was signaled in December 2013, when, after an almost 40-year hiatus, it took mankind back to the Moon, with the landing of the Yutu (Jade Rabbit) rover. This accomplishment was part of its expressed mission of exploring, developing, and mining the Moon, with a special emphasis placed on the ultimate use of the chemical isotope helium-3, which is found in relative abundance in the Moon's regolith (lunar soil). The commitment to develop helium-3 as a fuel for thermonuclear fusion-powered energy production—the next frontier in technology revolutions—reflects the level of scientific vision required for the necessary new Renaissance in human civilization, now long overdue.

The Chinese leadership's science-driver approach holds the key for transforming its economic and foreign policies as well. To accomplish its goals, the government has adopted a long-term credit policy, aimed at the promotion of front-end technology-driven, large-scale infrastructure projects, both at home and in its trading partners. These policies are also reflected in its foreign policy outreach, which has greatly expanded its economic cooperation and investment on virtually every continent. In sum, China is proceeding from the standpoint that economic development is the sine qua non for achieving peace.

Preparing for Take-Off

This current upshift toward the achievement of fusion power and industrial development of the Moon by China comes in the con-

China's moon rover, the Yutu (Jade Rabbit), taken from the lander soon after its landing in December 2013.

text of its having already completed, in the past two decades, an extraordinary infrastructure-development program, of which the following are notable examples:

• The largest dam in the world, the Three Gorges Dam, opened in 2008, with the last turbine becoming fully operational on July 4, 2012, with a total electricity generating capacity of 22,500 megawatts. The dam also provides flood control for the lower Yangtze, thus saving the lives of hundreds of thousands who would otherwise die in periodic deluges.

• The third largest dam in the world, the Xiluodo on the upper stretch of the Yangtze, opened in June 2014.

• The sixth largest dam in the world, the Xiangjiaba, also on the upper stretch of the Yangtze, opened in July 2014. Altogether China has more large dams than the rest of the world combined.

• The world's largest electricity generating capacity, increasing at 7.5% per year (building more than the UK's total generating capacity every year).

• The construction of 11,028 km of high-speed rail track since 2000, by far the most in the world, with plans to extend this to 18,000 km by the end of 2015. This includes the longest high-speed rail line in the world, the Beijing-Guangzhou Railway, at 2,398 km.

• The only commercial maglev rail line in the world, connecting Shanghai's Pudong district to the Shanghai International airport.

• The greatest water-transfer program in the world in the South-to-North Water Diversion project. With three routes designed to carry water from the Yangtze River basin north to the arid northern regions, the Eastern Route opened in November 2013, while the Central Route is opening in late 2014. The Western Route is in the planning stage. Altogether, the project will deliver 44.8 billion cubic meters of freshwater annually to the industrial and agricultural regions in the north.

• An expansion of the manufacturing workforce from 85.9 million in 2002 to 105.9 million in 2012—a total three times that of the United States, Germany, and Japan combined, while the manufacturing workforce in those three industrial nations declined during the same period by 10% to 32.9 million.

• The building of hundreds of new cities, with a commitment to build hundreds more by decade's end.

• The construction of 21 nuclear power stations (the first went on-line in December 1991), with 28 more currently under construction, and plans to more than triple its nuclear capacity by 2020. Although that will only bring the percentage of national power provided by

Academician Ouyang Ziyuan, "father of the Chinese lunar program," has been promoting the Chinese lunar exploration program since the 1990s. Here he is shown at a forum in 2012.

nuclear to 6% (compared to the United States at 20% and France at 74%), nonetheless China at the current pace will be the world's leading nuclear power producer within a decade.

• The beginning of construction of the world's longest quantum optics–based communications line, from Beijing to Shanghai, gives further credence to its place as a world technology leader.

All together, these accomplishments are establishing a new technological platform for the Chinese economy, one defined by an increased energy flux density of power production, and increased productive powers of labor—the very opposite of the direction of the ideology being imposed by the dying Trans-Atlantic financial system. The one significant discordant note in this policy is the continued acquiescence of the Chinese government to the international anti-growth environmental movement, which specifically attacks nuclear development, and promotes scientifically wasteful and incompetent energy sources like solar power, wind, and biomass. Actually sustainable development comes from increasingly higher energy-flux-density energy sources, not low-density "green" sources such as the sun.

In the follow-up article below, we elaborate in more detail on China's signature transportation program, which President Xi Jinping has dubbed the Silk Road Economic Belt, and which has become a leading facet of its cooperation with other nations, particularly those among the BRICS nations. But to understand the crucial characteristic that makes such cooperation possible, you have to start from the top: China's policy toward space and energy production.

China's Moon Program and Helium-3

China's perspective toward space and energy is unified from the top. The government has appointed a sole individual, Xu Dazhe, as, simultaneously, the director of three agencies—the China National Space Administration (CNSA); the China Atomic Energy Authority; and the State Administration for Science, Technology, and Industry for National Defense.

China's intentions for the future have been elaborated in discussions among people close to the space-science community, particularly remarks made since China's successful completion of its first soft landing on the Moon on Dec. 14, 2013, with the Chang'e-3 space capsule, and the deployment of its Yutu Moon rover. Famed Apollo 17 astronaut and former U.S. Senator Harrison Schmitt, following the Chinese Moon landing, said, "China has made no secret of their interest in lunar Helium-3 fusion resources.... In fact, I would assume that this mission is both a geopolitical statement and a test of some hardware and software related to mining and processing of the lunar regolith." This is an area that Schmitt knows well, having penned numerous papers and books on the prospect of lunar development and helium-3 mining, and having worked closely with the group at the University of Wisconsin that is developing helium-3 fusion technologies.

The truth of Schmitt's assessment is evident in the words of the "father of the Chinese lunar program," Ouyang Ziyuan, who began lob-

An artist's sketch of mining helium-3 on the lunar surface.

bying the Chinese government for a Moon program in the 1990s, and was finally rewarded in 2004 with the announcement of the China Lunar Exploration Project (CLEP), called the Chang'e Project, of which he became the first chief scientist. Ouyang said as early as 2006: "The Moon has huge reserves of metals such as iron," and "helium-3, an isotope of the element helium, is an ideal fuel for nuclear fusion power, the next generation of nuclear power. It is estimated that reserves of helium-3 across Earth amount to just 15 tons, while 100 tons of helium-3 will be needed each year if nuclear fusion technology is applied to meet global energy demands. The Moon, on the other hand, has reserves estimated at between one and five million tons."

In addition, Ouyang stated, in a BBC interview of Nov. 29, 2013, that "The Moon is full of resources—mainly rare Earth elements, titanium, and uranium, which the Earth is really short of, and these resources can be used without limitation.... There are so many potential developments—it's beautiful—so we hope we can fully utilize the Moon to support sustainable development for humans and society."

Ouyang identified three motivations for going to the Moon: "First, to develop our technology, because lunar exploration requires many types of technology, including communications, computers, all kinds of IT [information technology] skills and the use of different kinds of materials. Second, in terms of the science, besides Earth we also need to know our brothers and sisters like the Moon, its origin and evolution, and then from that we can know about our Earth. Third, in terms of the talents, China needs its own intellectual team who can explore the whole lunar and solar system—that is also our main purpose."

These stated motivations underscore the recognition on the part of China of the role that science-driver programs play in expanding the technology and growth of the nation as a whole. This is something that was once better understood by government layers in the United States, who would often quote the fact that the Apollo missions yielded a 10-to-1 return on investment, from technological spin-offs and increased production capabilities, as well as establishing many high-tech industrial firms, which prospered long after the Moon missions were complete.

Preparing for a Manned Mission

Rightfully declaring the Chang'e-3 mission a success, despite a glitch in the Moon rover Yutu's circuitry, China is currently putting forward a clear statement of the next steps in its Lunar Exploration Program.

Chinese scientists and engineers are working on designs for a lunar base that will include "new energy development and living space expansion," according to a manager of the Chang'e-3 spacecraft, speaking at the Shanghai Science Communication Forum, as reported in *Peoples' Daily* on Jan. 8, 2014. Zhang Yuhua affirmed that China's lunar sample return mission, Chang'e-5, is now scheduled for 2017, an acceleration of the original timetable, because of the success of the current mission. He stated that the interim launch of Chang'e-4 will not be a repeat of the current mission, but will incorporate some of the new technologies needed for the highly complex later sample return. Returning the planned five pounds of samples of lunar soil and rocks will allow a detailed analysis of the Moon's

China Space Timeline

China's National Space Administration was established by the government in 1992 as China's first civilian space agency, in order to implement national civilian space development plans and engage in international cooperative programs. The end of the Cold War, with the demise of the Soviet Union, led the Russian Federation, at the same time, to establish a civilian space agency.

Space Technology Milestones
- 1970: China launches its first Earth-orbital satellite, the Dongfanghong-1.
- 1984: The first geosynchronous communications satellite is launched.
- 2000: The first satellite in the Beidou experimental satellite navigation system becomes operational.
- 2003: A three-satellite experimental navigation system goes into operation with limited service to Chinese receivers.
- 2007: The first satellite launch of the second-generation Beidou system is carried out, to eventually include 35 satellites for global coverage by 2020.
- 2011: The first 10 Beidou-2 satellites are operational, providing service to China and the region.

Manned Program
- 1992: The government approves a manned space program.
- 1999: The Shenzhou-1 mission carries out an unmanned test of a manned space capsule.
- 2003: The Shenzhou-5 mission is China's first manned mission, with Yang Liwei carrying out 14 orbits of the Earth.
- 2005: Shenzhou-6 carries a two-man crew for a five-day mission.
- 2008: Shenzhou-7 achieves the first spacewalk, performed by two members of its three-man crew.
- 2011: The unmanned Tiangong-1 space module is launched.
- 2011: The unmanned Shenzhou-8 space capsule docks automatically with Tiangong-1.
- 2012: Shenzhou-9 docks with the Tiangong-1 module, and the crew includes the first Chinese female astronaut, in a crew of three.
- 2013: Shenzhou-10 has a three-man crew, with the second female astronaut, who teaches lessons from the Tiangong-1 module.

Lunar Program
- 2004: The government approves a three-phase lunar exploration program.
- 2007: The Chang'e-1 spacecraft orbits the Moon, mapping its surface.
- 2010: Chang'e-2 is launched, tasked with mapping the Moon's surface in greater detail. It later leaves lunar orbit for deep space, and flies by asteroid 4179 Toutatis in 2012.
- 2013: The Chang'e-3 lander sets down on the lunar surface and deploys the Yutu (Jade Rabbit) rover.

Marsha Freeman

minerals, chemistry, and other characteristics, which is a necessary step to precede sending people there. Zhang described the activity of a lunar base as setting up agricultural and industrial production, producing medicines in the vacuum environment, and "energy reconnaissance."

Add to this, the long-term intentions of the lunar program as stated by Luan Enjie, a senior advisor to China's lunar program, who told state media in December 2013 that the ultimate aim was to use the Moon as a "springboard" for deep space exploration, which many experts acknowledge would require a base on the lunar surface.

While it has not been made official by the government, it is clear to many that a manned landing on the Moon will one day appear on the horizon for the Chinese, as part of their continued expansion of manned space exploration. In a Jan. 13, 2014 interview with *Science*, Chinese premier Li Keqiang spoke about the manned program, saying, "China's manned space and lunar probe missions have a twofold purpose: First, to explore the origin of the universe and mystery of human life; and second, to make peaceful use of outer space.... Peaceful use of outer space is conducive to China's development. China's manned space program has proceeded to the stage of building a space station, and will move forward step by step.... As human life is precious, we will start with robotic exploration before gradually expanding manned space exploration. Space is all too mysterious. We need to take risks, but not at the cost of human life when conditions are not yet right."

More recently, China has made a number of announcements which concern both near-term and longer term plans for the Moon. For example, China will conduct simulation tests on the return to Earth of the Chang'e-5 lunar probe at the end of this year.

The longer term goals are made clear by the report of the completion of a 105-day test of the "Lunar Palace 1" (Permanent Astrobase Life-Support Artificial Closed Ecosystem), a facility created to test the requirements for a life-supporting Moon base.

As reported on us.news.cn on June 26, "The three 'Moon dwellers' drank recycled purified water, ate worms and food they grew themselves, conducted experiments, and chatted with their family on the Internet in the enclosed capsule from February 3 to May 20." Chief designer and lead scientist Liu Hong commented, "Lunar Palace 1 is different from Biosphere 2," an Earth systems science research facility in the United States. "Biosphere 2 is a duplication of the living environment on Earth, which is a failure we did not want to repeat. The system we made was directed toward the needs of humans. We carefully chose what plants, animals, and micro-organisms would be best included in the ecosystem.

" 'Many foreign experts think building a space base cannot be achieved in the near future, so they do not put many resources into research in this field,' says Liu, 'But the length of time needed to understand the complexity of an eco-system is why scientists should start experimenting now.' Liu says, 'It is necessary to build two mini Lunar Palace 1 systems—a monitoring station on the Moon and one on Earth—so the two sets of data can be compared,'" us.news.cn reported.

China is also deep into the development phase of a mission to return samples from Mars, with plans to land a probe on the Red Planet in 2020, and to return with samples in 2030. China's "Mars-Plus" plan was further elaborated by Ouyang Ziyuan, at the opening ceremony of the 22nd International Planetarium Society Conference held in Beijing on June 24, where he stated, "China's goal for space exploration is the Solar System." He added that future exploration in the Solar System will include the search for extraterrestrial life, the origin and evolution of the Solar System, solar eruptions, and other phenomena. Ouyang told *People's Daily* on June 26 that another important goal of the Mars mission is to detect solar systems beyond Earth's reach, and to compare the origins of Earth-like planets with the formation of our Solar System. The most ambitious project of the Chinese space agency is that they hope to "re-create" a planet, based on information obtained through exploration.

International Collaboration: China and Russia

In the wake of a May 2014 landmark deal signed between Russia and China for the export of $400 billion in natural gas to China, there has been an intensification of collaboration between the two countries in numerous areas, including space science. As reported by RIA Novosti on June 30, 2014, at the First Russia-China Expo, Russian Deputy Prime Minister Dmitri Rogozin said that Russia is ready to work with China to explore the Moon and Mars. "If we talk about manned space flights and exploration of outer space, as well as joint exploration of the Solar System, primarily the Moon and Mars, we are ready to go forth with our Chinese friends, hand in hand," he said. Rogozin believes that Russia and China could work together to create spacecraft, "a joint base of radio components independent from anyone," as well as cooperate in cartography and communication.

The Russian Federal Space Agency (Roscosmos) and its Chinese counterparts also signed a memorandum of understanding "on cooperation in global navigation satellite systems." Rogozin said that the Russian navigation system GLONASS and the Chinese Beidou will complement each other. China is also moving forward on space collaboration with India and the European Space Agency.

China's Fusion Program

At the top of China's long-view intention is the development of thermonuclear fusion power. Not only is China a contributor to the International Tokamak Experimental Reactor (ITER) project, but it is pushing ahead with its own tokamak reactor program. In fact, the Chinese have, with their Experimental Advanced Superconducting Tokamak (EAST) reactor, a higher-level tokamak than any found in the United States. As it is, the EAST uses superconductive magnets (which

China's Experimental Advanced Superconducting Tokamak (EAST), above, was the first fully superconducting tokamak in the world, and is today, a higher-level tokamak than any found in the United States.

no reactor in the United States currently does), which gives it a superior capability in magnetic confinement strength.

Additionally, in May 2014, Chinese scientists at the Institute of Plasma Physics completed a 20-month upgrade to their superconducting EAST tokamak, and will soon begin their 2014 experiments. The goal of the experiments will be to extend the duration of fusion production to more than 400 seconds, working toward steady-state operation of a fusion machine.

China is also well on its way to developing an inertial confinement fusion (ICF) facility comparable to the laser fusion facility at the National Ignition Facility (NIF) at Livermore, California. Named Divine Light 3 (SG-III), this facility is designed to use 48 lasers to compress an isotope fuel pellet to ignite fusion reactions. Although the facility is currently only in the target design experimental phase, the next phase, Divine Light 4, is scheduled to be running by 2020, with the intention of going for ignition of actual fuel.

In conjunction with this, in the process of building up its laser arsenal, China is developing a top-of-the-line, automated robotic machining capability for the construction of such lasers, which will give them a mass laser production capability that can be applied to other sectors of the productive economy.

Furthermore, China announced in May 2011 a policy to graduate 2,000 new fusion scientists and engineers by decade's end.

U.S.-China Collaboration

In a classic case of self-destruction, the U.S. Congress has banned any U.S. collaboration with China on manned space efforts. Nonetheless, U.S.-China collaboration in fusion research currently expresses best what the future might hold, were there to be a true cooperation among nations.

A number of the most progressive breakthroughs in the fusion world have come about as a direct result of U.S.-China collaboration. These include the record-setting achievement of a pulse-length confinement time of 30 seconds for an H-mode plasma at the Chinese EAST reactor in November 2013. This record result was achieved by Chinese scientists beaming a microwave frequency into the plasma, which reshaped the magnetic field lines that confine the plasma, thus reducing the instabilities. This new technique was combined with one developed by fusion scientists at Princeton Plasma Physics Laboratory, which is to coat the plasma-facing wall of the tokamak with a lithium metal, which absorbs the stray particles of plasma, stopping them from disrupting the fusion process.

Another collaborative advance was achieved with the Doublet III-D (DIII-D) tokamak at General Atomics in California, where U.S. scientists, along with visiting plasma scientists from the Chinese EAST project, conducted experiments with the DIII-D tokamak that demonstrated that the plasma itself generated a current that was more than 85% of the current in the plasma. This large "bootstrap" current will significantly reduce the amount of external power required for confinement of the plasma. It is likely that insights gained from this experiment with the DIII-D tokamak represent part of what contributed to the previously mentioned breakthrough at the more powerful EAST reactor in China. These breakthroughs and results are likewise likely to contribute further toward the larger international collaboration being carried out with the ITER.

Ironically, much of the high-end technology being developed by China, and other nations for that matter, has its origin in the United States. The bulk of that technology comes out of the scientific and productive potential built up under the leadership of Franklin Roosevelt, and in the echoes of his legacy, as under the Eisenhower and Kennedy administrations, with, for example, the Manhattan Project, Atoms for Peace, and the Apollo missions. It was out of these projects that the world got nuclear power, atomic medicine, and deep-space exploration capabilities, with leading scientists from around the world emigrating to America to participate in the advancement of human progress.

Likewise, the next generation of infrastructure technology, in particular, transportation and energy, has also been heavily borrowed from the United States. For example, China is the only nation with a functioning commercial magnetic levitation (maglev) train, but it is based on a technology originally developed by U.S. and German engineers. The latest design, using maglev trains traveling inside evacuated tubes that can reach speeds of up to 4,000 mph, and which was recently tested at China's Applied Superconductivity Laboratory of Southwest Jiaotong University, originates from the Brookhaven National Laboratory in New York, where it was patented in the 1960s.

Also, China's record-setting superconducting tokamak reactor EAST is based on designs (and scientists) from the Princeton Plasma Physics Laboratory, specifically the Tokamak Physics Experiment (TPX), which was designed to be the next generation of tokamak experiments, but was subsequently dropped in the United States due to lack of funding.

The Future Is in the Stars

With the development of space and fusion power driving the process, China sees its future in the stars. It is this which has propelled China to become the model the world must follow today, to create the new paradigm for the survival of civilization.

In a profound and ironical sense, China has taken up the mission of creative scientific development and progress that the West, and particularly the United States, has left behind. China has adopted the mission of a "knowledge-based society," and is working systematically to increase the skill level of its population. Simultaneously, China is increasing its investments in Research and Development. According to a December 2013 article in *R&D Magazine*, China has increased its R&D investments by 12% to 20% annually for each of the past 20 years, while U.S. spending has increased at less than half those rates.

Most importantly, the Chinese leadership sees accomplishing its development goals in cooperation with other major powers, including the United States and Europe, as well as the BRICS and allied countries. That will become evident, as we review the dramatic progress of the New Silk Road.

Marsha Freeman and Richard Freeman contributed to this report.

FOR FURTHER READING

"Dr. Yuanxi Wan: China's Ambitious Path to Fusion Power," *EIR*, March 11, 2011.
"China in Space: A Look at China's Ambitious Space Program," *21st Century Science & Technology*, Fall-Winter 2006.
"Mining the Moon to Power the Earth," *EIR*, January 24, 2014.

China's New Silk Road: Changing the Paradigm Toward Global Development

by William C. Jones and Michael Billington

October 2014

On September 7, 2013, the world changed. On that day Chinese President Xi Jinping, speaking at Nazarbayev University in Astana, Kazakhstan, called for the development of a "Silk Road Economic Belt" (SREB) stretching "from the Pacific Ocean to the Baltic Sea." "We must expand the development of Eurasia," Xi said, "creating an economic belt along the Silk Road."

The idea of the Silk Road hearkens back to a period 2,000 years ago, in the Han Dynasty, when Zhang Qian, an envoy of the Han Emperor, was sent on a visit to Central Asia in order to establish trade among nations of the region stretching all the way to Europe and to the Middle East and Africa. "More than 20 years ago relations between China and Central Asia began to take off," Xi said. "The old Silk Road began to radiate with a new vitality." President Xi was decidedly intent on creating a "new vitality in a world economy today" that was quickly self-destructing. "Developing friendly relations with the countries of Central Asia has now become a priority for China's foreign policy," Xi said. "We should have wider aspirations, broaden our field of vision of regional cooperation, and together create new brilliance in the region."

One month later, during a visit to Indonesia, President Xi announced a similar Maritime Silk Road, also referring to Chinese history, specifically when Chinese Admiral Zheng He in the 1400s conducted a series of maritime voyages to Southeast Asia, South Asia, and Africa, creating a network of economic and cultural ties between the nations along his route.

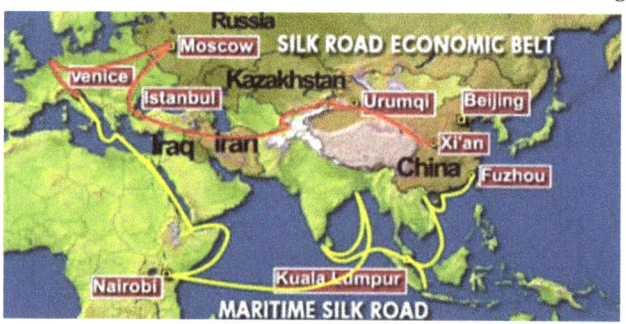
China State Council Information Office

What gives these proposals such power, of course, is the rapid development of the Chinese economy over the past four decades. Not only is this major diplomatic initiative of the "Two Silk Roads" a serious attempt to create a peaceful and prosperous neighborhood for China—an FDR "Good Neighbor Policy" in Eurasia—but also it provides a paradigm to bring prosperity to the world as a whole (see illustration).

This was underlined by the visits of President Xi and Premier Li Keqiang to Europe this past Spring.

The New Silk Road Becomes the World Land-Bridge

President Xi celebrated the ties between China and Germany during his March visit to Duisburg, Germany, the western terminus of the Silk Road, declaring, "Greater integration of our two economies, or cooperation between strong growth poles in Asia and Europe, will greatly promote the formation of a big Asia-Europe market and the growth of the entire Eurasian continent, and will have a far-reaching impact on the world economy and the world trade structure. Closer cooperation between China, a country that is committed to the path of peaceful development, and Germany will be in the interest of the formation of a multi-polar world and of world peace, stability, and prosperity."

Premier Li Keqiang, during his visit to Eastern Europe and Greece in June, proposed to bring the Silk Road Economic Belt to them, through rail, port, and related development projects. His proposals came as a clear contrast to the devastation that the collapse of the New York-London financial system has had on their countries.

Now, several months later, and in the wake of the BRICS process taking hold internationally, the Silk Road Economic Belt paradigm is being integrated with a global development perspective, in which increasing numbers of nations are joining together for great projects in space, nuclear power, water projects, rail, and in-depth development—a true new world economic order.

The Progress of the New Silk Road

China's 2013 Silk Road Economic Belt and Maritime announcements hearken back directly to the principle of development posed in the early 1990s, after the collapse of the Soviet Union, by Lyndon LaRouche and his wife, Helga Zepp-LaRouche, in their program for the "Eurasian Land-Bridge"—corridors spanning the Atlantic to the Pacific.

Their discussions with Chinese and Russian colleagues led to a conference on the topic of the "Eurasian Land-Bridge" in May 1996 in Beijing, sponsored by the Chinese Ministry of Science and Technology, at which Helga Zepp-LaRouche was a major speaker. From that conference, the LaRouche movement and *EIR* launched an international campaign for the Eurasian Land-Bridge/New Silk Road perspective, publishing an extensive special report, and organizing conferences throughout the world.

In 1998, Mrs. LaRouche returned to China, where she keynoted another conference dealing with the Silk Road perspective, entitled "Asia-Europe Economic and Trade Relations in the 21st Century and the Second Eurasian Land-Bridge." By that time, she had earned the title "The Silk Road Lady," due to her indefatigable battle for the peace-through-development perspective. That year also saw the beginnings of tripartite cooperation between China, India, and Russia, with the highly significant proposal by then-Russian Premier Yevgeni Primakov during a state visit to India in December, for the formation of a "Strategic Triangle" that would contribute to "peace and stability" in the Asia-Pacific region and the world.

But the 1997-98 "Asia Crisis," brought on by the raid on the Asian currencies by the hedge funds (which nearly brought down the world financial system), put a damper on the momentum for the Eurasian Land-Bridge project, although China proceeded at a slower pace to construct the primary rail route through China, Kazakhstan, Russia, and into Western Europe. In January 2008, the first pilot container train left Beijing for Hamburg, Germany, through Kazakhstan, Russia, Belarus, and Poland. By 2011 service was begun between China and Duisburg, carrying electronic goods and textiles from China to Germany, and industrial goods and machinery from Germany to the central Chinese industrial region.

Now the Silk Road is a reality. New lines, logistics, and infrastructure are being mapped out, built, and completed every month. At the time of President Xi's November 2013 speech calling for the SREB, representatives from 24 cities in eight countries along its route formally signed an agreement, committing to mutual development and prosperity. The SREB corridors cut across 18 Asian and European nations directly, but affect 40 nations, with a total population of three billion people.

One of the most dramatic manifestations of the process was the opening in June this year of the high-speed rail line, in the middle stretch of the SREB, in western China. It runs from Lanzhou, in Gansu Province, westward across Xinjiang, to Urumqi, and thence to either the northern corridor of the SREB across Kazakhstan into Russia and beyond, or arcing to Southwest Asia, to southern Europe and Africa.

On June 3, 2014, the first high-speed test train traveled this new Lanzhou-Urumqi High-Speed Railway route. Full commercial service is to start by the end of the year. Running for 1,776 km (1,104 mi), the line is exceeded in length only by the 2,298 km (1,428 mi) Beijing to Guangzhou high-speed route, and it travels through harsh western conditions, with an operating speed up to 350 km/h (217 mph). To counter the fierce desert winds, one part of the track is protected by a 67 km (42 mi)-long wind-break structure. The route also boasts the highest high-speed rail tunnel in the world, where, near Qilianshan, Tunnel No. 2 is at 3,607 m (11,834 ft) above sea level.

When Helga Zepp-LaRouche returned to China in late August 2014, she visited Lanzhou and Beijing, and encountered a pervasive spirit of optimism around the Silk Road Economic Belt perspective. It was also evident, as expressed at a Sept. 5 government-sponsored conference in Beijing on the SREB policy, that the conceptual input of the LaRouche movement, especially from her and her husband Lyndon LaRouche, was warmly appreciated (see speech at the conclusion of this article).

High-Speed Rail—Driver of Productivity

Join the three central lines by means of the fourth, and decide if, in ten years, a revolution will not have occurred in Peru, a revolution at once both physical and moral, because the locomotive—which, like magic, changes the face of the country through which it passes—also civilizes. And that is

perhaps its main advantage: populations are put into contact. It does more than civilize; it educates. All the primary schools of Peru could not teach in a century, what the locomotive could teach them in ten years.
—Manuel Pardo, President of Peru, 1872-76

The railroad is like a leaven, which creates a cultural fermentation among the population. Even if it passed through an absolutely wild people along its way, it would raise them in a short time to the level requisite for its operation.
—Count Sergei Witte, Prime Minister of Russia under Czar Nicholas II, 1905-06

As the two leaders cited above state, the construction of rail lines, especially high-speed ones, accompanied by development corridors, serves to lift up a population to a higher platform of productivity, thus better preparing citizens for participation in higher national missions based on a science-driver program.

The remarkable new Lanzhou-Urumqi high-speed span in the Silk Road Economic Belt, underscores this general point. **Figure 1** is a map from China Railways, showing the rail grid as of September 2014, with high-speed rail routes designated by travel speed.

China's total rail network has a route-length of more than 100,000 km (62,140 mi), the second largest after the United States, whose total land area is larger. But China's high-speed rail network has a total length of more than 11,028 km (6,852 mi) as of December 2013, by far the world's longest. It is expected to reach 18,000 km (11,000 mi) at the end of 2015, and is projected for 50,000 km (31,070 mi) by 2020.

In 2000, China had no high-speed service at all! But in the 1990s, plans had been laid for the HSR (high-speed rail) project, and in just over a decade, the biggest high-speed rail system in the world was built, and has been expanding rapidly since.

The principle involved, right from the start, was to increase the productivity of surface transportation, by separating passenger from freight haulage, and building key corridors. This has multiple gains, lifting the productivity up to new levels nationwide. This was spelled out emphatically by Dr. Sergei Sazonov, senior researcher at the Russian Academy of Sciences Institute of Far East Studies, in the newspaper of the Russian Railways, *Gudok,* June 24, 2014 ("The Development of High-Speed Mainlines Is a Stimulus for National Economic Development"). A few items in Sazonov's list of China's HSR hallmark successes are given under the railway map in **Figure 1**. His summary of the productivity impact of HSR is impressive:

"Key achievements of the Chinese railway sector's reform [upgrade] have been the ability of the railway complex to be specialized, allowing a radical increase in speeds, and a simultaneous increase in the carrying capacity of conventional-speed railroads, reducing the cost of bulk freight carriage on those older lines. According to analysis done by the Chinese Ministry of Railways, removing just one train from a mixed-use mainline (that is, building a special, dedicated HSR

FIGURE 1
China Railway Map-High Speed (CRH) and Other Services, September 2014

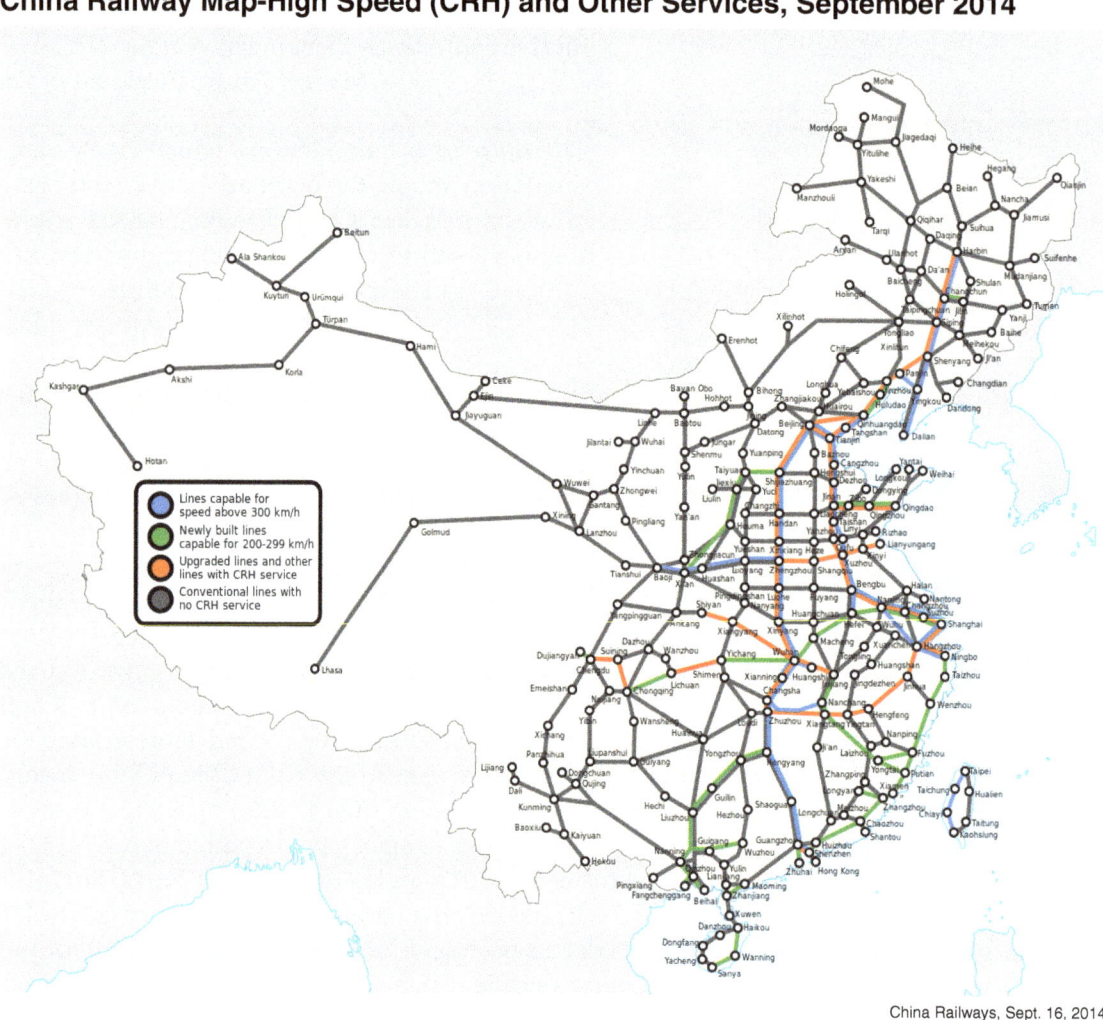

China Railways, Sept. 16, 2014

for the passenger traffic) raises the carrying capacity of the mainline by 1.5 to 2 freight trains daily.

"After the Beijing-Shanghai [passenger] HSR began operating, the daily carrying capacity of freight trains on this route rose by 140,000 metric tons, or 50 million metric tons per year.

"The development of HSR in China produces a significant multiplier effect within the industry. High-speed rail 'compresses' the great distances within China, not only connecting various cities, but driving domestic demand, and it is becoming a strategic sector of Chinese industry, promoting the development of related high-technology industries, as well."

In the forefront of new surface transportation technologies is magnetic levitation, where again, China leads the world. The maglev train between Pudong and the Shanghai International Airport, the only commercially functioning maglev in the world, can reach 430 km/h (268 mph) on its 30 km (18.6 mi) route. It opened in January 2004. Additional projects have been under consideration, including an

High-Speed Rail Drives Chinese Economic Development

From Dr. Sergei Sazonov:
- In 2006, the highest altitude railway in the world started operating between Qinghai and Tibet, with a high-speed segment, Golmud-Lhasa. Construction began in 2008 on the 1,318 km (819 mi) Beijing-Shanghai HSR, which had been in the planning for 18 years. December 2012 saw operations start on the Harbin-Dalian (904 km or 561.7 mi) and Beijing-Guangzhou (2,298 km or 1,428 mi) HSR; the latter cut the travel time between the country's two largest megalopolises from 22 hours down to eight.
- In June 2013, trains began to move on two HSR segments from Hangzhou: a 249 km (154.7 mi) line to Nanjing and a 150 km (93.2 mi) section to Ningbo.
- Traffic on the Nanning-Wuzhou HSR started moving on April 18, 2014. Construction was completed December 28, 2013 on the 1,249 km (776.1 mi) Beijing-Harbin HSR; in all, that month, seven new HSR segments, with a total length of 2,285 km (1,420 mi), began operating.
- Thus, in 10 years, as of the beginning of 2014, the biggest HSR system in the world had been created, with a total length of 10,463 km (6,214 mi), of which around 7,000 km (4,350 mi) is in the interior regions of China. As of early 2014, high-speed express trains are running on 34 dedicated HSR lines. Around 60% of the Chinese HSR network's trains run at 200-250 km/h, while the rest have speeds of 300 km/h or more.
- HSR has substantially increased the mobility of the Chinese population: as of the beginning of 2014, around 25% of all passenger rail carriage is on high-speed lines. The number of passengers on the Beijing-Tianjin HSR has grown 20% annually since it opened, while the Beijing-Shanghai HSR line's passenger growth has been 40% each year. High-speed trains on these routes now depart every four or five minutes.... In its first year of operation, the longest HSR in the world, the Beijing-Guangzhou route, carried 100 million passengers."

Source: *Gudok*, the newspaper of the Russian Railways, June 24, 2014, interview with Dr. Sergei Sazonov, senior researcher, Russian Academy of Sciences Institute of Far East Studies, "The Development of High-Speed Mainlines Is a Stimulus for National Economic Development."

---

extension from Shanghai/Pudong, southwest by 210 km (130.5 mi) to Hangzhou.

The Tibet rail line ("rail in the sky") is another demonstration of new technologies. The Qinghai-Tibet Railway is the first ever to connect the Tibet Autonomous Region, a land of extreme height and rough terrain, to anywhere else. The first 815 km (506 mi) section was constructed in 1984 to connect Golmud with Xining, in Qinghai Province. Then in 2006, the spectacular 1,142 km (710 mi) stretch was completed between Golmud and Llasa, the capital of Tibet, overcoming extreme conditions in ways that will be invaluable for building rail lines across the tundra of the far north of Eurasia and North America. For example, about 550 km (340 mi) of the main Golmud-Llasa line run on permafrost. There are 675 bridges.

Many new records were set. The line has the world's highest track route and railway station, in the Tanggula Pass, which is 5,072 m (16,640 ft) above sea level. The Fenghuoshan Tunnel is the world's highest, at 4,905 m (16,093 ft) above sea level. The trains provide oxygen contingency services for passengers.

In August 2014, the third stretch of the Qinghai-Tibet Railway was completed, a 253 km (157.2 mi) line linking Llasa with Shigatse,

FIGURE 2
Sun Yat-Sen's Vision of a China Rail Network
(At the time, China's borders included modern Mongolia)

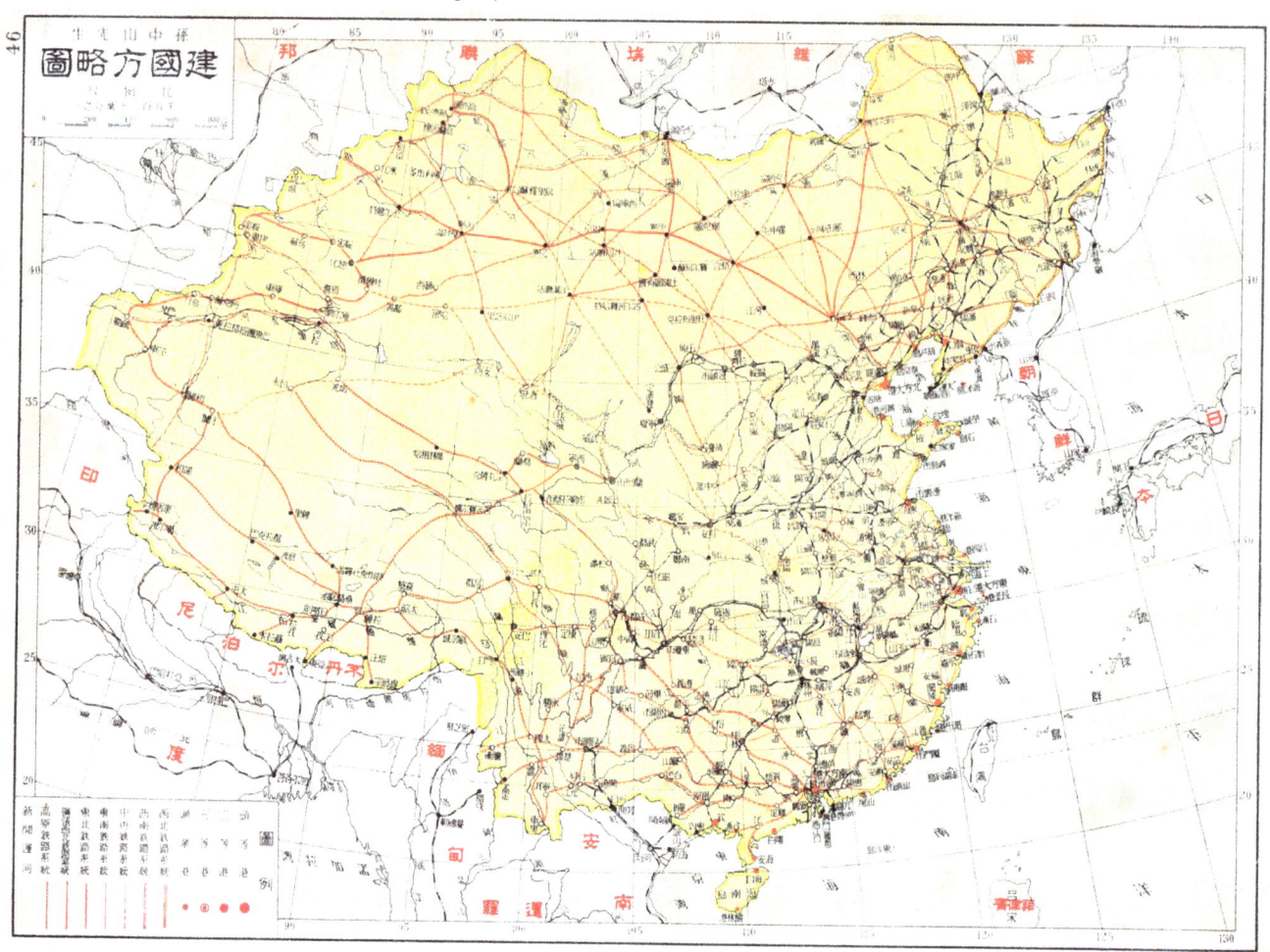

Tibet's second city. This puts the line 540 km (335.5 mi) from Nepal's border, and plans for a rail extension on this span were officially raised at the Nepal-Tibet Trade Facilitation Committee's fifth meeting, in Llasa, September 2014.

Thus, the drive by China for advanced rail service is contributing to the connectivity on many fronts in Asia. China's extensive and high-technology rail networks hearken back to the original nation-building railroad plan put forward at the beginning of the last century by the founding father of the Republic of China, Sun Yat-sen (**Figure 2**).

An International Mandate

Today China is in high demand internationally, to build high-speed or conventional modern rail routes in Africa, across South America, and other key spans, as well as the Silk Road proper. This derives from a conscious policy enunciated in November 2013 by Professor Wang Mengshu, a professor of engineering at Beijing Jiaotong (Communications) University, and a member of the National People's Congress of China. Wang told *Global Times* reporters that China is clearly a world leader in high-speed rail: "When people talk of watches, they think of Switzerland. When they think of small electronics, they think of Japan. When they think of space, they think of America, and talking about machinery, they think of Germany. Now when they think of high-speed rail, China becomes the brand name."

Under the strategy Wang described, China has completed a high-speed rail line for Turkey, connecting Istanbul and Ankara, the capital. China firms are on stand-by for track construction and/or train sets in many other locales, for example, if Kazakhstan decides to proceed to connect Almaty with its capital Astana by high-speed rail.

Wang Mengshu also envisions a line going through northeast China, and then through Siberia to Chukotka, where it would meet with the planned rail line to the Bering Strait tunnel.

In October 2014, at the time of a visit between Premier Li Keqiang and President Vladimir Putin, a memorandum of understanding was signed between rail firms and transport officials of China and Russia, on building a 803 km (499 mi) high-speed rail line from Moscow to Kazan, in Tatarstan, which, it is understood, could be the first stage of high-speed rail service to Beijing. The full project would be more than 7,000 km (4,350 mi) to connect the two capitals by fast train, making it more than three times the length of the world's current longest high-speed line, Beijing to Guangzhou.

In the Americas, a Chinese consortium is reportedly the sole bidder for a tender to build a 210 km (130.5 mi) high-speed line from Mexico City to Querétaro, in the north central region. In Venezuela, China Railway Group has a project underway to link the cities of Tinaco and Anaco, on the edge of the great central plains, to further inland economic activity and population growth potential. In the United States, Chinese manufacturers have tendered official submissions of interest to supply train sets to the California High Speed Railway Authority, if the future 1,287 km (800 mi) route from San Francisco to Los Angeles finally comes into being.

Xinjiang—'Traffic Hub' for the SREB

For China, the western Xinjiang region is intended as the jumping-off point for the Silk Road Economic Belt to multiple points westward and to the south. China is planning three corridors to run through Xinjiang. The northern-most route (not shown), not built yet, is a corridor to proceed from Beijing, via Hohhot, into Kazakhstan, north of Urumqi, thence to Russia and beyond. The other corridors are shown on the map in **Figure 3**, from north to south: (1) the current route going through Urumqi, then branching off into two routes westward across the Kazakhstan border; (2) an intended route from Kashgar into Kyrgyzstan and westward; and (3) the proposed southern route, for a rail line southward through Pakistan to Gwadar on the Arabian Sea. China has begun a study of this 1,800 km (1,118.5 mi) route, which presents stiff construction challenges to traverse the

FIGURE 3
Xinjiang—'Traffic Hub' of Rail Routes, Existing and Proposed for Silk Road Economic Belt

Shown are key rail routes through Xinjiang, either completed, e.g., Urumqi to Lanzhou (high speed); or intended, e.g., Golmud to Holan.

Karakoram Mountains and Pamir Plateau. In May 2013, Pakistan awarded control of Gwadar Port to the China Overseas Ports Holding Co., transferring operational rights from the Singapore Port Authority.

There are exciting new achievements, and visionary plans on all these routes. In July 2014, Zhao Shiaoyang, the Chairman of the Chinese Railway Company CSR, spoke of an option for China to build a high-speed line from China to Turkey through Kyrgyzstan, Uzbekistan, Turkmenistan, and Iran. This could then be integrated into Turkey's own Chinese-built high-speed line from Ankara to Istanbul.

Xinjiang, comparable in size to Alaska, borders eight countries, with its 5,700 km (3,542 mi) boundary. It is home to the Takla Makan and other deserts, mountain ranges, and continental temperature extremes, but is now on the front line as one of the world's most exciting development frontiers. Remote sensing analysis suggests groundwater resources in the Takla Makan Basin. The intention is for the new rail corridors, and new industrial and commercial establishments, to be built up in the politically sensitive Xinjiang region in order to raise the standard of living for its 22 million inhabitants.

China has long had a policy of developing its central and western regions, and the launching of the New Silk Road Economic Belt has given new life to this commitment. The key rail lines between the east coast of China and the western regions have been replaced by high-speed rail. Connectivity is being built up in Xinjiang itself. The largest expressway project now under construction in China is in southern Xinjiang. The highway stretches 428.5 km (266.3 mi) from Aksu (in the far west of the province), southwesterly to Kashgar. The multi-lane highway will be opened by the end of 2014, and developed as an economic corridor. Xinjiang now has 16 airports, with four to six new airports in the planning stages, and some others to be renovated or expanded in the coming three years.

More plans for Xinjiang were announced June 26-27, 2014 at a forum in Urumqi. Zhang Chunlin, Director of the Xinjiang Development and Reform Commission, said that the province will become a "traffic hub" for the Silk Road Economic Belt. It will "make full use of its geographical and cultural advantages, [to] further open up and make efforts to act as the main force and vanguard in building the Silk Road Economic Belt." He spoke of a range of economic activities, including developing oil and gas, mining coal and minerals, providing new medical services, and establishing science and education centers.

There are currently 17 land ports in Xinjiang, and plans are underway to expand the number of border railway ports, bonded zones, and other facilities to serve all aspects of domestic and international requirements for the SREB trunk lines and localities. Xinjiang is in the geographic heartland of Eurasia, and its commitment to systems of logistics lays a foundation for early negotiations for a China-Central Asia Trade Zone.

Research contributed by Mary Burdman, Rachel Douglas, and Marcia Merry Baker

---

FOR FURTHER READING

"China Builds Sun Yat-Sen's Great National Rail Project," by Mary Burdman, *EIR*, Jan. 29, 2010

"Ideas for Cooperation Along the Silk Road," a speech by Shi Ze, Director for International Energy Strategy Studies and Senior Fellow, China Institute of International Studies, given at the October 18-19, 2014 Schiller Institute conference in Frankfurt, Germany. See www.newparadigm.schillerinstitute.com

APPENDIX

China's Silk Road: Pathway to a New Human Civilization

Bao Shixiu, Professor (Emeritus) of Military Science, People's Liberation Army (PLA) Academy of Military Science, gave this speech at the conference "One Belt, One Road," held in Beijing on September 5, 2014. Mrs. LaRouche also addressed this conference.

I would like to extend my greetings and my gratitude to Dr. Xiao Jingqiu for inviting me here. At the same time, I have to say that studies concerning the Silk Road are a very big issue, and I'm not the top-notch researcher in this field. So I'd just like to share with you my very shallow thoughts about it, and please feel free to contribute all your insights and criticisms to me.

The topic of my speech is the realization of the new Chinese Dream by the New Silk Road. Two thousand years ago: The ancient Silk Road can be dated to this era, which had started a political and economic dialogue between the East and the West. Many different objects of the arts and of technology have been promoted through the process, which has attained great fame, and been transcribed in the annals of history.

Today, in the 21st Century, China is trying to build a New Silk Road. The world economic map is looking forward to some new dynamics and the realization of the new Chinese Dream. How can we make the two things promote each other? And how can we achieve the Chinese Dream? I think these are the challenges that have been put forward to all of us by the director of *China Investment* magazine. So taking into consideration the geopolitical situation, as well as economic conditions, and other factors, we do have the necessity of talking about this issue.

This is a long and very dynamic Silk Road. The concept of the New Silk Road was put forward by the Chinese leader in 2013, and now we have had one year full of the development of the dynamics, and I believe that this is a new concept that will make a great contribution to global governance, or the new world order. And it also concerns the role of China. What role does China want to play in the world?

A New Type of Thinking

I would like to talk about two points. The first is that the concept of the New Silk Road is making a contribution to global governance theory. This concept has reflected the canon of a new global governance theory, and it has provided tangible theoretical support for a

Helga Zepp-LaRouche and Col. Bao Shixiu, PLA officer and translator, in Beijing, September 2014.

new type of world order. Actually, we are looking forward to a new type of thinking mode, when we are dealing with international or global governance.

After the Second World War, we have measured almost a century, and what is the economic situation of today's world with multipolarization and economic globalization? Countries in the world have become increasingly interconnected and dependent on one another on a daily basis. Many developing countries around the world and billions of people are working toward modernization per se. An era of cooperation, collaboration, and win-win situation is continuing to mount.

However, we still face the problem of development. The world economy has suffered from the financial crisis and stagnation in the process of recovery. There are still many potential risks in the field of international finance, and many macro-regulation organizations in many countries are facing challenges and difficulties. The global financial crisis has reflected the systematic failure of the financial system, as well as challenges and crises such as climate change, food safety, security, and many other issues, which have reflected the fact that today's global governance system still has its weaknesses, and is in need of improvement.

In such an era of great change, we are all waiting for all sorts of upgrades, of positive reforms. So the demand for a new type of concept is very strong at the moment. Therefore, we have the "One Belt, One Road" concept, which has been welcomed and well received by many countries in the world because it is based on mutual respect, friendly relationships, a win-win situation, and cooperation. Therefore it created a new sort of atmosphere, which reflects the actual interests of the relevant countries, as well as a new dynamic in the global governance of the 21st Century. So this is a very good raw material, so to speak, for the development of the new global governance system.

In the Autumn of 2013, we had some movements from President Xi Jinping's visit to Kazakhstan and to other countries, and I think the core concept of this "One Belt, One Road" is a type of concept in which China is seeking the common interests of many countries, instead of the interests of itself alone. And President Xi Jinping has said that China will enhance its friendly relations with Central and Eastern Asian countries and work together with relevant countries to make contributions to the world. He also believes that as long as we adopt a rule of mutual respect, as well as a cooperative perspective, countries of different cultural backgrounds and ideologies can share prosperity and peace.

While visiting Indonesia, President Xi Jinping came up with the idea that the host and guest countries should become each other's good neighbors and good partners. And work together to build a China-ASEAN common destiny, actually, the concept of a New Silk Road as regional, innovative economic cooperation, which builds a platform for such cooperation and East-West cultural integration. The development of this new concept will refine the radiance of the Silk Road and make the East Asian and Central Asian economies more integrated than ever and leave a deep influence on the world.

The Role of Helga and Lyndon LaRouche

The new concept of the Silk Road has been given very high praise by many intellectual leaders. And this new concept is trying to absorb new contributions and insights from scholars and from people in academia—actually, many people of great knowledge have made contributions. But I think it is very important for us to mention the dean of the Schiller Institute in the United States, Mrs. Helga Zepp-LaRouche, and her husband Lyndon LaRouche.

To change the decades-long irrational global governance system, and to make the global governance system and the global order more sound and healthy, the couple, as early as the 1990s, had come up with a new idea about building a tunnel under the Bering Strait, as well as establishing a Eurasian Land-Bridge to connect the world, so that people of all countries and continents can benefit from this new connection. So common prosperity is the basis for a new global governance system.

These two dignitaries, who have been making contributions to the establishment of a new global order and a governance system, have paid special attention to the role of China and Asia in establishing this kind of new order. Mrs. LaRouche, as early as 1997, published an article about the Eurasian Land-Bridge as the most important geopolitical issue in the world, and has made a great effort in introducing China to the world.

When, in the Fall of last year, she heard the news of President Xi Jinping's visit in Indonesia and Kazakhstan, she was thrilled. She thought this new idea promoted by President Xi Jinping would actually produce prosperity in this part of the world and improve people's living standards. Now we have a common consensus in the world, which is that the New Silk Road is only the first step of economic integration of the world and the first light in the darkness toward a new human civilization.

Refuting the Critics

The second question is how to support this concept. It is a very important theoretical question for scholars. In the world of academia, it is very important for any idea or concept to be promoted or brought up by dignitaries of very famous names, which is quite normal. However, the problem is that out of the common interest of some political blocs or because of some ideological bias, many people hold a very critical and very arbitrary attitude to this new concept of the Silk Road, which is really hard for us to accept.

All of these theories criticizing the New Silk Road and tarnishing the New Silk Road must be clarified and corrected. The demand for actually supporting this theory is also a very important task for all the scholars in this field in China.

I just have two examples to show you. The first regards some people who wrote an article saying that this new kind of concept is very dangerous, and when the CICA [Conference on Interaction and Confidence-Building Measures in Asia] conference was initiated in Shanghai this year, people had discussions about why China was having this new policy, and a very famous magazine from Australia, *The Diplomat*, published an article that claimed the New Silk Road is not symbolic, but rather a diplomatic approach by China to establish a new economic and political order in East Asia and in Central Asia, which meant that China is intending to establish a new economic order, instead of fostering friendly and cultural communication and cooperation. And it also illustrated that China would like to become the core of this type of cooperation so that it could reflect its geopolitical importance.

That article reflected the belief that China's political ambition was to establish a transcontinental FTA [Free Trade Agreement] in trade.

And at that time, there were also many Western think-tanks that held a skeptical attitude toward the concept of the New Silk Road. Actually, there are also some who claim that the New Silk Road is a new form of Monroe Doctrine. The Monroe Doctrine was put forward by the fifth President of the United States, James Monroe, and was a very important symbol of the United States expansion in the world. It was the United States warning the European powers not to interfere in the American continent, namely, into the affairs of Mexico and other Latin American countries. And that the United States would remain neutral in the wars and conflicts happening in Europe, and the United States would always uphold its interests.

In recent years, with China's increasing flexing of its muscle in terms of safeguarding its legitimate territorial and maritime rights, many scholars have put forward the idea that China is pursuing a sort of "Monroeism." In 2012, James Holmes, from the United States Naval College, said that China and its South China Sea is just like the United States and the Gulf of Mexico. So he said that China is trying to build a new economic and political order in the region.

Another example is the very famous professor John Mearsheimer, from the University of Chicago, who gave a lecture not long ago, in which he said that if China continues to develop, it will push the United States out of Asia and pursue its own Monroe Doctrine.

And a Japanese scholar said he believed that China is trying to play the role of a regional leader and the leader of all of Asia.

This type of China-threat theory has been accepted by many people who hold a skeptical attitude toward China. But at the CICA conference held in Shanghai this year, President Xi Jinping had made it very clear that the security and peace of Asia should be safeguarded by its people.

According to the above-mentioned discussion, I think it is not hard for us to see that those studies based on history and the new

Asian diplomatic approach of China have nothing to do with the concept of a Monroe Doctrine, or the East Asia Prosperity Sphere promoted by the fascist Japanese Imperial Government. This is not an appropriate approach.

The new concept of the Silk Road of China has nothing to do with those old and absolute concepts. China's approach is based on open and friendly cooperation, and China is focusing on deepening cooperation in terms of security and economic development. China is expecting the benefits of the Silk Road to be shared, so that the prosperity of the entire region can be promoted.

And this is actually far from the concept of Monroeism, which is the United States trying to be the policeman of the American continent, as well as the fascist concept of the East Asia Co-Prosperity Zone promoted by Japan. President Xi Jinping, with his new concept of the Silk Road, represents China's image in the world and China's attitude toward the world.

We think that we should uphold this new concept, and its status in the field of academia, so that we can promote this new concept and safeguard China's legitimate rights, as well as promote a healthy development of the global governance system. Therefore, I say this is a very important theoretical project for all of us to accomplish.

PART 4
Russia's Mission in North Central Eurasia and the Arctic

Russia, Eurasia's Keystone Economy, Looks East

By Rachel Douglas

October 2014

The Eurasian Land-Bridge in Russia

The 1997 EIR Special Report *The Eurasian Land-Bridge: The 'New Silk Road'—Locomotive for Worldwide Economic Development* (ELB) was enthusiastically received and made a lasting impact in Russia. That is only natural, for Russia is "Eurasia's Keystone Economy," as Lyndon LaRouche titled a 1998 article.[1]

Like the pre-1917 Russian Empire and the Soviet Union of 1922-1991, today's Russian Federation extends across nine time zones, from the Baltic and Black Seas in the west, to Sakhalin Island, Kamchatka Peninsula, and the Bering Strait in the east (**Figure 1**). Its extreme southern cities on the Caspian Sea and the Sea of Japan are up to 4,200 km (2,610 mi) south of the northernmost reaches of the 5,000-km-long (3,100 mi) (east-west) Russian Arctic tundra zone. Russia above the Arctic circle, and other parts of Siberia, contain some of Earth's last undeveloped territories.

For centuries, efforts to master the Siberian frontier have been a driver of Russia's economic and cultural advances. Grand Prince Ivan III of Muscovy, the "gatherer of Russian lands" who consolidated the foundations of modern Russia after two and a half centuries of servitude to Tatar-Mongol invaders, in 1499 sent a force eastward from Moscow to the Ural Mountains, the divide between Europe and Asia, to establish a foothold in what the historian Nikolai Karamzin would later describe as "a terrible wilderness, bare cliffs, rushing streams, sad cedars, and white merlins, ... where, beneath the mossy granite, rich veins of metals and colored precious stones lie hidden." By the late 18th Century, iron mills in the Urals had helped to form the basis of Russian industrial growth. The publisher of the 1807 Russian edition of Alexander Hamilton's *Report on Manufactures* wrote in an introduction that, because of similarities between Russia and the USA "in the expanse of land, climate and natural conditions, [and] in the size of population disproportionate to the space, ... all the rules, remarks and

---

1. Lyndon H. LaRouche, Jr., "Russia Is Eurasia's Keystone Economy," *EIR*, March 27, 1998.

FIGURE 1
North Central Eurasia

means proposed herein are suitable for our country."

A century later (1891-1916), Russia firmly established itself as a continental power by building the more than 9,000-km Trans-Siberian Railway (TSR), after the precedent of the U.S. Transcontinental Railroad. The Soviet Union's eastward evacuation of entire factories to new locations behind the Urals, in the late 1930s and following the Nazi invasion in 1941, was crucial to its fighting capability, while permanently enhancing the industrial power of the Ural cities of Yekaterinburg and Chelyabinsk. Another Siberian center of industry, Novosibirsk, founded in 1893 where the TSR would cross the Ob River, was chosen in 1957 as the home of the Siberian Branch of the Soviet (now Russian) Academy of Sciences and its "science town" of Akademgorodok, which would pioneer regional development initiatives and new technologies for the Siberian frontier.

When the ELB report came out, however, the physical-economic development of the Eurasian continent was far from the minds of those holding power in Moscow. Post-Soviet Russia was five years into the "shock therapy" of deregulated prices and crash privatization, carried out by a clique of London- and Chicago-trained neoliberal economists, operating within the regime of President Boris Yeltsin. One of these so-called "young reformers," Anatoli Chubais, was finance minister and deputy prime minister in Prime Minister Victor Chernomyrdin's government, after Yeltsin's manipulated and fraud-ridden re-election as President in 1996. On Chubais's watch, the greatest innovation in Russian economic policy during 1996 and 1997, the ELB report's year of publication, was to ease access for foreign investors to exchange-trading of Russian government bonds, known as GKO. This was an invitation for hot money from around the world to pour into Moscow, building the financial pyramid that one year later, in August 1998, would collapse in default, nearly triggering global financial meltdown. The crisis year of 1998 spelled the end of the Chernomyrdin government, and opened a crack in the young reformers' control of economic policy.

As these events unfolded, attention to LaRouche's "physical economy" ideas, and the Schiller Institute's programs for revolutions in nuclear power and infrastructure, soared in Russian political opposition and academic circles. Russian-language summaries of the Institute's Paris-Berlin-Vienna Productive Triangle plan circulated, with its call to activate machine production, infrastructure capacities, and skilled labor in the historical heart of Europe, as an engine for the economic recovery of the former socialist bloc. The message was that the Cold War could be followed by an era of peace and prosperity, only if nations ended the reign of speculative looting in international finance, and put in place a new system to fund real economic development. In 1995, LaRouche had submitted a memorandum to the Russian State Duma on "Prospects for Russian Economic Revival,"[2] in which he introduced the concept of development corridors, and suggested that the industrial and technological capabilities of the former Soviet Union should not be scrapped, as was then happening, but rather exploited to produce new, more advanced infrastructure and other capital goods.

From discussions of the Productive Triangle with Russians and many others, the land-bridge concept emerged: the prospect of cooperative projects along infrastructure corridors, extending as arms from the Productive Triangle, into Eurasia. Visiting Moscow in April 1996, LaRouche had keynoted a seminar chaired by Academicians Leonid Abalkin and Gennadi Osipov, at which he proposed to revive Franklin Delano Roosevelt's vision of the United States, Russia, and China assuming joint leadership for the economic development of a post-colonial world. "The job before us sometimes … comes down to diplomats and elected government officials," LaRouche concluded his remarks at that seminar, "but governments cannot act on ideas, unless those ideas are established in some influential circles. My concern is to broaden and deepen the discussion of precisely this, among intellectual layers, which are influential in shaping the thinking of governments."[3]

Lyndon LaRouche (second from left) faces Academician Leonid Abalkin (fourth from right) at the April 24, 1996 seminar in Moscow on "Russia, the U.S.A., and the Global Financial Crisis." To Abalkin's left is economist Valentin Pavlov, former prime minister of the Soviet Union.

---

2. LaRouche, "Prospects for Russian Economic Revival," *EIR*, March 17, 1995.

3. "Russia, the U.S.A., and the Global Financial Crisis," seminar transcript, *EIR*, May 31, 1996.

FIGURE 2

The online archive of Metals of Eurasia *displays Academician Vladimir Myasnikov's article, "The Continental Bridge: a Project for the 21st Century," in the magazine's issue #3 for 1997. It is a précis and review of EIR's Special Report,* The European Land-Bridge, *published in January of that year.*

FIGURE 3

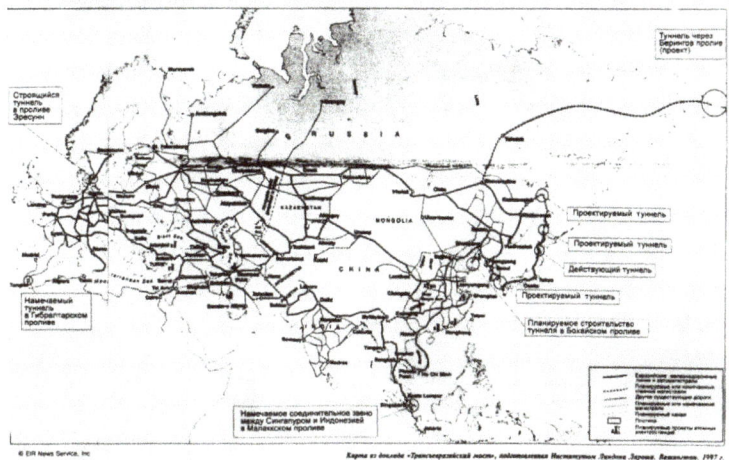

This reproduction of the main map from EIR's Special Report The European Land-Bridge *illustrated a two-page centerfold by Academician Sergei Rogov in* Nezavisimaya Gazeta-Tsenarii, *#3, 1998, titled "The Contours of a New Russian Strategy: Only Its Central Location on the Geoeconomic Map of Eurasia Can Save the Country."*

In Russia, the influential layers stirred by the Eurasian land-bridge idea included institutes of the Russian Academy of Sciences (RAS), where LaRouche had been a guest for seminars and private discussions in 1994, 1995, and 1996. Academician Vladimir Myasnikov, a prominent specialist in Asian affairs who was then deputy head of the RAS Institute for Far East Studies (IFES), favorably reviewed *EIR*'s ELB report in the May-June 1997 issue of the industry magazine, *Metally Yevrazii* (*Metals of Eurasia*) (**Figure 2**). In 1998, an article by Academician Sergei Rogov of the RAS USA-Canada Institute appeared in a supplement to the popular daily *Nezavisimaya Gazeta*, under the headline "The Contours of a New Russian Strategy: Only Its Central Location on the Geoeconomic Map of Eurasia Can Save the Country." Illustrating the two-page newspaper spread was a reproduction of the main Eurasian Land-Bridge map from the 1997 report, properly credited to *EIR* and LaRouche (**Figure 3**).

Interviewed by *EIR* on the sidelines of a May 2001 Schiller Institute conference, Russian economist Sergei Glazyev replied to a question about Eurasia.[4] At the time, Dr. Glazyev was chairman of the Russian State Duma Committee on Economic Policy; today he is a full member of the RAS and advisor to President Vladimir Putin on Eurasian integration.

---

4. Sergei Glazyev (interview), "How Can the World Get out of This Crisis?", *EIR*, May 25, 2001.

EIR: How do you and others in Russia view the infrastructure proposal of the Schiller Institute and Mr. LaRouche on the Eurasian land-bridge?

Sergei Glazyev: I think that Russia really could and should play a very important role in the establishment of a new world architecture. If we speak not about virtual economic activity, but about real economic growth, we should think about how economic development could be promoted, not only in Russia or in other countries, but for the whole world. At the moment, the whole world financial system is going into trouble. A lot of countries now lack capital.... They do not have credit to develop infrastructure, and to develop their products.

For this reason, at this conference, and especially in the work conducted by the Schiller Institute and Mr. LaRouche, we see a lot of very fruitful ideas, concerning the promotion of economic growth. One of those ideas, the establishment of large infrastructure projects, is one of the most actively discussed in Russia. We are speaking about a new stage of life for the Trans-Siberian Railroad, which could connect our Far East with Western Europe.... The Baikal-Amur Mainline is a part of this Trans-Siberian corridor. It is being finished now. Just a few months ago, the last important tunnel [the Severomuysky Tunnel] was built, and now we are ready to put this railroad into full-scale operation. This means that various goods could be delivered from the Far East to Eastern Europe and Western Europe in two weeks....

In Eurasia, ... we must think about how to improve south-north links.... Recently, during the visit of the President of Iran to Russia, we agreed to open another very important link and development channel, from south to north: from Iran, through the Caspian Sea and the Volga River, to the northern part of Russia and the Baltic Sea.

EIR: Iran has negotiated also with India on the completion of this north-south route.

Glazyev: Yes, and on this route you can use the already existing waterways that connect the Caspian Sea with the Baltic Sea. It makes the transportation of goods in large quantities easier and cheaper, through these transnational routes.... The formation of these development corridors, which will be based on modern transportation technologies, is a very important device to stimulate economic growth and economic development.

I would like to stress that at this conference, we are thinking about how to combine the competitive advantages of various countries in this very large region, to push them into mutually beneficial cooperation—how to combine efforts in order to achieve higher rates of economic growth.

Primakov's Legacy

Academician Yevgeni Primakov and defense-industry expert Yuri Maslyukov, as prime minister and first deputy prime minister, respectively, formed a government in the wake of the August 1998 default. In power for only eight months, before Primakov was forced out in the midst of the global crisis over NATO's Spring 1999 bombing of Serbia, they took emergency action to stabilize Russian industry and internal financial flows. The results of the Primakov-Maslyukov government's measures to salvage Russia's real economy were inherited by Vladimir Putin, upon Yeltsin's appointment of him as prime minister (August 1999) and then acting president (December 1999; he was elected in his own right in March 2000). They created a framework, in which decisions in favor of Eurasian continental development might be seriously considered.

Also of strategic importance was the outstanding diplomatic engagement of this government: Primakov's December 1998 visit to India, during which he proposed the formation of a "strategic triangle" among Russia, India, and China. The collaboration of these Eurasian powers subsequently came to life through a years-long sequence of three-way academic and diplomatic meetings; after many turns in the road, the "RIC" combination today is the core of the alliance called the BRICS.

There was to be no smooth sailing toward Eurasian development for Putin, however. Upon taking office, he faced the possible break-up of the Russian Federation, amid insurgencies in the North Caucasus, which grew into the Second Chechen War. Then, as the first foreign leader to phone President George W. Bush during the 9/11 attack on the United States in 2001, Putin offered strategic cooperation. Nonetheless, three months later, the USA withdrew from the 1972 Anti-Ballistic Missile Treaty, and the U.S. war party, then led by Vice-President Dick Cheney, launched the attempt build a global ballistic missile defense (BMD) system, a screen behind which nuclear-missile attacks on Russia and China might ultimately be attempted. The Russian President was therefore preoccupied with defense and security issues, a priority expressed in his speech at the February 2007 Munich Conference on Security Policy, warning that "the process of NATO expansion is not at all related... to ensuring security in Europe. On the contrary, it is a serious provocation, which reduces the level of mutual trust."

At the same time, the bitter economic legacy of the 1990s within Russia was not easily overcome. Much of Russian industry was now owned by newly minted "oligarchs," who registered the companies abroad and kept their money in offshore tax havens. Ex-young reformers remained in charge of government economic ministries and institutions. The same monetarist dogmas that have wrecked the economies of the trans-Atlantic sector continued to hold sway in Russia, again and again impeding the implementation of infrastructure and industrial development plans, up to the present. The events of 2014, centered on the BRICS alliance, present an opportunity for Russia to jettison monetarist practice and take leadership in Eurasian development.

The Vernadsky Strategy

During the first years of Putin's presidency, LaRouche wrote extensively for his friends in the Russian intelligentsia, on Russia's challenge to find its mission in the world. In essays such as "The Vernadsky Strategy" and "The Spirit of Russia's Science," he invoked the noösphere concept of Academician Vladimir Vernadsky (1863-1945), as the proper guide in transforming Eurasia.[5] LaRouche visited Moscow to discuss these ideas in person, including to testify at June 2001 State Duma hearings, convened by Glazyev, on national economic security in times of financial crisis. There he said:

> Those regions, within and among nations, which can generate "fountains" of scientific and technological output to regions which are deficient in their available supply of such technology, must be envisaged as the suppliers not of money-loans, but of long-term purchasing credit, at nominal borrowing-costs. Continental Eurasia should be the center of such global economic recovery and growth, but all the world will benefit through participation as partners in that effort.[6]

The Vernadsky strategy for Eurasia was the topic of LaRouche's speech to the International Symposium on Space and Time in the Evolution of the Global System Nature-Society-Man, held in Moscow on Dec. 14, 2001,[7] excerpted here:

Lyndon LaRouche speaks in Moscow on the "Vernadsky strategy" for Eurasia, at the Dec. 14, 2001 International Symposium on Space and Time in the Evolution of the Global System Nature-Society-Man, dedicated to the memory of the scientist Pobisk G. Kuznetsov.

> If the world is to come out of this great financial, and monetary, and economic crisis successfully, Russia, as a Eurasian nation, must play a very crucial, central role.
>
> Looking from the Atlantic Ocean to the Pacific Ocean, across Eurasia, we see countries, such as China, India, and Southeast Asia, and other countries, which are in great deficit in respect to the amount of technology they have, and can supply, to meet the urgent needs of their populations, as a whole. So the nations, such as China, Southeast Asia, and India, must now catch up with the technology they have not had and have not assimilated, or have not developed, over the recent century. To a certain degree, India has a significant scientific community.... China has significant technology.

---

5. LaRouche, "The Vernadsky Strategy," *EIR*, May 4, 2001; "The Legacy of Mendeleyev and Vernadsky: The Spirit of Russia's Science," *EIR*, Dec. 7, 2001.

6. LaRouche, "Policy Changes Needed to Overcome the Collapse," *EIR*, July 6, 2001.

7. LaRouche, "Russia's Crucial Role in Solving the Global Crisis," *EIR*, Dec. 28, 2001.

But Chinese technology is far less than the urgent needs of China, as a nation, as a whole. The sources of this technology available within Eurasia, include Japan, Russia, and, mostly, Western Europe. As we can observe today,... that scientific potential in Russia, has been sleeping for a while, without work.

While related problems exist in other parts of the world, we can concentrate upon the Eurasian continent and the islands associated with it, as the typical center of the world's problem today.

This brings us to Vernadsky. One of the greatest concentrations of mineral and related resources in the world today, is an area, which includes Central and North Asia, including the tundra areas of Russia. Of course, it is possible to loot some of these resources, and ship them abroad at cheap prices. That would be a tragedy for Russia, and a betrayal of the interests of Eurasia, as a whole. So, I have proposed, that we must develop development corridors, superseding the Trans-Siberian Railroad, across Eurasia. Through large-scale water management, improved transportation, power generation, and other infrastructure, including human support infrastructure, in these regions, we can transform these areas of Asia.

To the west of Russia, in Europe, we have bankrupt nations: Germany, France, Italy, other nations. They are bankrupt, presently—nations which are traditionally producers of modern technology. So, there's a natural market for these parts of Europe—as for Japan—in Asia, if the proper system of economic development is organized. And Russia and Kazakhstan represent the principal conveyor belt of development, and other things, necessary to tie the potentials of Europe with those of various parts of Asia. *This would require, and would mean, the greatest transformation in the biosphere, in the history of humanity.*

Now, obviously, we can not do the kinds of things we've often done, in looting the biosphere. Often, at present, through looting policies, we degrade the biosphere more rapidly than we extract useful results from it; for example: mineral resources.

So therefore, when we are going to transform the biosphere, by means of a policy action, we must consider the implications of what we're doing, and approach the problem in a way which becomes, then, a net improvement in the biosphere, as the basis for man's activity. This forces us to think in terms of all modern economy from the standpoint of Vernadsky....

This also involves, how we look at man's relationship to the Solar System and beyond. This means that space exploration and space science become an integral part of developing life on Earth.

The Bering Strait Crossing

The idea of a connection between Eurasia and North America across the Bering Strait has captured imaginations for the past 150 years. Even before Russia sold Alaska to the United States in 1867, there was support in the U.S. Congress for a telegraph line across the Bering Strait to Russia, a U.S. ally. In 1890, Governor William Gilpin of the Colorado Territory promoted a "grand scheme of a Cosmopolitan railway," reaching "north and west across the Strait of Bering; and across Siberia, to connect with the railways of Europe and all of the world" (**Figure 4**). There were several initiatives to launch the project in subsequent decades, both before and after the Russian revolutions of 1917.

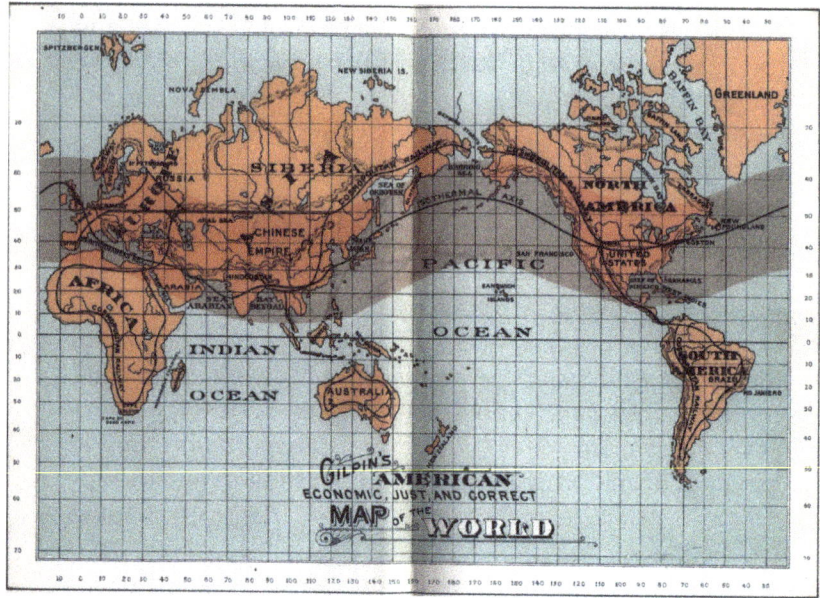

FIGURE 4

This "American Economic, Just, and Correct Map of the World," including a Cosmopolitan Railway across the Bering Strait, was issued in 1890 by Governor William Gilpin of the Colorado Territory. He advocated extending the Transcontinental Railroad "north and west across the Strait of Bering; and across Siberia, to connect with the railways of Europe and all of the world."

LaRouche's movement has campaigned for the Bering Strait connection since 1978. In 1995, *EIR* published an overview, written by American consulting engineer Hal B.H. Cooper, Jr. and Professor Sergei A. Bykadorov of the Siberian State Academy of Transport in Novosibirsk, of railroads in Siberia and the Russian Far East: existing lines, those started but not completed, and an ambitious vision of a far denser network to support the economic development of the continent and its resources.[8] **Figure 5** shows the Cooper-Bykadorov design for a web of the Trans-Siberian Railway, the Baikal-Amur Mainline, completion of the Near-Polar Mainline, and construction of a Bering Strait Connector, a North Siberian Mainline, and a Cross-Siberian Connector.

Within Russia, the Council for the Study of Productive Forces (SOPS, from its Russian acronym) took the lead in promoting the Bering Strait connection. A joint organization of the RAS and the Ministry of Economics, SOPS has its roots in Vernadsky's KEPS organization, the Commission for the Study of Natural Productive Forces, of 1915-1930. From 1992 until his death in 2010, the head of SOPS was Novosibirsk-based Academician Alexander Granberg, Russia's leading specialist on regional development. In November 2009, Granberg endorsed a public call to "Put the 'LaRouche Plan to Save the World Economy' on the Agenda."

On April 24, 2007, SOPS co-sponsored a conference in Moscow called Megaprojects of Russian East: An Intercontinental Eurasia-

The late Academician Alexander Granberg, long-time head of Russia's Council for the Study of Productive Forces (SOPS).

---

8. Hal B. H. Cooper, Jr., Sergei A. Bykadorov, "North Eurasian Rail Systems and Their Impact on Siberian Economic Growth," *EIR*, May 19, 1995.

FIGURE 5
Proposed Expansion of North Eurasian Rail Systems

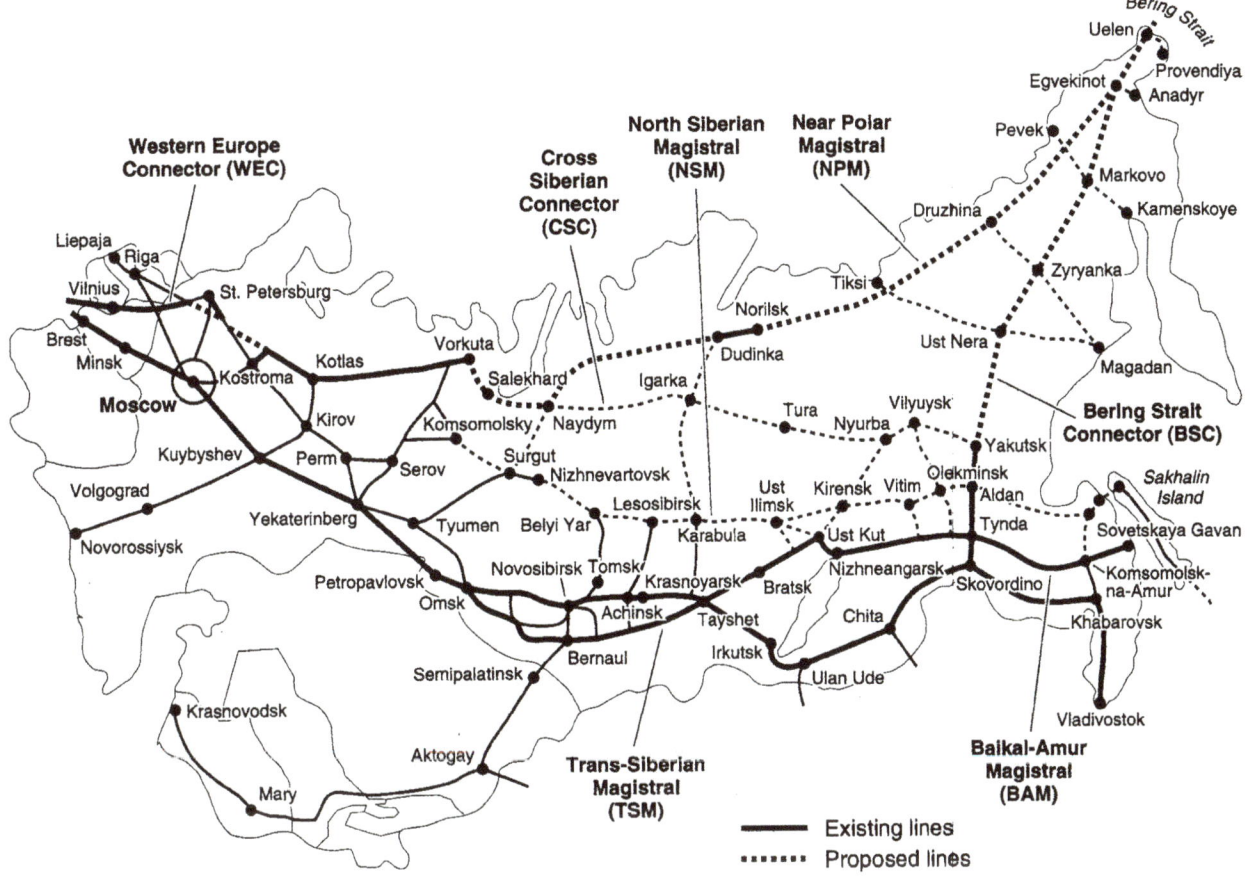

This 1995 map of existing and proposed rail lines, by Hal H. B. Cooper and Sergei Bykadorov, encompasses several countries adjacent to Russia: Estonia, Latvia, Lithuania and Belarus in the west, and Kazakhstan, Uzbekistan, Turkmenistan, Kyrgyzstan and Tajikistan to the south. Not shown are features built since 1995, such as Kazakhstan's new capital city of Astana and several railways in Central Asia.

America Transport Link via the Bering Strait.[9] Granberg told the meeting that Russia's leaders saw transportation infrastructure as essential for uplifting the country's vast outlying regions. He cited a presentation that Vladimir Yakunin, CEO of the state-owned Russian Railways company, had made earlier that month at a meeting chaired by Putin on rail transport, favoring construction of a 2,500- to 3,000-km railway from the Lena River to the Bering Strait. Later in 2007, that railway to the coastal village of Uelen, with a spur to the gold-mining center of Magadan on the Sea of Okhotsk, would be incorporated into the Russian Railways Strategy for 2030 (**Figure 6**), as a "new railroad of strategic importance."

SOPS invited LaRouche to address the Megaprojects conference. Although he was unable to attend, his contribution was read to the gathering: "The World's Political Map Changes: Mendeleyev Would Have Agreed!"

---

9. Rachel Douglas, "Russian-American Team: World Needs Bering Strait Tunnel!", *EIR*, May 4, 2007. The conference proceedings were published in the bilingual English-Russian Moscow periodical FORUM International #7 (2007).

FIGURE 6
The 2007 Russian Railways Strategy for 2030

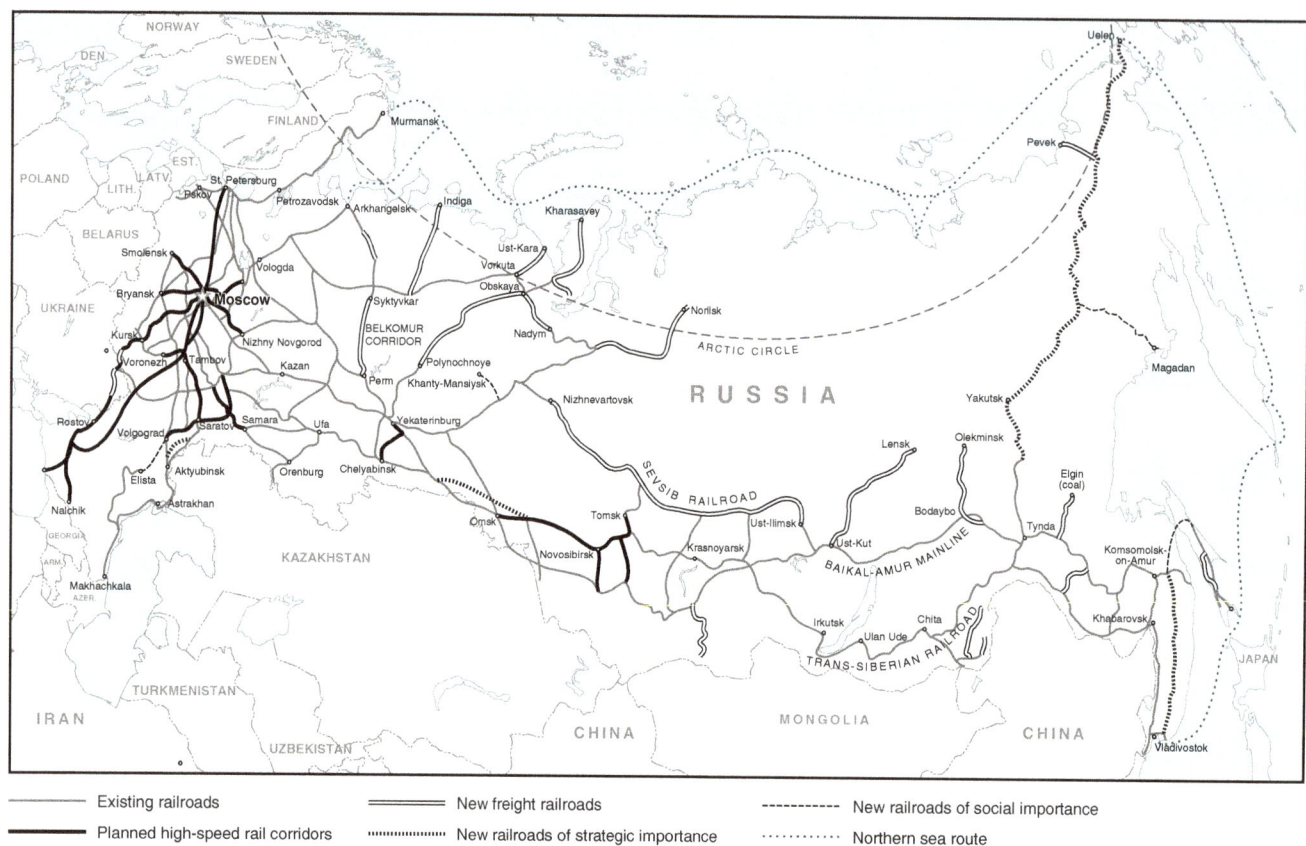

— Existing railroads
═ New freight railroads
---- New railroads of social importance
— Planned high-speed rail corridors
⋯⋯ New railroads of strategic importance
⋯⋯ Northern sea route

This depiction of existing and planned railways in Russia was created by EIR in 2007 from the Russian Railways map of the Strategy to 2030, adopted that year. Not all existing railroads are marked, and many lines connect to the rail systems of adjacent countries, not shown.

In eastern Siberia, the Berkakit – Yakutsk segment of the south-north Amur-Yakutsk Mainline (AYaM) has since been completed to Nizhny Bestyakh, across the Lena River from Yakutsk. The AYaM is the first leg of the "railroad of strategic importance," to the Bering Strait.

In 2007, high-speed rail (HSR) was not planned between Khabarovsk and Vladivostok in the Far East, but two years later this route was named as a possible Chinese-Russian HSR project. High-speed trains, sharing tracks with conventional trains, now travel between Nizhny Novgorod, Moscow, St. Petersburg, and Helsinki, Finland. Not shown as of 2007 is the now-proposed HSR extension past Nizhny Novgorod to Kazan. No work has been done on projected HSR in the Urals (Chelyabinsk – Yekaterinburg) and Siberia (Omsk – Novosibirsk – Tomsk, plus spurs), but that may change in view of an Oct. 2014 Chinese-Russian memorandum of understanding on building the Moscow – Kazan HSR as one segment of a Moscow – Beijing HSR corridor.

The planned new freight railroads at upper left center include the Belkomur Corridor (Perm – Arkhangelsk) and,

northeast of it, the Barentskomur Corridor (Urals – Indiga). They are elements of the now-stalled Industrial Urals–Arctic Urals project. Viewed together, the Belkomur Corridor, SevSib ("North Siberian") freight railroad in central Siberia, and Baikal-Amur Mainline (BAM) to the Pacific follow a 1928 plan for a Russian Great Northern Railway, from the White Sea in the West to the Sea of Japan in the East.

The main, 525-km Obskaya – Bovanenkovo section of the Yamal Peninsula freight railway toward Cape Kharasovay has been built, opening in 2010 as the northernmost railroad in the world. Some freight spurs from the BAM have been built (the 321-km line from the BAM to the Elga coking coal deposits was finished in 2012 by the Mechel company, as a private railroad), while others have not. In 2014, for example, construction material for the new Power of Siberia gas pipeline, a Russian-Chinese project from the Chayanda gas fields near Lensk in Yakutia, was shipped from the BAM to Lensk by barge on the Lena River, since the Ust-Kut – Lensk railroad has not come into being.

On this map, the Northern Sea Route (NSR) dips to the Arctic coast in two places: the mouth of the Yenisei River northwest of the mining center of Norilsk, and, 1500 km farther east, at the port of Tiksi in the mouth of the Lena River. Leaders of the Sakha Republic-Yakutia want to develop Yakutsk – Tiksi water transport via the Lena to the NSR.

ARTIST'S CONCEPTUAL VIEW OF THE NORTH AMERICAN ENTRANCE TO THE BERING STRAIT RAIL TUNNEL

BERING STRAIT RAILWAY TUNNEL
Between
Wales, Alaska, USA
and
Uelen, Chukotka, Russia

© J. Craig Thorpe, commissioned by Cooper Consulting Co.

The SOPS Plan for the Bering Strait

In September 2007, Dr. Victor Razbegin, deputy head of SOPS, contributed a presentation of the Bering Strait megaproject to a Schiller Institute conference in Germany.[10] The abridged excerpts below include key parameters of the plan.

> The need for a combined, multimodal transport corridor linking four of the six continents is obvious to everyone. Scientists have already solved practically all the technical tasks. A detailed look at the preliminary design reveals that the proposed route is neither longer, nor much more complex, than some other transport arteries that are already operational.
>
> The permafrost and harsh conditions of the extreme North are not an obstacle for the builders, since Russia has vast construction experience in similar climatic zones. Though laying a tunnel under the Bering Strait will require complex engineering solutions, it is also quite possible. World experience in recent decades demonstrates that such routes can be successfully operated, even in countries with high levels of seismic activity.
>
> On Sept. 6, 2007, the Russian government approved the Strategy for Railroad Development in Russia to 2030. This plan includes the line from Yakutsk (right branch of the Lena River) to Uelen, coming out at the Bering Strait, as a priority project of strategic significance.
>
> The Intercontinental Link will be a multimodal corridor, including:
>
> • A double-tracked, fully electrified, high-speed rail

Gazprom

Russia has construction experience in arctic climate zones. Here the offshore ice-resistant stationary gas extraction platform "Prirazlomnaya" is delivered from from the White Sea port of Severodvinsk, to Murmansk on the Barents Sea, January 18, 2011.

---

10. Victor N. Razbegin, "Eurasia-North America Multimodal Transport," *EIR*, Sept. 28, 2007.

mainline Yakutsk-Zyryanka-Uelen-Fort Nelson (Canada), total length 6,000 km;

• An electric power transmission line, with up to 1,500 kV direct current, and capacity of 12,000-15,000 MW;

• Fiber-optics telecommunications lines.

The Intercontinental Link Project will unite continental transportation lines in a single global network, create an international transport corridor, and make it possible to organize large-scale freight transport between Eurasia and America. This will accelerate global economic integration, opening up new opportunities for development. In particular, it will be possible to develop the northern regions of Russia, the U.S.A., and Canada, linking their enormous natural resources to world markets. The Intercontinental Link across the Bering Strait is the missing element in the global transportation network (**Figure 7**). This 6,000-km rail line could potentially carry about 500 billion ton-kilometers annually, or 3% of world rail cargo flows.

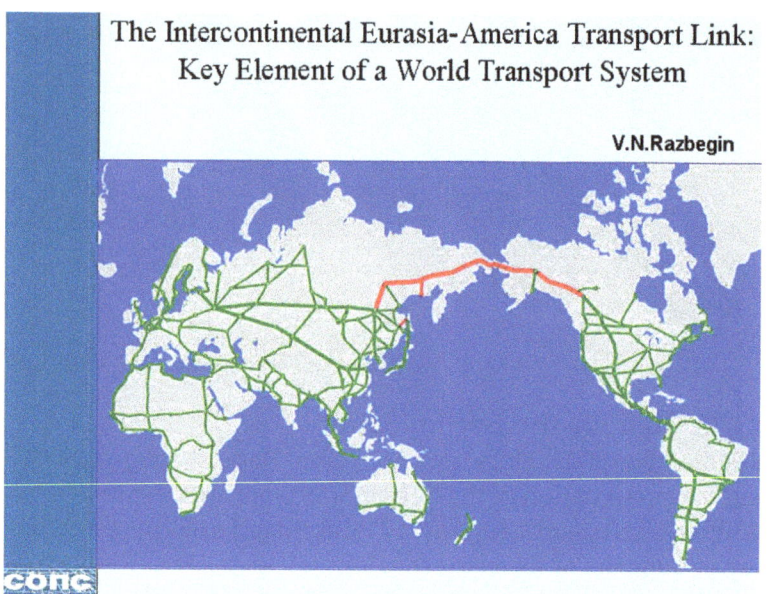

FIGURE 7

Dr. Victor Razbegin of Russia's Council for the Study of Productive Forces (SOPS – its Russian logo is at bottom left) adapted EIR's World Land-Bridge map, to illustrate the Bering Strait crossing as "a key element of a world transport system."

Razbegin stated the following costs of the Bering Strait, estimated by SOPS.

TABLE 1
Bering Strait Intercontinental Link investment requirements

	Billions of 2005 U.S. dollars
Railway component, total	12–15
Yakutsk-Uelen (Russia) railway	(9.5–11.5)
Wales (Alaska)-Fort Nelson (British Columbia)	(2.5–3.5)
Tunnel construction	10–12
Electric power industry, including intercontinental transmission line	23–25
Other (social infrastructure, fiber optics lines, etc.)	10–15
Total	**55–67**

Such investment would be recouped in 13 to 15 years, according to Razbegin, through development of the regions around the railway and their natural resources ($25-30 billion), freight transit revenue ($8-10 billion per annum), and other effects. He summed up:

The Intercontinental Link is important as a national project for Russia. It will give Russia a greater geopolitical presence in the Asia-Pacific Region and an improved position in world transportation services markets, as well as energy and industrial markets. It will be an important link in Russia's own transportation network, linking northeastern Russia to international transportation corridors.

Construction of the corridor will also be a precondition for the intensive economic development and settlement of northeastern Russia, providing year-round transportation access, reduced transportation costs, and competitive advantages for key manufacturing. It can improve living standards, create new jobs, and reverse out-migration from the region.

At the same time, the Transcontinental Link is a project of worldwide importance. The project will give the U.S.A., Canada, and the nations of South America direct access to China, Southeast Asia, Central and South Asia and beyond, for their products and technologies. At the same time, the Asia-Pacific Region will gain regular and mutually beneficial access to Siberia's resources. The project can bring about a shift toward civilian industrial production, as against military.

As a transnational project, the Intercontinental Link can improve international relations. It is a project that can change the world. It pulls together creative energies. Instead of putting up BMD systems, we can create a zone of international cooperation. Transnational infrastructure projects are the only real alternative to confrontation, including military confrontation, between nation-states and peoples.

SOPS took the Bering Strait multimodal tunnel project to the Shanghai World Expo-2010, where it won a Grand Prize for innovation. As the Silk Road Economic Belt policy began to take shape after President Xi Jinping's late-2013 announcement, respected rail expert Academician Wang Mengshu voiced China's keen interest in the scheme. In May 2014, Wang told the Beijing Times that Chinese high-speed rail (HSR) plans now encompass the Bering Strait tunnel, which was being discussed with Russia. With HSR technology, Wang forecast, a passenger would be able to travel between China and North America by train in only two days. High-speed trains are already operating on the Beijing-Shenyang-Harbin line in northeast China. Greater Tumen Initiative Corridor 4 (See following article) is intended to connect this route across the Amur River to the Trans-Siberian Railway, whence the Amur-Yakutsk Mainline and planned Bering Strait Connector lead to the Bering Strait (**Figure 6**).

Dr. Vladimir Petrovsky of the RAS IFES proposed in 2011, that Chinese involvement could be crucial in making the Bering Strait project happen. He cited not only China's experience with undersea tunneling projects, completed in Shandong and Fujian Provinces, but also the scope of its national railway program. "China may become a serious stakeholder in the Bering Strait project in the foreseeable future," wrote the Russian expert. "It would be difficult to take this proposition seriously, if China were not already half way through construction of the world's largest high-speed rail network. But the most populated country on Earth has shown no deficit of skill recently in undertaking massive public works projects, and its ambitions—and willingness to finance them—show no sign of slowing."[11]

Siberia Needs More People!

At the April 2007 meeting where Russian Railways announced that the Bering Strait Connector railway was in its plans for 2030, President Putin spoke of providing "transportation access to sparsely inhabited regions of the country and promising industrial zones," allowing such regions to be developed. The physical economy of the task starts with a look at the geography of the Siberian frontier.

Russia from the Ural Mountains to the Pacific is 13.1 million square kilometers (km^2) in area. Three of the 12 longest rivers in the world run through Siberia: the Ob-Irtysh system in western Siberia, the Yenisei-Angara system in central Siberia, and the Lena River in eastern Siberia (**Figure 1**). Lake Baikal, in southeastern Siberia, contains one-fifth of the Earth's fresh water. Besides the north-south ridge of the Urals, Siberia's western boundary, there are mountain ranges in eastern Siberia and the Russian Far East. Most of the region lies in the taiga climate and vegetation zone, with swampy coniferous forests. There is a narrower strip of tundra in the far north, along the Arctic coast. Permafrost, ground that remains continuously frozen for two or more years, covers approximately 10 million km^2 of Russia, most of it in Siberia and the Far East.

In a paper for the September 2007 Schiller Institute conference,[12] Dr. Sergei Cherkasov and Academician Dmitri Rundqvist of the RAS Vernadsky State Geological Museum summarized the physical and geological riches of Russia, and the rigors of the future Bering Strait Connector zone. Russia has 20.5% of the world's land area, 3% of its population, 22% of the forests, 30% of the total continental shelf area, and 16% of all mineral resources, they reported. Besides oil and natural gas, Russia has high percentages of the world's rare earth elements, as well as agrochemical ores such as potassium salts, apatite, and phosphorite. Its diamond resources are the largest in the world, and it ranks third in gold. Russia has up to 50% of the planet's reserves of some metals, but its share of their production and consumption is far smaller (**Figure 8**), except for platinum group elements (PGE) and nickel (**Figure 9**).

The geologists documented the wholesale looting, under economic deregulation and the collapse of Russian industry in the 1990s. In 1999, for example, Russia exported 57.3% of the oil it produced and 32% of the natural gas, while the export levels of key metals were:

---

11. Vladimir Petrovsky, "Bering Strait Project—Russian Perspective Update," Universal Peace Federation, Oct. 4, 2011.

12. Sergei Cherkasov, Dmitri Rundqvist, "Raw Materials and Russian Infrastructure," *EIR*, Sept. 28, 2007.

FIGURE 8
Russian Resources: Reserves, Production and Consumption

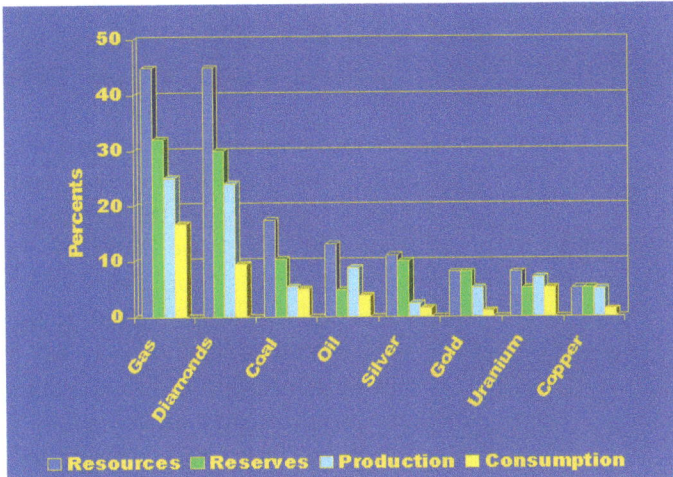

This 2007 Vernadsky State Geological Museum chart compares Russia's percentage of world deposits and proven reserves of selected subsoil resources, with its lower share of world production and consumption of them. It reflects Russia's export potential, as well as the need to boost its own manufacturing industry.

FIGURE 9
Russian Nickel and Platinum Group Elements

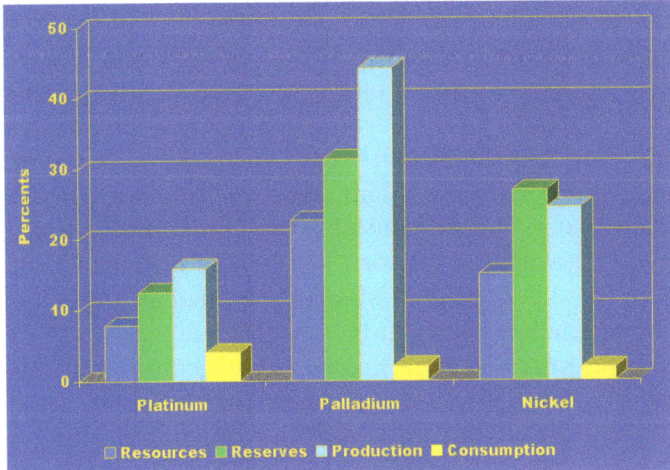

Russia is a relatively big producer of nickel and platinum group metals, mined around the city of Norilsk above the Arctic Circle, but domestic consumption remains miniscule.

copper, 85% of production; nickel, 91%; and tungsten, 96%. In 1996, Russia exported an amount of uranium equivalent to almost 417% of that year's production. Another year, 356% of the amount of molybdenum produced was exported.

Although geological exploration had declined drastically after the break-up of the Soviet Union (a still ongoing problem), the Geological Museum team presented results from an RAS study, showing previously undeveloped fields of gold, silver, copper, and molybdenum in northeastern Siberia (**Figure 10**). "The spectrum of metals in use is changing," Dr. Cherkasov told the conference, while Russia has also discovered new types of deposits, such as oil-bearing sands consisting largely of ilmenite, the most important ore of titanium.

These resources must be mined in a harsh climate, as in the case of the nickel and platinum group elements near which the industrial city of Norilsk was built, above the Arctic Circle. Cherkasov and Rundqvist described the rugged climate and topography of the Bering Strait corridor route through the Sakha Republic-Yakutia. Houses there are built on stilts, because of the permafrost. Yakutia's Popigay diamond deposit, a meteorite impact crater, contains more industrial diamonds than all other known resources in the world, they reported, "but it is not being developed, because there is not *any* infrastructure in that place. Nobody lives within 200 km." Developing the Yana-Kolyma gold and silver zones, charted by the Russian Geological Survey as comparable in size to Germany, would take the creation of approximately 300,000 jobs, "yet, in this area we have a population of maybe 10,000 people."

In 2001, responding to a warning by Academician Dmitri Lvov, that Russia was becoming a "raw-materials appendage" of Western finance, LaRouche wrote about the threatened loss of Russia's, and the world's, even more precious asset: skilled and creative human beings, the outcome of whose productive activity is not "objects," but the continued life of the human species.[13]

Siberia demonstrates better than most places, that the world needs more people, especially more of the well-educated and often daring intellects who made Soviet science a

---

13. LaRouche, "On Academician Lvov's Warning: What Is 'Primitive Accumulation'?", *EIR*, Aug. 17, 2001.

FIGURE 10

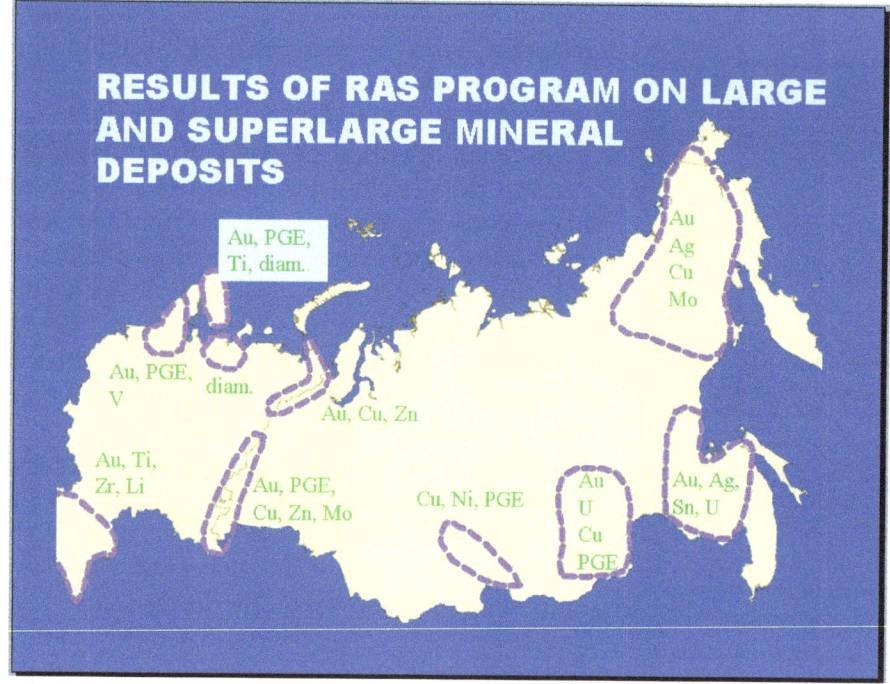

The 2007 map from the Vernadsky State Geological Museum summarizes results from a Russian Academy of Sciences survey program. Not shown are previously confirmed and exploited resources, such as the nickel and platinum group elements (PGE) deposits around Norilsk in northwestern Siberia and the famous diamond fields of the Sakha Republic-Yakutia in the East.

Courtesy of Dr. Sergei Cherkasov, RAS Vernadsky State Geological Museum.

Norilsk (top), population 175,000, is above the Arctic Circle in Russia's Krasnoyarsk Territory. Nearly 25% of the world's nickel and 44% of the palladium are produced here. In the lower picture: an oilfield snowstorm in Siberia.

formidable power in the 20th Century. In his last book, written in 1906, the universal genius Dmitri Mendeleyev said that Russia should have 500 million people by the year 2000, in order to progress. When the Soviet Union broke up in 1991, after revolutions, civil war, two world wars, and all the other traumas of those intervening 85 years, its population had reached 293 million; that of the Russian Federation was 148.6 million. As of 2013, Russia's population was down to 143.5 million, but had rebounded slightly from a low of 141.9 million in 2009.[14] For nearly 20 years, Russia lost population because of an excess of deaths over births, despite net immigration as Russians moved there from other former Soviet republics.

Nationwide, Russia suffered a brain drain in the 1990s, as scientists were hit simultaneously with an abrupt drop of funding for their institutes, and allurements to emigrate; many younger educated people likewise sought greener pastures abroad. Equally, if not more devastating has been an internal brain drain, namely, the massive 1990s shift of university-age youth into fields like business education, while a new generation of scientists and engineers was not trained. Veteran cosmonaut Georgi Grechko bluntly stated the result, in terms of its impact on the Russian space sector: "What is our main trouble?" he asked, after Russia's Phobos Grunt Mars probe failed in 2011. "It is an open secret for specialists. The staff members are either over 60 or under 30. There is no intermediate age group. A generation was lost for the space industry, when it was struggling to survive. People, most of them young, energetic, and talented, would seek higher earnings in other places."

The Far East Federal District (FD) was especially hard hit, in terms of absolute population, and the quality of employment. The area has nearly 2 million fewer people than in 1989. In the Far East and eastern Siberia combined, the population fell by 21% in 1989-2012. Net losses of 300,000 people from Primorsky Territory (the Maritime Territory, or "Primorye"), and 230,000 from the Amur Region bordering China,

---

14. With the annexation of the Crimean peninsula in May 2014, Russia gained 2.5 million people.

FIGURE 11
The Federal Districts and Selected Regions and Cities of the Russian Federation

The Siberia/Far East frontier is spread across three of Russia's Federal Districts: the Ural, Siberian, and the Far East FDs. Their combined area is 13.1 million km², with a population of just under 38 million counted in the 2010 Census (2.8 people per km²). The two vast easternmost FDs, without the Urals, are the most sparsely populated, with 25.6 million inhabitants. Only 6.3 million people live in the Far East FD, on 6.2 million km² of land—1 person per square kilometer. In the Siberian FD, 19.3 million people live on 5.1 million km², an average density of 3.7 per km²; 3.5 million of them reside in three cities: Omsk (1.1 million), Novosibirsk (1.5 million), and Krasnoyarsk (1 million).

The Northwest FD, where the ports of Arkhangelsk and Murmansk are situated, is also sparsely populated. Setting aside the 5 million population of St. Petersburg, 8 million people live on the remainder of its 1.7 million km² (4.7 per km²).

are typical (**Figure 11**). The drastic curtailment of social and cultural infrastructure, previously financed by the Soviet government or state-owned companies, and the end of affordable transportation, including subsidized fares for travel to other parts of Russia, were factors in people's desire to leave.

Much of the Soviet Far East's industry had produced for the military, and was drastically taken down in the 1990s. Shipbuilding, for the Pacific fishing fleet as well as the Navy, was slashed. Director Stanislav Govorukhin's 1994 film *The Great Criminal Revolution* documented Russia's border-areas near China in the first two years after the break-up of the Soviet Union: truckers driving loads of metal products, pilfered from industrial plants, into China for sale; unemployed men and bands of small children robbing trains; and the contrast between depressed Primorye towns and booming cities like Suifenhe, in Heilongjiang Province across the border. Fifteen years on, in 2009, the area's economy was still depressed—an estimated 100,000 of the nearly 600,000 people of Vladivostok were employed in the city's biggest line of business: importing and servicing Japanese and Korean used cars.

Many Chinese cross-border street traders, called shuttle traders, worked in Russian Far East cities in the 1990s. More than 130,000 Chinese workers from Heilongjiang Province began traveling to Russia as contract workers in construction or agriculture. Despite sensational media stories about "millions" of Chinese pouring into the Far East and, by one wild-eyed account, living in "autonomous communities deep in the forest," qualified Russian estimates are that 200,000-400,000 Chinese citizens currently live in all of Russia.

For Chinese and Russians in the Far East, crossing the border is nothing new. The southern part of today's Far East FD was Chinese territory until the mid-19th Century, while Russia later dominated northeastern China during construction of the Chinese Eastern Railway (Chita-Harbin-Vladivostok), in a corridor leased from China beginning in 1898. Russians were the largest ethnic group in Harbin, China in 1913, and still comprised one-third of its population in 1921. Conversely, the Soviet 1926 Census recorded that the population of Vladivostok was 22% Chinese.

Russian Orthodox Church of the Holy Iveron Icon, Harbin, China, c. 1900.

Today, Chinese people living in the Far East are outnumbered by those from formerly Soviet Central Asia, who enjoy visa-free entry into Russia, are more likely to remain in the country illegally, and occupy many unskilled, low-income jobs. Former Prime Minister Primakov, at a March 2014 conference, noted that 60% of the 12.5 million immigrants currently in Russia are there illegally; the Russian economy, he said, loses 117 billion rubles ($3.6 to 4 billion, in recent years) annually from the effects of Russian businesses hiring these undocumented workers, as the businesses evade taxes and the workers are paid a pittance. Despite limits on hiring foreign guest workers, imposed in 2003 and toughened in 2007, the Federal Migration Service reported a 50% increase in migrants entering Russia in 2011-2013, with the greatest numbers arriving from Kazakhstan, Uzbekistan, and Ukraine, even before civil war in Ukraine drove hundreds of thousands more to flee to Russia.

Even as the role of immigration in meeting labor needs in Siberia and the Far East continues to be hotly debated, the composition of the Chinese migrant population in Russia has shifted. A former official of the Russian Federal Migration Service's Far East office, criticizing Russian internal opposition to Chinese immigration, reported: "In the mid-1990s when we first opened up Vladivostok—and China was less economically developed—we had lots of Chinese [working] in border bazaars and sleeping in train cars. Nowadays you have to pay around a thousand dollars monthly to attract a specialist from China. God gave us a peaceful, hard-working neighbor, eager to engage in mutually beneficial economic activity... and [yet] we live in fear of China!"[15]

For those who are serious about the reindustrialization of Russia and development of its frontiers, there is no substitute for economic policies to promote family-formation and a positive demographic trend in the Russian Federation itself. But in the short term, immigrants are needed. Mikhail Vinokurov, rector of Baikal State University of Economics and Law (Irkutsk), called in 2013 for a goal of doubling Russia's population, to 300 million, within the next 100 years. For the immediate future, he said, "the right thing to do is to bring 40 to 50 million people into the country to work," under a well-regulated migration program. The distinguished director of the RAS IFES, Academician Mikhail Titarenko, concurs. Speaking at the same March 2014 conference as Primakov, he said, "We need immigrants, if we want to develop the economy of Siberia and the Far East," and called for the Federal Migration Service to concentrate less on punishing illegals and more on "creating the conditions for education and assimilation."

Two contrasting polls dramatize that the future of Russia's eastern frontier depends on policy decisions, not on so-called objective factors. In Summer 2012, many were dismayed at a survey by the VTsIOM polling agency, which found that 39% of Siberia and Far East residents would leave those regions if they could. The reasons cited were low wages, lack of career growth prospects, inability to obtain their own housing, and lack of educational and other opportunities for their children. Prof. Yuri Tavrovsky, however, last year reported findings pointing in the opposite direction: "Polls have shown that over one-third of the inhabitants of the European part of Russia would readily move to the East, if a major national project were launched there, and high wages offered, with the opportunity to purchase housing."

The Corridor Solution

Russian Finance Minister Count Sergei Witte (1849-1915), organizer of the Trans-Siberian Railway, knew the principle expressed in that optimistic poll. He stated it in his famous dictum about railroads creating "cultural fermentation" (see Part 3, China's New Silk Road) Witte also knew that development along a railway, concentrated in a corridor on both sides of the track, would transform that region and the whole country. In an English-language book prepared for the 1893 World's Columbian Exposition in Chicago, Witte's Ministry of Finance staff wrote that the TSR, "intersecting the whole of Siberia for a distance of 7,112 versts,[16] embraces a very wide zone, which cannot be taken at less than 100 versts on either side of the line, or about one million and a half square versts"—larger in area than any nation of

U.S. Library of Congress, Proskudin-Gorskii Collection.

An ethnic Buryat Mongol switchman on the Trans-Siberian Railway, 1910.

---

15. Artem Zagorodnov, "Is Russia's Far East Overcrowded by Chinese Immigrants?", interview with Sergei Pushkarev, former head of the Federal Migration Service office in Vladivostok, *RBTH*, April 28, 2012.

16. 7,539 km (1 verst = 1.0668 kilometers). This is the length of the TSR measured not from Moscow, but eastward from Chelyabinsk, where new construction began.

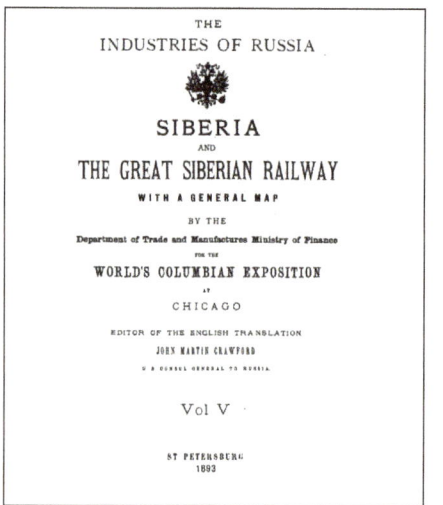

Prepared by Count Witte's staff at the Russian Ministry of Finance for the 1893 World's Columbian Exposition in Chicago, the book Siberia and the Great Siberian Railway *said that the benefits of a development corridor along the Trans-Siberian would be impossible to measure in monetary terms.*

Europe. They promised that the benefits from the Siberian railroad would be impossible to measure in monetary terms, for it "probably for a long time will not prove remunerative in the strict sense of the word," but would have "numerous advantages not subject to arithmetical computation." Its "powerful impetus to the whole economic development of the country, [would] call into existence many new branches of industrial activity."[17]

Omsk, Novosibirsk, and Krasnoyarsk, three boomtowns on the TSR of a hundred years ago, today have populations of more than 1 million each. In a study of the history of trans-Siberian railroad projects, economist Victor Suslov of the RAS Siberian Division's Institute of Economy and the Organization of Industrial Production (SD IEOIP), a colleague of the late Academician Granberg, described the multiplier effects of the TSR throughout southeastern Russia as "unexpectedly great." Wherever the TSR intersected major north-south waterways in Russian Eurasia, he wrote in a 2008 article, "its beneficent influence reached 600-800 km in both directions."

The TSR itself, naturally, figures in some of the most advanced thinking about development corridors in Russia today. One example is Sibstream (**Figure 12**), a high-speed rail and city-building concept developed in the 2000s by Prof. Ilya Lezhava, head of the Department of

---

17. Department of Trade and Manufactures Imperial Ministry of Finance, *Siberia and the Great Siberian Railway*, St. Petersburg, 1893.

FIGURE 12
Sibstream "Linear City" To Revitalize Siberia's Science Centers

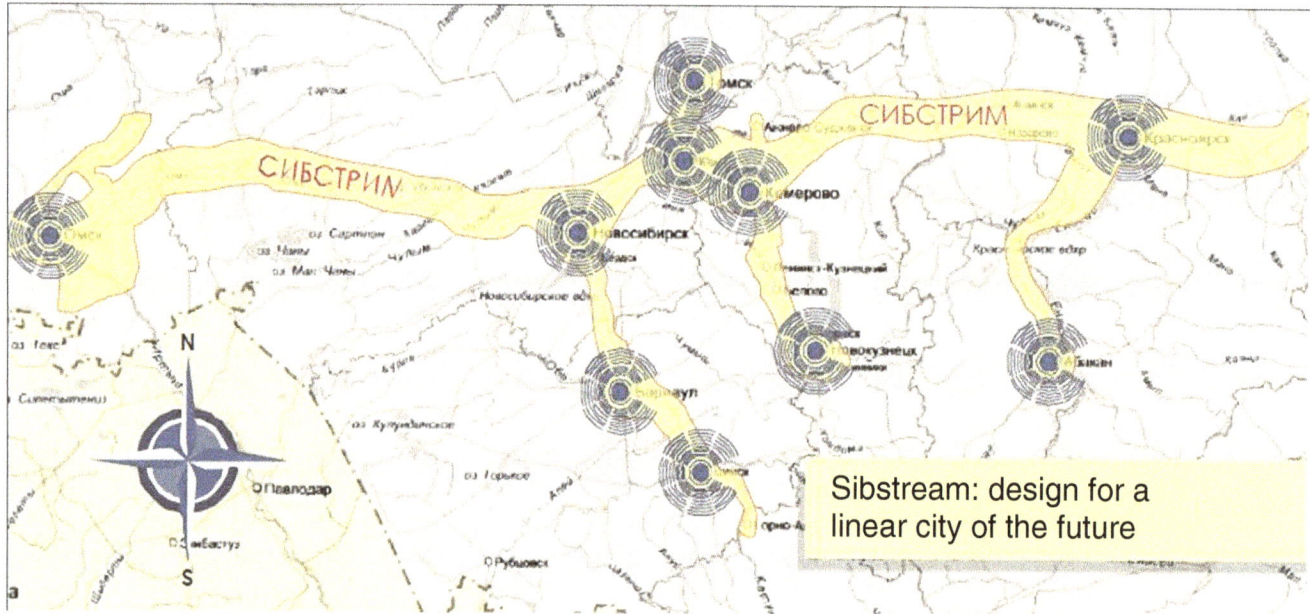

Siberia: The New Central Russia, or How the South of Western Siberia Will Become an Economic Center of the Planet, IDMRD, 2012

The schematic map shows the proposed stretched-out city called Sibstream, with a high-speed railway in the Trans-Siberian Railway (TSR) right-of-way as its spine, designed by Russian architect Ilya Lezhava. In this diagram, Sibstream is labeled with red letters in the wide, pale-yellow east-west corridor. Existing cities shown by the circular symbols, west to east along the TSR, are Omsk, Novosibirsk, Yurga, Kemerovo, and Krasnoyarsk. Tomsk, north of Yurga, and industrial towns south of the TSR—Barnaul (machinery, coal), Biysk (defense industry), Novokuznetsk (coal), and Abakan (light industry)—would be redeveloped as science and new-technology centers

City-Building at the Moscow Architectural Institute (State Academy). Its implications are discussed in a 2012 report from the Institute for Demography, Migration, and Regional Development (IDMRD), titled *Siberia: The New Central Russia, or How the South of Western Siberia Will Become an Economic Center of the Planet*. Like all the best designs for Siberian development, this one emphasizes that the region's infrastructure must be far more than only pipelines for hydrocarbon exports or railway spurs for shipping out minerals: people must be able to live happy and productive lives in these areas.

Sibstream: a new, linear city along the Trans-Siberian Railway 2.0

It is appropriate to build an original, linear city from Omsk to Novosibirsk to Krasnoyarsk, centered on a high-speed rail line, as a flourishing geocultural and economic space, on a firm infrastructure platform. This would be the city of Sibstream, developed by a team under Academician [of the Academy of Architecture] Ilya Lezhava.

Around Sibstream, in convenient locations, what should be done is not only to build up the urban "fabric" along the main line, but also to carry out massive reconstruction of existing and construction of new traditional ("focal point") cities of all sorts: technopolises, university cities, agropolises, science cities, and industrial cities, as well as recreational, sanatorium-and-spa, tourist, and nature preserve areas. A stretched-out megacity could be an alternative to the typical form of urbanization in a megalopolis; it not only does not preclude, but actually demands urbanization based on low-rise dwellings with land attached, which should become the model for all Russia. This will promote the formation of a geographically dispersed, but nonetheless consolidated and unified community of leaders and specialists—a "development class." At least a million new, free such homes should be built in this region, to attract specialists from throughout Russia and accelerate the formation of this social layer, aiming to increase the population of the area to 12-14 million by 2030.

Since the "load-bearing" transportation backbone of Sibstream will be a high-speed railway along the Trans-Sib, implementation of this project will also serve as an important launch mechanism for the modernization of the Trans-Sib. Sibstream will open the way to developing and implementing a Russian city-building doctrine, based on the deliberate planning of economic and demographic development around a new philosophy, system, or, as Lezhava puts it, "channel" of settlement.

The IDMRD's New Central Russia program proposes to rebuild the traditional industrial and scientific strengths of west-central Siberia, as an engine for Russia's own reindustrialization with advanced tech-

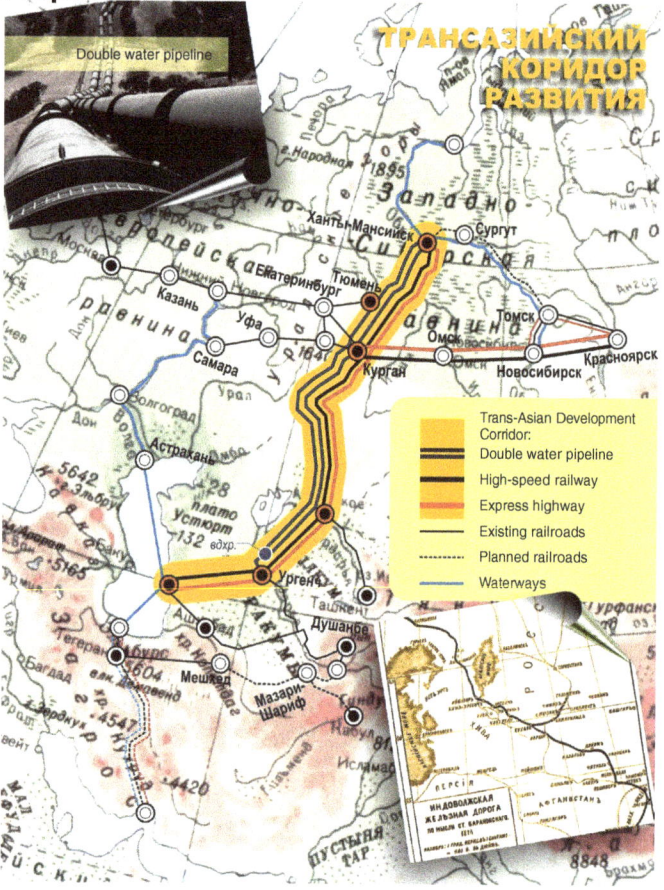

FIGURE 13
North-South Trans-Asian Development Corridor Proposal

The map is titled at the upper right, "Trans-Asian Development Corridor." This north-south infrastructure scheme, reaching from the science and industry centers of West Siberia, through Kazakhstan and into Central Asia, is proposed by the Institute for Demography, Migration and Regional Development (Russia). Bodies of water shown are the Persian Gulf (bottom left), Black Sea (far left), Caspian Sea (center-left), the currently desiccated Aral Sea (center), and the Barents Sea and Kara Sea (top) off Russia's Arctic coast. The proposed east-west high-speed rail and super-highway (black and red horizontal lines) run along a segment of the Trans-Siberian Railway in west-central Siberia (Omsk, Novosibirsk, Krasnoyarsk). The southern terminus of the corridor is at Turkmenistan's port of Turkmenbashi on the Caspian Sea.

nologies, as well as to produce capital goods and infrastructure for the development of countries to the south—the Central Asian nations and Afghanistan (see Appendix, Part 5). **Figure 13** illustrates the IDMRD concept of a north-south Trans-Asian Development Corridor from this western Siberian resource and industry zone to the Caspian Sea.

In a March 2007 report on the major scientific work done by the Russian Academy of Sciences during the previous year, then-RAS President Academician Yuri Osipov highlighted an industry, science, and technology plan drawn up by the SD IEOIP for the same west-central Siberian region (**Figure 14**).

Russian Railways CEO Vladimir Yakunin, who continues to advocate building the Bering Strait railway crossing, is the author of another programmatic outline for a corridor, which he calls the Trans-Eurasian Development Belt. It is centered on building high-speed rail along the TSR and BAM, integrated with energy and telecommunications systems, new towns, and "10 to 15 new industries," emphasizing the economic ripple effect of rail construction through the creation of demand for capital goods. In March 2014, Yakunin's draft, with two senior academicians of the RAS as co-authors, was favorably received by the Presidium of the RAS after he presented it. On October 29, 2014, RAS President Academician Vladimir Fortov briefed President Putin on the Yakunin plan, as one of four RAS proposals for jump-starting the Russian economy. Although Yakunin's report in March had barely alluded to "a possible Eurasian-American development belt, linking Alaska with Chukotka," Fortov now emphasized this aspect, according to Russian media reports, saying that "the idea is to build a railway alongside the Trans-Sib, based on new high-speed rail technologies and possibly going on to Chukotka, across the Bering Strait, and to the American continent," and proposing that this would make it possible to settle and industrialize Siberia.

The State Geological Museum's Cherkasov and Rundqvist likewise present Siberia's potential in terms of development corridors, as the key to the proper exploitation of new mineral resources. The mineral zones mapped by the RAS survey (**Figure 10**) coincide with the routes of existing railroads and projected ones, like the Bering Strait Connector and railways in the Urals. The Industrial Urals-Arctic Urals megaproject, they noted, involves a $2.4 billion, 1,000-km railroad running north from Yekaterinburg, and $3.5 billion for energy infrastructure.

FIGURE 14
Russian Academy of Sciences Graphic: Strategic Projects for Western Siberia and Krasnoyarsk Territory (2006)

RAS media release

*This 2006 graphic, made by an institute of the RAS Siberian Division, shows projects that could be launched immediately in west-central Siberia, given federal government commitment. The differently colored areas are regions in the Siberian Federal District. Parts of the Trans-Siberian Railway (TSR) and the Baikal-Amur Mainline (BAM) are shown by the yellow and black line, while the planned North Siberian or SevSib Railway (red and black) runs southeast from the Ural Mountains, following the curve of the Angara River to connect with the BAM. The red dots on the TSR mark the cities of Omsk, Novosibirsk, and Krasnoyarsk (west to east), followed by Tayshet, where the TSR and the BAM split, and the East Siberia-Pacific Ocean oil pipeline starts. As in the IDMRD's version of Sibstream (**Fig. 12**), future technology, university, resource-development, and energy facilities are indicated for the area south of the TSR, while an Angara Industrial Complex north of the SevSib (broken-line ellipses, upper right) will be based on natural resources there. The broken double lines mark oil and gas pipeline routes.*

This Industrial Urals-Arctic Urals scheme, called by its Russian acronym, UP-UP, was designed by Granberg and a SOPS team for the industrial development of the northern Ural Mountains. Its main components are 60 new mines and ore-processing facilities, power plants, and new railways. The concept was that subsoil resources from these northern latitudes would not only be exported, but also serve as raw materials for revitalized and new industries in the southern Urals. The plan was lobbied for in 2006-2007 by Pyotr Latyshev, then-Presidential representative in the Ural Federal District, which extends from the Arctic down to the million-population cities of Yekaterinburg and Chelyabinsk, and encompasses the West Siberian oilfields. Pointing out that little of the district's northern reaches had been thoroughly studied by geologists, Latyshev and the UP-UP company, formed to oversee the project, anticipated finding deposits of coal, iron ore, titanium and manganese ores, gold, platinum, quartz, copper, chromium, alumina, tantalum-niobium ore, bentonite, and precious gemstones.[18]

The main railway for the UP-UP project, an extension of the Yekaterinburg-Polunochnoye line north to Obskaya (on the west bank of the Ob River, opposite Salekhard on the east side, and at the base of the natural gas-rich Yamal Peninsula), was included in the Russian Railways Strategy for 2030, as were two other railroads involved in UP-UP. Those are the Belkomur and Barentskomur corridors, named after their routes from the Urals, through the Komi Republic on the west side of the Urals, and on to Arkhangelsk on the White Sea and a future expanded port at Indiga on the Barents Sea, respectively (**Figure 6**). Those routes would provide export outlets westward through Scandinavia, as well as onward to the British Isles and even North America, and eastward to Asia, via the Northern Sea Route along Russia's Arctic coast.

---

18. Hopes for speedy implementation of the plan faded, however, with the deaths of Latyshev in 2008 and Granberg in 2010, and the inability of the UP-UP company to attract either state funding or international investors during the aggravated global crisis, beginning in 2008.

The routes of both the Belkomur corridor and the planned SevSib industrial railway, the latter running east from the Urals, proceeding along the Angara River, and connecting to the Baikal-Amur Mainline (BAM), have been on the drawing board for a long time. If Belkomur, SevSib, and BAM are connected end to end, as shown schematically on our World Land-Bridge Network map (see Part 1), they trace the entire route of the Russian Great Northern Railway (or North Siberian Mainline), a diagonal northwest to southeast trans-Eurasian railway proposed in 1928 by the visionary Alexander Borisov (1866-1934). An artist, who had sailed in Arctic waters with Count Witte in 1894, painted Arctic landscapes, and explored Russia's Arctic islands, Borisov spent his later life designing railways. The Russian Great Northern was to extend more than 10,000 km from Murmansk on the Kola Peninsula in far northwest Russia, to the Tatar Strait opposite Sakhalin Island on the Pacific Ocean.

A Corridor for a Cosmodrome?

In late 2007, President Putin gave the go-ahead to build a new space launch site in the Amur Region. Cosmodrome Vostochny (meaning "East") will be the first facility on Russian territory that can handle satellite launches to geostationary orbit, and other space missions for which Russia still uses the former Soviet Baikonur Cosmodrome, located in Kazakhstan. As prime minister in 2008-2012, Putin continued to push for Cosmodrome Vostochny in the face of budget-cutting pressures.

At the height of debate over whether to build the new cosmodrome at all, IDMRD proposed to make it the anchor of a "space industry cluster" and a high-technology industrial corridor from the town of Uglegorsk to Komsomolsk-on-Amur, a manufacturing city near the eastern end of the BAM (**Figure 15**). This cosmodrome corridor concept was presented at the September 2007 Schiller Institute conference, in a paper by IDMRD Supervisory Board Chairman Yuri Krupnov, then serving as an advisor to Far East government officials.[19]

Adjacent to Uglegorsk, the small military Cosmodrome Svobodny was being mothballed at the time. IDMRD called for retaining the skilled military personnel who

FIGURE 15
Proposed Uglegorsk to Komsomolsk-on-Amur Development Corridor

As the decision was being made in 2007 to build Cosmodrome Vostochny (East) at the site of the former military launch site, Cosmodrome Svobodny, the Institute for Demography, Migration and Regional Development proposed a development corridor from the existing military town of Uglegorsk, to Komsomolsk-on-Amur, a city which is home to major aircraft and shipbuilding plants, near the terminus of the Baikal-Amur Mainline. The map shows the corridor's location with respect to China and to the ten provinces and main cities of the Far East Federal District. (Russian "Krai" in place names means "Territory," "Oblast" means "Region.")

---

19. Yuri Krupnov, Ilnur Batyrshin, "Svobodny Cosmodrome: a Potential Space Industry Cluster and Development Corridor in Russia's Amur Region," *EIR*, Sept. 28, 2007.

worked at Cosmodrome Svobodny and lived in Uglegorsk, to be the core of the regional workforce, while also encouraging "internal immigration"—an influx of people from western Russia and former Soviet republics. Thus the project would be a model for the repopulation of Siberia and the Far East, and creation of skilled manufacturing jobs. Specific goals included:

- Increasing the population of the Amur Region by 175,000 within five years, reaching 1 million inhabitants and raising the population density from 2.4 persons per km^2, to 2.8 or 3 persons per km^2, compared with 80 per km^2 across the Amur River in China's Heilongjiang Province.
- Creating tens of thousands of new high-skill jobs in economic clusters centered on the space exploration industry, machine-building, the timber industry, soybeans, and transport and energy infrastructure.
- Generating more than 30,000 skilled and at least 60,000 semi-skilled jobs at the Komsomolsk Aircraft Company and at shipyards in the same city, to support full-cycle space industry production and services around the cosmodrome. These workers and their families would nearly double the population of Komsomolsk-on-Amur, to 500,000.

The IDMRD report defined the corridor as "a platform for the industrial development of the Far East on the basis of innovation." Because the Uglegorsk to Komsomolsk-on-Amur route runs toward Sakhalin Island and the Pacific Ocean, it would be an incentive to modernize the BAM, and its natural continuation would be a bridge or tunnel across the Tatar Strait to Sakhalin Island. The planned industrial growth in the Amur Region and Khabarovsk Territory would create demand for new steel and other metallurgical plants, for which the Garin iron ore deposit in the Amur Region could be developed to provide raw materials.

IDMRD described the Svobodny project as well-suited to "play a crucial role for the development of the whole region, and therefore for the Eurasian Land-Bridge," also because of its proximity to existing east-west infrastructure. The Chita-Khabarovsk federal highway passes a few hundred meters from Uglegorsk and the TSR is 1 km away. The BAM is 220 km to the north. Also in the vicinity are three airports, the new East Siberia-Pacific Ocean oil pipeline, transmission lines from two hydroelectric power plants, and a shipping terminal on the Zeya River, providing access to Blagoveshchensk and other Amur River ports.

Under the project actually adopted, on which construction began in 2011, Cosmodrome Vostochny is to be completed in 2016-2018, with the first launches taking place in 2015. After a September 2014 inspection visit, Putin put Deputy Prime Minister Dmitri Rogozin in charge of making up a two-month lag behind the construction schedule. He also mandated doubling the 6,000-person workforce. The Vostochny project includes production facilities for the launch base itself, such as plants to

The late Gen. Vladimir Popovkin, then head of the Russian Federal Space Agency (Roscosmos), briefs President Putin during an April 12, 2013 visit to the site of Cosmodrome Vostochny, under construction in the Amur Region. Left to right: then-Minister of Far East Development Victor Ishayev, Defense Minister Sergei Shoygu, Amur Governor Oleg Kozhemyako, Popovkin, Putin, and Deputy Prime Minister Dmitri Rogozin.

produce liquid rocket fuels. A new city, for at least 30,000 people, is to be built, expanding around Uglegorsk; it will be named Tsiolkovsky, after the Russian space pioneer. It is planned to build research centers, an academy for young scientists, and a cosmonaut training center.

So far, the IDMRD idea of a corridor reaching to Komsomolsk-on-Amur has not been adopted. Employment at the Komsomolsk-on-Amur Aircraft Production Association, which makes Sukhoi fighter jets and remains Russia's premiere aircraft plant, dropped from 17,000 in 2007 to 13,500 in 2011. The famous Amur Shipbuilding Plant (former Lenin Komsomol Shipyard), which used to build nuclear submarines, nearly went bankrupt in 2009 after a fatal industrial accident and a financial fraud investigation of its former director.

The "Turn to the East"

Russia's "turn to the East" began long before the U.S. and EU sanctions of 2014. Vladimir Putin's February 2012 Presidential campaign article on foreign policy, with "The Growing Role of the Asia-Pacific Region" as one subhead, and his October 2011 article in *Izvestia*, "A New Integration Project for Eurasia," announcing plans for a Eurasian Economic Union, were milestones in post-Soviet Russian foreign policy.[20] Much earlier, on the eve of the November 2000 Asia-Pacific Economic Cooperation (APEC) meeting in Brunei, the first APEC summit after his election as President, Putin set forth this theme in an article published throughout the region, titled "Russia Is a Eurasian Country." At that APEC forum, Putin called for worldwide efforts to overcome the North-South economic development gap. He emphasized the potential for Asian investment in infrastructure projects in Russia, especially Siberia and the Far East, and Russia's ability to export capital goods, not only raw materials. Russian science will promote Eurasian development, Putin declared, pointing to potential cooperation on space-launch services, nuclear reactor technologies, and a secure nuclear fuel cycle.[21]

Putin's personal diplomacy, with summit visits to India in 2000, China in 2001, and both countries again in 2002, also carried forward the idea of Eurasia's "strategic triangle," enunciated by Primakov in New Delhi in 1998.

---

20. Vladimir Putin, "Russia and the Changing World," *Moskovskiye Novosti*, Feb. 27, 2012; "A New Integration Project for Eurasia: The Future in the Making," *Izvestia*, Oct. 3, 2011.

21. Douglas, "Russia's Eurasian Diplomacy Might Spark Real Economy," *EIR*, Nov. 24, 2000, with excerpts from Putin's pre-summit article.

In domestic policy, Putin and a succession of prime ministers attempted to shift the 1990s pattern, whereby "growth"—measured in monetary terms—had been overwhelmingly concentrated in the capital, Moscow, while outlying regions deteriorated. A Federal Targeted Program for the Economic and Social Development of the Trans-Baikal Region and the Far East, already on the books for several years, was placed under the oversight of a reorganized Ministry for Regional Development in 2004. In 2007, Russia formed a state commission on developing these areas. Its chairman, then-Prime Minister Mikhail Fradkov, toured the eastern provinces, promoting the strategic importance of building up their transportation infrastructure. He undertook diplomacy with Japan and other neighbors, seeking an expansion of investment and trade with the Far East.

2007 saw a joint commitment by China and Russia to cooperate in their Far East border regions. The decision to proceed with Cosmodrome Vostochny also came that year. Construction of the East Siberia-Pacific Ocean (ESPO) oil pipeline started in 2006; this 42.8 billion ruble (nearly $1.5 billion) project created more than 6,000 jobs at its peak, before completion of the main route to a port near Vladivostok in 2010, and a spur to northeastern China in 2012. ESPO initially delivered 15 million metric tons (mmt) of oil (300,000 barrels per day) to China annually; the overall capacity of the pipeline is planned to reach 1 million bpd by 2016 and 1.6 million bpd in 2025.

The new, cable-stayed bridge from the city of Vladivostok to Russky Island was dedicated in July 2012, in time for the APEC summit, hosted by Russia on the island.

The cosmodrome and ESPO are on the official list of major investment projects, achieved in the Russian Far East and Baikal region during the first dozen years of the 21st Century:

- Construction of the two ESPO pipelines.
- Infrastructure for the 2012 APEC summit in Vladivostok, including the first bridge to Russky Island off the tip of the Vladivostok peninsula, where the conference was held. The new conference facilities were subsequently transferred to the Far East Federal University as a campus.
- Paving of the "Amur" stretch of the Trans-Siberian Highway, between Chita and Khabarovsk.
- Production of the Sukhoi Superjet-100 medium-range aircraft at the Sukhoi plant in Komsomolsk-on-Amur (start of prototype production, 2004; first commercial flight, 2011).
- Breaking ground for Cosmodrome Vostochny.

Then-Prime Minister Vladimir Putin drives a Lada car on the Amur Highway in 2010. The segment of the Trans-Siberian Highway between Khabarovsk and Chita had just been paved for the first time. Construction of Cosmodrome Vostochny, just off this highway, began the next year. The road goes through Belogorsk, a Trans-Siberian Railway station that is the northern point on Asian Highway Network route AH31 and Greater Tumen Initiative Corridor 4, both running 1,437 km south to the Chinese port of Dalian on the Yellow Sea.

Another Far East energy project is the Sakhalin-Khabarovsk-Vladivostok natural gas pipeline, which opened in September 2011. Owned and operated by the giant state company Gazprom, it supplies gas for domestic consumption, with plans to add exports, including to Japan and South Korea. Since 2005, Gazprom has been working with the Japanese Ministry of Economy, Trade, and Industry (METI) to boost Russian liquefied natural gas (LNG) deliveries to Japan, the world's largest LNG importer, by building an LNG plant and export terminal at Vladivostok. The prospect of increasing Gazprom's LNG market share through greater sales to Japan became even more irresistible, with the tendency toward a disastrous "non-nuclear Japan" policy, after the 2011 tsunami damaged the Fukushima nuclear plant. In June 2012, the Russian Energy Ministry and METI signed a memorandum of understanding on the Vladivostok project, which is to produce 15 mmt of LNG (20.7 billion cubic meters of natural gas) annually.

Mining enterprises in Siberia and the Far East have been investment targets of London- and British Commonwealth-centered metals cartels, ever since the 1990s. Most of their spending for infrastructure to exploit Russian mineral resources, strictly for export, follows the "looting and shipping abroad at cheap prices" model that LaRouche warned about in 2001, as what a "Vernadsky strategy" should prevent. The Russian gold industry, for example, is 15% to 20% foreign-controlled, with cartel majors like Kinross Gold Corp. (Canada) among the active players. The CEO of Trans-Siberian Gold Ltd, a smaller operator that is 31% owned by the giant AngloGold Ashanti (an offshoot of mining interests close to the British Crown), has boasted that his company "came in under the radar" after the August 1998 crash, snapped up properties in Kamchatka and on the Yenisey Ridge near Krasnoyarsk, and has done quite nicely on its Siberian goldfields.

While some resources have thus come under direct foreign control, there are also nominally Russian companies making money in mining, which are part of the Pirates of the Caribbean phenomenon that arose during privatization—the overnight billionaires who acquired former Soviet industries and registered them in offshore tax havens. Polyus Gold, Russia's biggest gold producer and among the top ten in the world, is registered in the Bailiwick of Jersey, one of the British Crown's Channel Islands, and listed on the London Stock Exchange, although it is under pressure to relist within Russia.[22]

A private steel company, Mechel, invested $2.5 billion in 2008-2011 to begin mining the giant Elga Coal Deposit in Yakutia, with financing from the state-owned VEB bank and the Eurasian Development Bank, founded by Russia and Kazakhstan in 2006. Mechel, which originated from the privatization and merger of the Southern Kuzbass Coal Company and the Chelyabinsk Steel Plant, is owned by Igor Zyuzin through three Cyprus-based holding companies. Mining of the high-quality coking coal began in 2011, and sales in 2012, after Mechel sank upwards of $1.25 billion into building a 321-km private railway to link Elga to the BAM. (As of 2014, the near-bankrupt company is attempting to sell the railroad to Russian Railways.)

Apart from these export-oriented energy and mining projects,

---

22. Polyus Gold was hived off from Norilsk Nickel in 2006. The latter, in 1995, had been the most notorious case of "loans for shares" privatization, under which upstart private Moscow banks loaned money to desperate Russian factories, devastated by 2,600% inflation in 1992 after "shock" price deregulation, in exchange for shares of the companies. The bankers ultimately retained ownership of those industrial facilities, when the latter defaulted.

Russian officials have continued to search for a Siberia-Far East development model that will work.

The long-time governor of Khabarovsk Territory, Victor Ishayev, became presidential envoy to the Far East FD in 2009. With training as an engineer, experience as an aluminum plant director, and the political style of a regional boss, Ishayev also has a doctorate in economics. He led the drafting of a report to the inaugural November 2000 meeting of the State Council (of regional governors), calling for dirigist measures to restart Russian industry after a decade of destruction.[23] Glazyev and Granberg were co-authors of the Ishayev Report.

In late 2011, Ishayev complained that federal programs in the Far East were receiving only 20% of their assigned government funding. The next year, he warned that federal investment in the region would soon dry up, with the completion of APEC-related projects and ESPO. He voiced concern about the economic decoupling of the Far East from Russia, noting that in just 20 years, the region had gone from delivering 75% of its output to other parts of Russia, to only 21%. In Ishayev's view, "The Far East can be developed only through federal investment and big projects."

Ishayev teamed up with Russian Railways CEO Yakunin, in 2010, to present a 400 billion ruble (~$13 billion) plan to upgrade and double-track the BAM. He pushed for the Sakhalin-Russian mainland link (see Part 1, World Land-Bridge Network), for which SOPS had drafted plans and estimates in 2008. In early 2013, Ishayev stated that construction of a bridge to Sakhalin could and should start in 2016.

Then-Prime Minister Putin, on a late-Summer 2010 tour of Siberia and the Far East to dedicate the Russia-China section of ESPO, drive on the brand-new Khabarovsk-Chita highway, and visit the Cosmodrome Vostochny site, told the country that economic life revolves around great projects.

In the first year of Putin's new Presidential term, 2012, the build-up to the September APEC summit in Vladivostok framed a national debate on how the big-project policy—or some alternative—for Siberia and the Far East should proceed. Even before the March election, government circles discussed an idea from then-Emergencies Minister Sergei Shoygu, Putin's ally and a native of the Tyva Republic on the border with Mongolia, to set up a State Far East Development Corporation. Language in the proposal about "unlocking the treasure chest of Siberia" alarmed many, leading to public debate of whether the new entity would imitate the looting practices of the British East India Company, or the nation-building of Franklin Roosevelt's Tennessee Valley Authority. In May 2012, Putin announced formation of a new government ministry, instead. Ishayev took charge as minister of Far East development, while remaining presidential representative. By December, however, Putin was criticizing the ministry for inaction and a lack of new ideas.

On the eve of the APEC summit, the prestigious Valdai Discussion

---

23. "Strategy for the Development of the State to 2010" (translated excerpts), *EIR*, March 2, 2001.

Club released a report titled "Toward the Great Ocean," using the historical Russian term for the Pacific.[24] It acknowledged "the shift, unprecedented in scale and speed, of the global economic and political center to 'the new Asia,' or more precisely to East and Southeast Asia and India." The report said that "Russia has so far been unable to participate" in the Asian development process, "being held back by the backwardness of its infrastructure, an underdeveloped economy and the demographic situation," as well as "obsolete Euro-centric" policies. At the same time, the Valdai Club set forth an argument that would soon come to a head, asserting that no "comprehensive development strategy for Siberia and the Russian Far East" could work, because such programs are "state-centric" and would inevitably be mired in corruption.

Devastating Amur River floods in Summer 2013 were followed by Ishayev's abrupt surrender of both his Far East portfolios. Alexander Galushka, a young businessman experienced chiefly in property appraisal, succeeded him in the ministerial post. In his 2013 year-end Address to the Federal Assembly, Putin declared Far East development a strategic priority for the 21st Century.

Russian Emergencies Ministry

Flooding of the Amur River in Summer 2013 forced 36,000 people to flee 8,000 flooded houses in Russia's Amur Region, while 120,000 in all were evacuated. Three thousand houses and apartment buildings, home to 35,000, were flooded in Khabarovsk Territory. The high water washed out bridges and highways, and destroyed crops. In China's three Amur basin provinces, 60,000 houses were destroyed, 840,000 people evacuated, and 787,000 hectares of cropland flooded.

Galushka began promoting a new approach, consistent with the Valdai Club reports, and dubbed "the Singapore model." It is proposed that special zones called TOR, standing for "priority development zones," will attract Russian and foreign investors with incentives such as government-built infrastructure and tax exemptions, including ten years without paying the 18% VAT. Their success, the argument goes, will pull the rest of eastern Russia along. A year later, the Ministry of Far East Development has drafted enabling legislation—not yet passed—for the TOR, and vetted a large number of proposals, from among which were selected 14 TOR, to be developed by 2020, and 18 ready-to-start projects. The TOR would aim to sell processed products, rather than completely raw materials, in Asia-Pacific markets. According to calculations Galushka presented to Putin in August 2014, government spending of 345 billion rubles[25] (89 billion on

---

24. Sergei Karaganov, et al., "Toward the Great Ocean, or the New Globalization of Russia," July 5, 2012; Igor Makarov, et al., "Toward the Great Ocean-2, or Russia's Breakthrough to Asia," Feb. 27, 2014. Both documents are available on the Valdai Discussion Club publications page. The Valdai Discussion Club hosts an annual get-together between Putin and foreign analysts. Its 2014 report details the physical-economic potential of Siberia and the Far East, as well as reviewing Russian economic relations with individual Asian countries. The flaws of these reports are chiefly connected with monetarist assumptions.

25. Approximately $8 billion in late-2014 dollar terms (cited for indicating approximate project size only), with the ruble ending October 2014 down 30% against the dollar since one year earlier.

the zones and 256 billion on the projects) is supposed to attract 1 trillion rubles of Russian and foreign corporate investment in the TOR as such, and another 2.5 trillion in the projects.

The TOR scheme has been presented to Chinese, Japanese, and other investors, throughout 2014. It has also raised serious questions at home and abroad, starting with the dissimilarity between Singapore and Russia's Far East: Singapore, a major international financial center, sits on the most heavily trafficked maritime trade route in the world, and has a population density of 7,615 per km^2, whereas the Russian Far East and Siberia have huge distances, and the population densities described above (**Figure 11**). Accordingly, seven of the planned TOR are in coastal areas, two more are in the large cities of Khabarovsk Territory, and only five are farther inland. The 18 priority projects are nine mining and minerals-processing complexes, four agro-industrial projects, two ports, two oil and gas projects, and one for chemical fertilizers. Identified as "key investors" in the projects are the state-owned oil company Rosneft, and several firms from the "oligarchical capitalism" layer created in the 1990s—the above-mentioned Polyus Gold and Mechel; Sibur (natural gas); Summa Group, the telecom and logistics holding company of Ziyavudin Magometov; and Intermetals, the mining management arm of Victor Vekselberg's Renova Group.[26] The projects reflect the corporate priorities of these large minerals-exporting companies, rather than being territorial development plans. The TOR are small: except for a 59 km^2 spread near Yakutsk, slated to host the Diamond Valley gem-cutting center and tourist park, they range in size from 1,200 hectares (12 km^2) down to 186 hectares.

Small and medium businesses in the Far East worry about the impact of the TOR, if the spots within them are won by big firms. Russian majors operating in the zones may be hard put to compete with Asian producers, but, with their costs lowered by 20-30% thanks to TOR incentives, they could run circles around local companies, warned Primorsky Territory Chamber of Trade and Industry executive Dmitri Tsaryov in a PrimaMedia.ru interview. He fears that separate regulations for the TOR will make them "states within the state," exempt from reasonable local regulation and oversight. Daisuke Saito, an economist at the Japanese Association for Trade with Russia and the Newly Independent States, who toured three prospective Primorye TOR sites in October 2014, told PrimaMedia.ru that Japanese companies believe that successful projects in Russia depend more on good relations with regional governments, than tax incentives.

TOR boosters cite the success of China's Special Economic Zones, besides Singapore. Deputy Minister of Far East Development Maxim Shereykin offers Chinese investors the chance to get in on the ground floor and help design the TOR infrastructure. But *Expert* magazine, Russia's leading economic weekly, observed that it makes no sense to

---

26. Stanislav M. Menshikov, *The Anatomy of Russian Capitalism* (EIR News Service: Washington, DC, 2007), profiles these Russian conglomerates and relates how they came to be. Renova is owned by Bahamas-registered Renova Holding Ltd and managed through TZ Columbus Services Ltd in the British Virgin Islands.

use the Chinese zones as a model, not only because of the orders-of-magnitude difference in population density, but in view of China itself having moved beyond that approach: "The experience of the Asia-Pacific countries that adopted the export model at a certain point, has shown that it leads to dependence on the patron countries. China had a downturn, because of dependence on the U.S. and European economies. As a result, China has now shifted the emphasis, to increasing domestic demand." Similarly, experts at the RAS IFES have concluded that China's stupendous high-speed rail program, rather than its special zones, is the greatest transformative force in the Chinese economy.

Cracking the "Money" Barrier

A greater obstacle to Siberian development than the taiga or blizzards or permafrost, is monetarism.

President Putin declares his intention to liberate the Russian economy from dependence on raw-materials exports, bring Russian companies home from offshore, open up new manufacturing, and develop the Siberian frontier through great projects. Yet these goals have been held hostage to rules and practices decreed by the now bankrupt trans-Atlantic financial world, to which Russia's main economic policy institutions have remained committed. With the imposition of sanctions against Russia in 2014, the deadlock caused by this hybrid policy could be set to break.

Starting in 1992, Russian neoliberal governments suppressed the money supply and credit, in the name of preventing inflation. Alexei Kudrin, as finance minister from 2000 to 2011, continued this practice. Beginning in 2004, tax revenue from the production and export of oil and gas, up to a certain percentage of Russia's GDP, was withheld from the economy in a Stabilization Fund, which was invested in foreign government bonds. The fund had built to 3.8 trillion rubles (then ~$157 billion) by 2008, when it was split into a Reserve Fund, continuing the same function, and a National Welfare Fund (NWF), earmarked to support the pension system.

When the global financial crisis hit Russia again in 2008, especially with crashing oil prices, the government mobilized the Reserve Fund and Russia's $600 billion in foreign currency reserves, to bail out public and private corporations, which were heavily indebted to foreign lenders. Kudrin then called for even tighter fiscal policies at home, along with imposition of a "global Maastricht" to prevent countries from "living beyond their means." Whereas the Maastricht accords had obligated EU countries to have budget deficits of no greater than 3% of GDP, Putin himself was sold on the notion that a *zero* federal budget deficit was a crucial goal to achieve by 2015. Kudrin left office in 2011, but in 2012 the Russian government adopted a new, strict Budget Rule, drafted under his successor, Finance Minister Anton Siluanov. Now, federal deficit spending is limited to 1% of GDP; in addition, oil and gas revenues, in excess of projections based on a five-year rolling average oil price, are diverted to the Reserve Fund, while any shortfall below such projections must be compensated through budget cuts.

Under these policies, big projects may be placed on the agenda, but are often taken off again. Currently on ice are regional development schemes like UP-UP, and the Tatar Strait bridge or tunnel to Sakhalin Island—the one Ishayev wanted to start in 2016.

A breakthrough occurred in June 2013, when Putin announced that NWF money would be released for priority infrastructure goals. Speaking at the St. Petersburg International Economic Forum (SPIEF), he said that "about half" the NWF—450 billion rubles ($15 billion) at that time—would "be invested in projects in Russia." Putin named three:

> What are these projects? The first is a high-speed train link between Moscow and Kazan. This will be a pilot project for the route that will eventually connect the Central, Volga, and Urals economic regions.
>
> Second, a central ring road will be built from scratch and run through the Moscow Region and the New Moscow area.... The Central Ring Road ... will transform transport logistics in the European part of Russia, connect the country's central regions, and open up new development opportunities....
>
> As a third project, we are going to significantly upgrade the Trans-Siberian Railway and expand its capacity.... A direct rail route across Eurasia will act as a key artery between Europe and the Asia-Pacific region. It will give a powerful impetus to the development of the Far East and Siberia.

By the end of 2013, the President's go-ahead for the Moscow-Kazan high-speed railway had been overridden by the Budget Rule. At a September budget meeting, Siluanov informed Putin that the Finance Ministry proposed to save 57.5 billion rubles by delaying federal funding for it until 2016. In November, the government postponed a tender for the design phase of the project, scheduled for the following month, despite preliminary applications from two European consortia.

Russian Railways, essential as it is to Russia's participation in the World Land-Bridge, is squeezed under this hybrid economic policy, which mixes great development intentions with monetarist rules. The company was formed in 2003, with Yakunin as CEO, out of the Russian section of the former Soviet railway system, a government ministry which had employed 2.5 million people in 1991. Like Gazprom, Russian Railways was protected from full and immediate privatization, as a state-owned so-called "natural monopoly," but a policy of eventually privatizing it remains in effect. In step with the British Commonwealth practice of hiving off units of major state infrastructure enterprises during privatization, Russian Railways had to sell its freight-carriage subsidiary, Freight One, in 2011-2012. A 75% stake went to Vladimir Lisin, a 1990s-vintage steel magnate who runs his biggest asset, the NLMK plant, through a Cyprus-based holding company, and his other companies through funds registered in the British Overseas Territory of Gibraltar. With his Freight One rolling stock, Lisin now controls one-fourth of the Russian freight market, a formerly profitable part of Russian Railways' business.

In 2012, Kremlin aide and soon-to-be Deputy Prime Minister Arkadi Dvorkovich included 25% + 1 share of Russian Railways itself on the list of state assets for sale in 2012-2013, in an intended Second Great Privatization. Yakunin resisted the timetable, and most of the sale roster was soon put on hold because of global economic and political conditions. Russian Railways has suffered a decline in the volume of freight carried because of the depression, as well as the defection of some shippers to long-haul trucking (on Russian highways not built for it) for reason of high rail freight rates. To keep fares from rising faster than allowed by law, Russian Railways passenger services depend on federal budget subsidies, which Yakunin this year said were at only half the needed level. In October 2014, Russian news agencies reported that Russian Railways may be forced to spend TSR/BAM modernization funds on fare subsidies.

These pressures constantly block Russian Railways from undertaking new projects. Precious little of the rail expansion in the company's 2007 Strategy has been done, as is detailed under **Figure 6**:

- High-speed service now connects Nizhny Novgorod, Moscow, St. Petersburg, and Helsinki, Finland, albeit on shared tracks, not dedicated lines.
- Two freight railways were funded by corporations, one by Gazprom and one, to a coal mine, by Mechel.
- The Amur-Yakutsk Mainline (AYaM) is marginally operational as of 2014, but reaches only to a terminal across the Lena River, not to the city of Yakutsk itself. Construction of a bridge over the Lena was cut from the federal budget once again in Autumn 2014.
- Upgrades of the TSR between Mezhdurechensk (central Siberian coalfields) and the branching point with the BAM at Tayshet are under way, but average Russian freight train speeds have fallen to less than 10 kph, or about 6 *(six)* mph (2012).

Sakha.gov.ru

Workers and officials in 2011 celebrate construction progress on the Berkakit – Tommot – Yakutsk (Nizhny Bestyakh) railway, part of the Amur-Yakutsk Mainline (AYaM), from which the Bering Strait Connector (BSC) will start. The "golden spike" ceremony was in November 2011; the first freight was carried in 2014. Like the Baikal-Amur Mainline, the AYaM is single-track and for diesel-powered trains only. No bridge has been built across the Lena to Yakutsk, population 270,000, the Sakha Republic-Yakutia capital. The most optimistic Russian plans call for the AYaM and BSC to be double-tracked, electrified, high-speed railways.

Vladimir Petrovsky of the RAS IFES warned in 2011 that at the average rate of construction of the AYaM, 12 to 15 km/year, it would take 200 years to reach the Bering Strait, although a hundred years ago Russia built the TSR at 560 km/year. In an open brawl at a September 2014 investment meeting chaired by Putin in Yakutsk, Yakunin fended off a demand by Galushka, to divert NWF money from the TSR/BAM upgrade, into setting up the Far East TOR zones.

Like many other Russian companies, Russian Railways gets around the dearth of funding and credit at home, by borrowing abroad. (As of 2014, total Russian foreign corporate debt has climbed to $650 billion, above its level in the 2008 crisis.) In mid-2014, the Russian Railways debt portfolio was 605.8 billion rubles (~$18 billion), 37% of which is foreign currency-denominated. Included is 175 billion rubles in special infrastructure bonds, which Russian Railways began to issue in 2013 after government approval.

Russia, and Russian Railways in particular, has been courted by foreign financial interests that promote the public-private partnership (PPP) model. Economists from the Long-Term Investors Club (LTIC), an organization formed by continental European megabanks and advocating a green agenda of low–energy-flux-density projects such as wind farms, have lectured in Russia on the need to assemble and invest pools of money from pension funds, sovereign wealth funds, and insurance assets, claiming that in the crisis-ridden, globalized economy, governments and commercial banks cannot extend long-term credit any more. A dead giveaway that the PPP model is worlds removed from the type of private-sector subcontracting practiced under, for example, the 1930s Reconstruction Finance Corporation in the United States, is that its promoters promise usurious rates of return on project bonds.[27] Such financial instruments may be associated with infrastructure projects, but they are designed chiefly to extract a monetary income stream from the infrastructure's users, with the added plus, for investors, of government guarantees: a transparent scam in which the "Public" assumes the risk and the "Private" takes the profit.

The main plenary of a Summer 2011 conference on the Bering Strait project, hosted in Yakutsk by the government of the Sakha Republic-Yakutia, was on PPPs as the way to obtain financing. Chaired by an official of McKinsey & Company, it focused on enabling legislation, like a planned regional Law on Public-Private Partnership, to create incentives for private investment. At the 2013 SPIEF, a Russian Railways-sponsored panel on port, rail, and road development heard from an LTIC economist and was keynoted by one of the world's leading private infrastructure managers, Damien Ronald Secen of the Macquarie Renaissance Infrastructure Fund.[28]

Russia does have alternative, sound proposals that could be adopted for financing its economic development. In Spring 2014, a conflict broke out within the government, between Siluanov's tough fiscal austerity line, and attempts by Minister of Economic Development

---

27. Project bond supporters advertise their "predictably high returns" for investors (PPP specialist Mark Hellowell in the London School of Economics European Politics and Policy blog, June 27, 2012); George Inderst, in "Infrastructure as an Asset Class," Pensions Institute EIB-sponsored Discussion Paper PI-1103, Vol. 15, No. 1, 2010, cited the ability of unlisted infrastructure funds to deliver returns of 14% or higher.

28. John Hoefle, "Macquarie Bank Takes the Low Road," *EIR*, July 28, 2006, profiles the Anglo-Australian giant Macquarie, the world's leading private infrastructure specialist.

Sergei Glazyev, economist and adviser to President Putin

Victor P. Ivanov, director of the Russian Federal Drug Control Service.

Alexei Ulyukayev, following up on Putin's 2013 decision on the use of NWF resources, to increase federal spending for economic projects. Against the backdrop of this clash between two varieties of monetarism, a proposal by Academician Glazyev surfaced in the financial daily *Vedomosti*, for measures to protect the Russian economy in the face of foreign sanctions. Glazyev's 15-point plan implies an escalation of Putin's announced policy of moving Russian finances out of offshore jurisdictions, and raises the possibility of new, state-guided credit generation for real economic investment.[29]

Another promising initiative is that of Victor Ivanov, director of the Federal Drug Control Service (FDCS). Charged with defending the Russian population against a flood of heroin, Ivanov has called in speeches delivered all over the world, for mounting an international drive for Glass-Steagall-type banking separation, as an essential component of a new global financial architecture that would stop illicit drug-money-laundering. "I am sure that the effectiveness of anti-drug policies would increase considerably, were the organization of economies for development adopted as the primary method," Ivanov said in Argentina in 2012. To ensure productive credit-generation, he asserted, nations must have "sovereign development" and be "endowed with financial and credit independence," whereas "the existing world monetary and financial system, built on the ruins of national economies and by sucking their resources, is the main reason for the spread of drug trafficking on a world scale."[30] In January 2012, Ivanov proposed to create a Russian Corporation for Central Asia Development Cooperation (See Appendix, Part 5), based at the state-owned Vnesheconombank (VEB), Russia's second-largest bank. In 2008, VEB bank handled the disbursement of government bail-out funds to Russian banks and corporations. In 2013, it became the designated agent for delivering NWF funding to selected infrastructure projects. By assuming the function proposed by Ivanov, VEB could follow in the footsteps of the Reconstruction Finance Corporation, which changed from a bail-out mechanism in the late Herbert Hoover administration, into a full-fledged economic development institution under Franklin Roosevelt.

The China Factor

The Moscow-Kazan high-speed railway (HSR), which in 2013 was put on the agenda by President Putin and taken off again by the Russian Finance Ministry, has made a comeback, now as part of a Beijing-Moscow HSR corridor. Alexander Misharin,[31] first vice-president of

---

29. Douglas, "Asymmetric Response to Sanctions: Russia Debates Dirigist Credit-Creation Plan," *EIR*, May 2, 2014, details the 15 points.

30. Victor Ivanov, "Drug Trafficking and the Financial Crisis," *EIR*, Dec. 2, 2011; Cynthia R. Rush, "Ivanov: Attack the Drug Trade with Glass Steagall," *EIR*, 2012.

31. In April 2007, as a deputy minister of transport, Misharin was the government's main speaker at the SOPS-sponsored conference on the future Bering Strait crossing. He discussed the strategic significance of a multimodal infrastructure connection between Eurasia and North America.

Russian Railways and managing director of its High-Speed Mainlines subsidiary, negotiated with the China Investment Corp. and the Chinese construction and engineering company CREC in Summer 2014. The relevant memorandum of understanding between Russian Railways, the Russian Ministry of Transport, and their Chinese partners was signed during Prime Minister Li Keqiang's visit to Moscow. These agreements exemplify how the obstacles to great projects may begin to be overcome, as Russia-China relations and BRICS financing develop.

Sinologist Tavrovsky, in a 2013 article on the need for Russia to emulate China's HSR program, pointed out that doing so would reciprocate the impetus Russia gave the Chinese economy a hundred years ago, when cities like Harbin and Dalian (then known as Port Arthur) boomed as a result of Russia's building the Chinese Eastern Railway to Vladivostok. There was another period of intense industrial cooperation in the 1950s, between the Soviet Union and the People's Republic of China, but during the subsequent Sino-Soviet split, relations deteriorated to the point of armed clashes on the Ussuri River in 1969.

Nauka Siberia (Siberian Science), weekly of the Siberian Division of the RAS

Chinese President Jiang Zemin meets with leaders of the Siberian Division of the Russian Academy of Sciences, November 24, 1998.

The split lasted almost until the demise of the Soviet Union in 1991. Diplomatic contacts were sporadic in the 1990s, but a visit to Russia by then-President Jiang Zemin in 1998 served as a keynote for coming changes.[32] Arriving three months after the August 1998 financial crash in Russia, Jiang not only visited Moscow, but stopped in Novosibirsk on the return trip. Meeting with RAS leaders at the science town of Akademgorodok, he delivered what LaRouche termed "a brilliant, carefully prepared intervention, laying out the principles of a policy." While trans-Atlantic sector financiers were quaking over the global effects of the Russian government bond default, brought on by policies they themselves had insinuated into Moscow, Jiang spoke to the real Russia. The speech, which deserves to be read in full, included these highlights:

> I have long heard about the Science City of Novosibirsk. But seeing is believing. During the visit, I have been deeply impressed by your scientific research capabilities and the explorative atmosphere....
>
> Russia is a scientific and technological power in the world. Russian scientists have made outstanding contributions to the progress of human civilization.... Even today, Russia leads the world in many key scientific and technological areas....
>
> The progress of human civilization has more and more convincingly proved that science and technology constitute a primary productive force and an important driving

---

32. Mary Burdman, "Jiang in Russia: a Speech That Can Change History," *EIR*, Dec. 4, 1998, with the full text of the speech.

force for economic development and social progress. None of the achievements mankind has scored in understanding and taking advantage of nature would have been possible without scientific and technological advancement. Human wisdom is inexhaustible. Science and technology are a shining beacon of this wisdom. A great many scientists, one after another, have kept scaling new heights in science and technology after overcoming numerous obstacles through arduous efforts....

In order to meet the challenge of rapid scientific and technological progress and the fast-rising knowledge economy, we must keep on creating and innovating. Creativity is the soul of a nation and an inexhaustible source of a country's prosperity. The key to creation and innovation lies in human resources, whose development depends on education. Only a well-developed education can sustain scientific and technological progress and economic development. Scientific and technological strength and the educational level of a nation have always been an important yardstick for measuring the overall national strength and the civilization of a society. Like indispensable wheels, they propel a country to prosperity.

Primakov's initiative for the Russia-India-China strategic triangle came one month later, and, within two years, the Russian-Chinese diplomacy of the Putin era began. In addition to bilateral and trilateral contacts, an important venue for Eurasian deliberations was the Shanghai Cooperation Organization (SCO), formalized in 2001. Starting as the Shanghai Five in 1996, when China and the four ex-Soviet countries with which it has borders—Kazakhstan, Kyrgyzstan, Russia, and Tajikistan—signed a Treaty on Deepening Military Trust in Border Regions, the SCO has expanded from security to economic issues, and added members.

As their trade began to grow, Putin and the next Chinese President, Hu Jintao, agreed in March 2007 that "the two countries should strengthen coordination in implementing their respective strategies to revitalize the old industrial bases in the northeastern region of China and to promote development in the far eastern and eastern Siberian regions of Russia, and work out a plan of cooperation in this area." That pledge resulted in a list of 205 joint projects in the border provinces on both sides, signed by Hu and Russia's new President Dmitri Medvedev in September 2009. Many of these involved raw materials and lumber, feeding China's construction boom, but there were also eight joint technology parks on the list, and a hefty section on transportation infrastructure, including rail links from China to the TSR. Contracts and other agreements signed when Prime Minister Putin visited Beijing the next month, touched on several high-tech areas. Putin admonished the media not to focus exclusively on natural gas price issues, but to pay attention to the agreements on Russian aid in expanding the Tianwan nuclear power plant in Lianyungang, and on the first-ever Russian export of two sodium-cooled breeder reactors to China. During that visit, Yakunin and Chinese Minister of Railways Liu signed the first memorandum of understanding on "organizing and developing high-speed rail service on the territory of the Russian Federation."

LaRouche hailed the October 2009 Chinese-Russian agreements, which he said could make China's U.S. dollar reserves worth something real, if they led to such funds being invested in infrastructure and other tangible production. If the Russia-China economic cooperation proceeded, involving India as well, LaRouche said, it would create an opportunity for the United States to change its own policies and join in economic cooperation with these other great powers.

China has surpassed Germany as Russia's largest trading partner, with bilateral turnover worth $95.6 billion in 2012 and $89 billion in 2013. With new currency and credit swaps arranged in Summer 2014 to finance trade, that level will surpass $100 billion (but, denominated in rubles and yuan). Speaking at the May 2014 SPIEF, Chinese Vice President Li Yuanchao reported that Chinese direct investment in Russia reached $4 billion in 2013, but added that this figure was minuscule, compared with China's total outward foreign direct investment of more than $100 billion per year, including cumulative tens of billions in some countries in South America. Japan and South Korea both have invested more in Russia than China has. To foster growth in this area, Russia and China set up the Russia-China Investment Fund (RCIF) in 2012, with initial commitments of $1 billion each from the Russian Direct Investment Fund and the Chinese Investment Corporation, and planned to raise as much again from private institutional investors. As of May 2014, the RCIF had made investments of nearly $1 billion. Additionally, a Russian-Chinese Investment Committee was established in 2014.

There was no progress on HSR cooperation after the 2009 memorandum, until 2014. Vladimir Remyga, head of the Russian-Chinese Center at the Russian Government Financial University, told an interviewer in July 2013, "The Chinese treated what we signed in 2009 seriously. They set up a large number of special research centers and concentrated their financial resources, with the Chinese State Development Bank taking charge of implementation; recently, it opened an office in Moscow. But from the Russian side, nothing happened. The [205-project] program successfully migrated from the [disbanded] Ministry of Regional Development to the Ministry of Far East Development, but there is no financing."

Now, Chinese involvement in the Russian economy is set to surge. The contract on building the 4,000-km Power of Siberia natural gas pipeline from eastern Siberia to Heilongjiang Province was finalized during Putin's visit to Shanghai in May 2014. It calls for China to invest about $20 billion in building the Chinese end of the pipeline, as well as to fund Russia's start-up investment by prepaying $25 billion for the natural gas it will receive, beginning in 2019, at the rate of 38 billion cubic meters a year. In addition, it was reported in August 2014 that the Russian Ministry of Economics is soliciting

Russian President Vladimir Putin and Chinese Vice Premier Zhang Gaoli autograph the Power of Siberia pipeline, at the launch of its construction in Sakha Republic-Yakutia, Sept. 1, 2014. Under the $400 billion deal, Power of Siberia will deliver 4 trillion cubic meters of gas to China over 30 years. Putin underscored that the pipeline will be the biggest construction project anywhere on the planet, over the next four years.

applications from Russian businesses, interested in receiving Chinese investments, up to the equivalent of $7.2 billion.

Bilateral cooperation deals include the following, besides the Power of Siberia pipeline, high-speed rail, and nuclear power projects mentioned above:

- In September 2014, state-owned Russian Technologies (Rostekh) and China's state-owned Shenhua Group, the largest coal producer in the world, signed a memorandum of understanding on a $10 billion project to develop coalfields in Siberia and the Far East, particularly the Ogodzhinskoye deposit in the Amur Region. The two firms will build a 20 mmt coal terminal at Port Vera, on the Ussuri Bay east of Vladivostok. The project includes a power plant and transmission lines, and pledges of social and transportation infrastructure improvements.
- In connection with Greater Tumen Initiative corridors 1 and 2, also called Primorye Corridors 1 and 2, China and Magometov's Summa Group are promoting possible construction of what would be Russia's largest port, at Zarubino on the Sea of Japan, southwest of Vladivostok. Summa and affiliated companies have already signed deals with China for a 5 mmt grain terminal at Zarubino, but Summa executives and the Chinese *People's Daily* both suggested in Summer 2014 that Zarubino could expand to handle 10 mmt of grain and container cargoes annually by 2018 and 60 mmt thereafter. There would also be a large "dry port" logistics terminal at Hunchun in Jilin Province. The border is only 18 km inland from Zarubino, in this region where China has no seacoast of its own. China is eager to acquire a modernized short route from heavily industrialized Jilin Province to seaports. Primorye officials, concerned about any tendency toward "export-only" investments, are simultaneously exploring denser, job-creating cooperation with Japan and Korea, as well, particularly in and around Vladivostok and Khasan, the city on Russia's Tumen (Tumannaya) River border with North Korea (see following article).

The Port of Zarubino, southeast of Vladivostok on the Sea of Japan coast of Russia's Primorsky Territory, is 40 km north of the North Korean border and 18 km from China's Jilin Province. With current capacity of 1.2 mmt, but actual turnover of only one-tenth that in recent years, Zarubino is already connected by rail into Jilin, as well as north to Vladivostok. As the eastern terminus of Greater Tumen Initiative Corridor 1, it is slated for expansion. A Russian private holding company, Summa Group, has reached agreement with Chinese agencies on building a 5 mmt grain terminal at Zarubino, with rumored expansion to 10 and then 60 mmt.

- Several more mining projects are under discussion with Chinese firms, including the development of Udokan in Transbaikal Territory, the third-largest known copper deposit in the world; lead and zinc deposits in Buryatia, just east of Lake Baikal; and polymetallic ores in the Tyva Republic.

In his April 2014 book *Russia and China: Strategic Partnership and the Challenges of Today*, Academician Titarenko wrote that the Russia-China strategic partnership is "a tectonic shift." Not everyone in Russia welcomes that; belief in the inevitability of geopolitical conflict is as tenacious as monetarism. The Institute for Contemporary Development (INSOR), a government-linked liberal think-tank, continued right up until the Ukraine crisis to promote Russian cooperation with NATO against alleged threats from China. Hence the importance of Chinese sensitivity to Russian sensitivities. In a 2013 article about the Silk Road Economic Belt's implications for Central Asia, as an example, Tavrovsky quoted Jiang Yi, a Chinese Academy of Social Sciences specialist on Russia: "The greatest challenge for China is to arrange our relations with Russia. We cannot permit our close ties with Central Asia to harm Chinese-Russian cooperation." Titarenko's assessment is that "Beijing has declared the goal of building a world in which all countries ... have equal rights to realize their legitimate interests. This fully coincides with the famous commandment of Confucius, 'Do not do unto others, what you would not wish upon yourself.'"

The Northern Sea Route and the Arctic

According to no lesser authority than Dmitri Mendeleyev, it was Russian Tsar Peter the Great (1672-1725) who commissioned Vitus Bering's expedition that discovered the strait between the Pacific Ocean and the Arctic, and sought the passage known today as the Northern Sea Route (NSR). Mendeleyev himself, in a November 1901 memorandum to Count Witte "On the Exploration of the Arctic Ocean" (see *EIR*, Jan. 6, 2012), proposed to lead an expedition, in hopes of breaking through the ice and establishing a passage. He considered the shortening of the sea lanes to be a means of advancing civilization and industry, including the development of Russia's Arctic Ocean coast.

In the space age, Mendeleyev's mission must lawfully be expanded to an even higher one, appropriate to mankind's role as a citizen of the solar system. In addition to linking the two hemispheres of the globe, the Arctic region is, and has been called, mankind's "window to space," a laboratory for exploration of the causes of climate and weather, and for devising means to meet the challenges of cosmic radiation in its dynamic interaction with our planet. Such a laboratory, like the ITER program on nuclear fusion, requires international collaboration among the scientists of all the world's major nations.

This approach contrasts sharply with the current dominant discussion of Arctic development, which tends to focus on the mining of natural resources and establishing geopolitical advantages against potential adversaries. The proper approach would require investments by all the

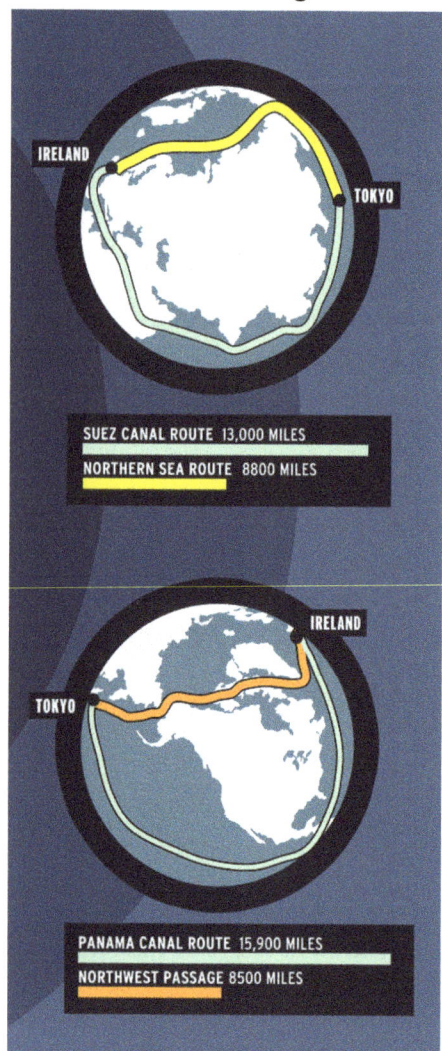

FIGURE 16
The Northern Sea Route and the Northwest Passage

LPAC/Chris Jadatz

The northern maritime routes in each of Earth's hemispheres are compared with the traditional passage through the Suez Canal and Panama Canal, respectively. The Northern Sea Route length shown here is longer than stated in the text, which concerns only the part along the Russian Arctic coast proper.

world's major powers around the common aim of making the Arctic a domain of the Vernadsky strategy for transforming the planet through the most advanced scientific research and development.[33]

The science of physical economy makes this entirely possible. Any effective recovery program must be a revolutionary one, driven by scientific work in the most challenging physical environments. Arctic development will leapfrog over the swamp of seeming intractable economic problems, whereas merely incremental increases in investment never would. With sovereign credit-creation, and cooperation among Eurasian and transpolar neighbors, it can be done.

The Northeast Passage

The NSR, also called the Northeast Passage, has just as long and dramatic a history, as does the elusive Northwest Passage in the opposite hemisphere—or more so, considering the argument by India's Bal Gangadhar Tilak, in *The Arctic Home in the Vedas* (1898), that Indo-European civilization began with people who lived and studied the stars along this coastline.

Under the World War II Lend-Lease program, 120 ships brought 450,000 mt of materiel from American West Coast ports to the Soviet Arctic via the NSR, for forwarding to the Eastern Front. Fifty-four of them docked in Tiksi at the mouth of the Lena River, 13 rounded the Taymyr Peninsula to reach Yenisei River ports, and one continued westward to Arkhangelsk on the White Sea. Of all these NSR ports, only Murmansk in the far northwest is classified as "ice-free" year-round. Reinforced ship hulls and icebreaker escorts are the rule.

The NSR went dormant after the break-up of the Soviet Union, as state-subsidized winter provisioning of its port towns ceased. At least one large icebreaker was stolen and fenced on the international market. As Sergei Glazyev told *EIR* in the interview quoted above, by 2001 Russia had plans to resuscitate the NSR. Greater Arctic ice-melt in some years encouraged these intentions, but their realization was not made dependent on climate; all policy declarations on the Arctic refer to developing new ice-breaker and underwater technologies.

Following the preliminary announcement of a new Arctic policy in 2001, a major Russian government report in 2009 termed the Arctic a "strategic resource base of the country," requiring upgrades to social and economic infrastructure, a heightened military presence, and cooperation among northern nations on utilizing the region's resources.

The Russian Geographical Society (RGS) is a leading institution for the Arctic. Since 2009, its president has been Sergei Shoygu, now also Russia's defense minister. The RGS initiated periodic Arctic—Territory of Dialogue conferences in 2010, attended by President Putin and leaders of other Arctic Ocean littoral countries. RGS First Vice President Artur

---

33. This section includes highlights of three in-depth *EIR* articles on Arctic development: Ulf Sandmark, "Arctic Development: Sweden and Finland Test the Waters, May Join Efforts of Norway, Russia," Jan. 6, 2012; Michelle Fuchs, Sky Shields, "Self-Developing Systems and Arctic Development," Jan. 6, 2012; William C. Jones, "Russia Prepares to Develop the Arctic as Earth's Next Great Project," Sept. 7, 2012.

Chilingarov, a polar explorer and former Russian Federation Council member, grabbed headlines in 2007 with his bathyscaphe descent near the North Pole, where he placed a Russian flag on the sea floor. Meanwhile, the Russian government filed claims with the UN Commission on the Limits of the Continental Shelf (UNCLOS), with evidence that the Lomonosov and Mendeleyev underwater ridges in the Arctic Ocean are natural extensions of the Eurasian continent and, therefore, part of Russia's shelf under the Law of the Sea. As of Autumn 2014, Russia is still assembling additional documentation at UNCLOS's request.

The coastal NSR, however, is undisputedly in Russian waters. Encompassing several shipping lanes, stretching distances ranging from 2,200 to 2,900 nautical miles (4,075-5,370 km), it is 6,400 km shorter for travel between Seoul and Rotterdam, than sailing south of Eurasia and through the Suez Canal (**Figure 16**). In 2011, 34 ships traversed the entire NSR, carrying 820,000 mt of cargo, compared with only four ships and 110,000 mt the previous year. NSR freight carriage surpassed 1 mmt in 2012; its Soviet-era peak had been 6.6 mmt in 1987. Also in 2011, the *Vladimir Tikhonov* became the first supertanker to sail the NSR end-to-end, with a cargo of gas condensate. 2010 saw the first shipment of iron ore from Norway to China via the NSR.

In China and northern Europe, there is great interest in expanding NSR operations and boosting rail access to its White Sea and Barents Sea ports: Murmansk, Arkhangelsk, and a future port at Indiga. In

Wikimedia Commons/Yaropolk
Railways approaching harbor facilities at the Murmansk Commercial Seaport.

Figure 17, Murmansk appears as the hub of the Northern East-West Freight Corridor trans-Eurasian shipment route, which will become even more direct if Russia completes the Belkomur or Barentskomur railway corridors northwest from the Urals.

To function as a full-fledged transport corridor, the NSR requires new oversight and guidance systems, an administrative fleet, and hydrographic and hydrometeorological support. Modernization and harbor expansion of several existing Arctic ports is planned, as well as new ports. New facilities at Murmansk, which obtained special economic zone status upgrades in 2007 and 2010, include oil, coal, and container terminals. In 2004-2009, €4.4 billion in investments boosted the port's capacity to 20 mmt, with 52 mmt projected by 2020.[34] The Russian

---

34. The Murmansk Commercial Seaport is wholly owned by Andrei Melnichenko's SUEK company, Russia's largest coal producer, which he controls through a holding company in Cyprus, and his Switzerland-headquartered EuroChem Group AG (Yevrokhim), producing fertilizers mainly in Russia. Melnichenko made his fortune through MDM Bank, which he co-founded in 1993, at age 21, out of a chain of currency-exchange street kiosks. Economist Stanislav Menshikov described its operations as of 2006: "MDM Bank caters to relatively young industrial magnates from Siberia and the Urals, among them [Oleg] Deripaska (aluminum), [Iskander] Makhmudov (copper), and [Alexander] Abramov (founder of Yevrazholding in the steel industry)." In 2009, Melnichenko sold out his bank shares to his business partner, Sergei Popov, while purchasing the latter's stakes in the MDM Group's chemicals and coal holdings. Menshikov (Note 26) documents how such new "oligarchs" acquired their properties during the 1990s fire sale of ex-Soviet industrial assets.

FIGURE 17
Northern East-West Freight Corridor with Hub at Murmansk

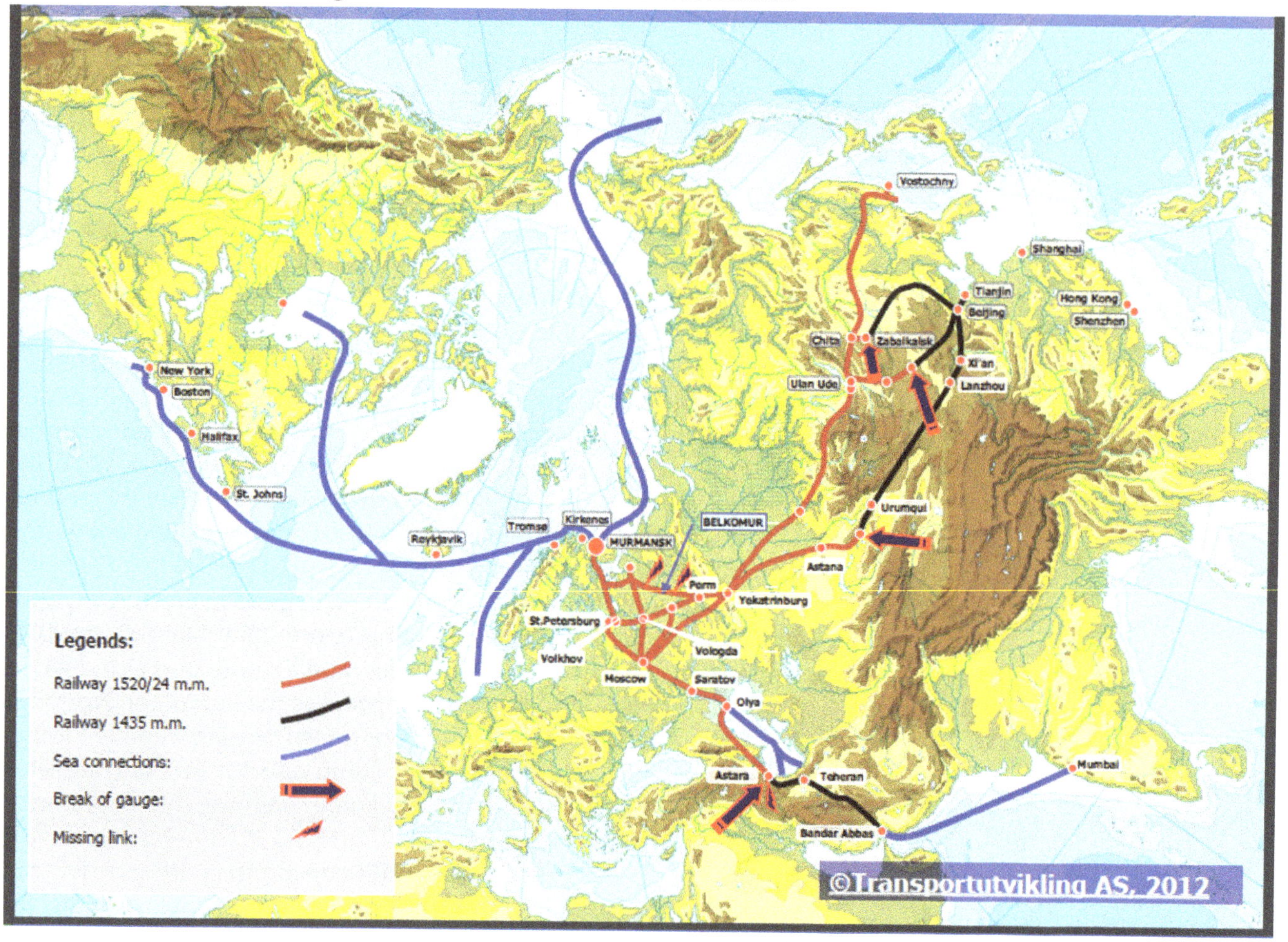

Courtesy of Stig Nerdal, Transportutvikling AS (www.transportutvikling.no)

This polar projection map by the Norwegian consulting firm Transportutvikling AS, transport and logistics specialists, highlights the role of Russia's Barents Sea port of Murmansk as a multimodal transport hub. The Northern Sea Route, the Trans-Siberian Railway (Pacific Ocean – Chita – Ulan Ude – Yekaterinburg – Perm – Moscow), Greater Tumen Initiative Corridors 4 and 2 (Beijing – Harbin – Zabaikalsk – Chita – TSR) and Corridor 8 (Tianjin – Beijing – Ulaanbaatar – Ulan Ude – TSR), the Silk Road Eurasian Continental Bridge (Beijing – Xian – Lanzhou – Urumqi – Astana – Yekaterinburg – TSR), and the International North-South Transport Corridor (Mumbai – Bandar Abbas – Tehran – Astara – Olya – Saratov – Moscow) all connect to Murmansk. At present, access is via the St. Petersburg – Murmansk railway, on which full electrification was completed in 2005. Proposed by Norway, the Northern East-West Freight Corridor (NEW), with outlets toward northern Europe and North America, has been endorsed by the International Union of Railways (UIC). With China interested in sending 50,000 containers a year on this route, initial plans for NEW to terminate at the Norwegian harbor town of Narvik were changed to emphasize Murmansk, because of iron-ore freight traffic congestion on the Scandinavian Peninsula and the need for an additional gauge-change when passing from Finland (whose 1524 mm broad gauge is compatible with the Russian 1520 mm) to Sweden. In the future, the Belkomur corridor (labeled) may allow transit directly from Perm to Arkhangelsk on the White Sea (shown but not labeled), whence an east-west rail connection to the Murmansk line and Finland has been completed.

Federal Space Agency and the Coast Guard are applying the space-based GLONASS system for Arctic navigation, environmental monitoring, and rescue services. In August 2012, Russian National Security Council Chairman Nikolai Patrushev announced the coming construction of ten sea and air stations along the NSR, at river mouths, ports, and railheads. China's Ministry of Transport said in 2014 that it will issue a complete navigation guide to the NSR.

The Russian Arctic fleet, including ships for sea and riverine use, dry and liquid cargo vessels, container ships, icebreaker-class tankers, fishing boats, and scientific research vessels, is being modernized. Russia is one of the few countries with an icebreaker fleet, for which the next generation of nuclear-powered icebreakers, including specialized icebreakers, reinforced icebreakers, and double-plated tankers, is under development. Six new icebreakers are planned or already under construction, three of them nuclear-powered. In November 2013, the keel was laid on the first of a new model of icebreaker, the largest ever, to be powered by two reactors. Plans are to extend the NSR operating season from seven months to year-round.

An Arctic Development Corridor

So far, freight transit and the oil and gas industry are the drivers of Russian Arctic development. A 2011 plan for continental shelf exploration until 2030, issued by the Russian Ministry of Natural Resources, projected that 8-16% of Russia's oil production and 32-35% of its gas production will come from the sea by that time, including Pacific Ocean fields and, primarily, the Barents Sea and the Arctic Ocean.

Hydrocarbon- and mineral-centered development implies the construction of mines and smelters, food-processing industries, and housing on the Arctic coast. As in every area of Siberian and Far East development, there also exist more ambitious ideas for the Arctic. One is to couple the NSR with an on-land rail transport route, parallel to the coast. This route was once planned for a railway, the Near-Polar Mainline, as Razbegin of SOPS reported: "In the Soviet period, in the 1930s and the 1950s, an Arctic Railroad was planned from Vorkuta in the northwest to Anadyr in the northeast, and 1,700 km of this railroad was built from the western end." That project used prison-camp labor and was abandoned after 1953, becoming known as the Dead Road. Today, it would be resurrected under more decent conditions, as a railway or, possibly, a system of industrial transport technologies suited to the polar climate. The 2012 IDMRD report on Siberian industrialization proposed that, with designs such as those depicted in **Figure 18** "An important component of the new [development] platform could be a near-polar mainline, using string transport and trolley principles, and paralleling the NSR as a kind of Northern Land Route."

FIGURE 18

A Russian artist's conception of innovative string-transport, maglev, and trestle transport systems for application in the Far North.

A train crosses the Shchuchya River bridge on the Obskaya – Bovanenkovo Railway, built by Gazprom on the Yamal Peninsula. Opened in 2010, it is the northernmost railroad in the world.

Chinese high-speed rail technologies, developed for the high-altitude Tibet railway and the relatively high-latitude lines in northeastern China, will be relevant. Russia itself opened the world's northernmost conventional railway, on the Yamal Peninsula, in 2010.

Russia also plans to build several floating nuclear power plants, each with two 35 MW nuclear reactors of the type used on icebreakers, for deployment on the Arctic coast. The keel of the prototype, the *Academician Lomonosov*, was laid in 2007; after several delays, it is scheduled for commissioning off Kamchatka Peninsula in 2016. **Figure 19** shows the top of our planet as a bustling zone of economic and scientific progress.

A Window to Space

In January 2012, SOPS issued a Development Strategy for the Arctic Zone of the Russian Federation and the Maintenance of National Defense to 2020. As with its Bering Strait project design and the UP-UP mining and industry scheme, SOPS looked beyond present budgeting, to present a broader vision. The package also included guidelines for a clean-up, now almost complete, of former Soviet military bases in the Franz Josef Land archipelago. Using ideas from the earlier UP-UP plan, SOPS proposes innovative mining technologies, to extract minerals above the Arctic circle without laying waste to the areas around the mines. The IDMRD program and SOPS both call for developing north-south river and rail corridors, to form multiple links between the Arctic zone and the existing trans-continental railways.

Proximity to the Earth's magnetic pole and the geomagnetic fields defines the Arctic as the junction between the Earth and galactic radiation, an ideal place for exploring the causes of climate and weather, and meeting the challenges of cosmic radiation in its dynamic interaction with our planet. The Arctic has been called a "window to space," because of this position as one of Earth's two invisible polar portals through the atmosphere, receiving an influx of extraterrestrial radiation of which the Aurora Borealis is only the visible, beautiful fringe.

Curtain-type Aurora Borealis, photographed in Wiseman, Alaska.

SOPS presents a science-vectored policy for transforming the Arctic. The Strategy includes plans for a new generation of scientific ships, to study the deep-sea environment with instruments adapted for polar conditions. A new satellite observation system called Arktika, utilizing GLONASS, is to support polar hydrography (dynamic processes in the Arctic Ocean), study of geophysical conditions, and hydrometeorology. Former Russian Federal Space Agency (Roscosmos) head Anatoli Perminov said in 2010 that the Arktika system would enable year-round monitoring of the Arctic shelf environment, water temperatures, ice floe thickness, and pollution levels, as well as ensuring safe and efficient exploration of the shelf.

FIGURE 19
Arctic Development on the World Land-Bridge and Northern Sea Passages

LPAC/Chris Jadatz

1. With construction of the Bering Strait tunnel, new cities in Alaska and western Canada will not be at the edge of civilization. They will be on the only overland route connecting Russia and the USA, or, by the World Land-Bridge, Argentina and South Africa, by maglev rail.
2. The Canadian North is another gateway to the Arctic. The Hudson Bay port of Churchill, Manitoba is the western hemisphere end of the Arctic Bridge shipping route, with has its other end at Murmansk, Russia.
3. Ireland can develop both by building a deepwater port in the Shannon Estuary, and by taking the lead in astrobiology through deep-sea exploration. In the coming Arctic renaissance, the great Harland & Wolff shipyard in Belfast (Northern Ireland), currently specializing in offshore wind-farm construction, might be revived to build nuclear-powered icebreakers.
4. Under-populated Siberia should be not only mined, but developed and settled, with new towns along the high-speed modernization of the existing Trans-Siberian Railway and related corridors, and future ones such as the Near-Polar Mainline, paralleling the Northern Sea Route.
5. The proposed self-contained city of Umka, on Kotelny Island off Russia's Arctic Coast near the mouth of the Lena River, will be perfectly situated at Earth's Arctic "window to space," for scientific study and the development of the technologies needed to conquer mankind's next frontiers, the Moon and Mars.

Russia is expanding the Northern Federal Arctic University, in Arkhangelsk, as a center for training Arctic development specialists. Establishment of an Arctic Research Center, to conduct interdisciplinary Arctic studies under RAS auspices, is also proposed for Arkhangelsk Region.

At the September 2011 Arctic—Territory of Dialogue conference, Russian architect Valeri Rzhevsky's plan for a new facet of the "window to space" was exhibited. Then-Prime Minister Putin viewed the three-dimensional design of a "wonder city" in the Arctic, which would pioneer ways for humans to live in environments far removed from those of Earth. The project is called Umka, a word suggesting "cleverness" (and the name of an enterprising polar bear character in a Soviet-era animated cartoon). The site of the domed, self-sustaining city is Kotelny Island, between the Laptev Sea and East Siberian Sea, some 400 km northeast of the Lena River delta. It is 1,500 km from the North Pole. The giant dome is designed to contain life-support systems, modeled on those of the International Space Station. With power from a floating nuclear power station, Umka will have a regulated temperate climate, in which air circulation and the biological cycles of all plants and creatures within are interdependent. Circulation of oxygen and plant growth-stimulating carbon dioxide will be contained under the dome, minimizing contact with the frigid outside environment. Fish farming, in waters slightly warmed by the nuclear power plant, will help to supply food. All waste will be recycled or reduced to less than ashes. Recreational facilities will lessen psychological strain on Umka's human residents, helping them adjust to living in an enclosure.

A cutaway design model of the domed Arctic city of Umka.

The Umka design has been submitted to all five Arctic littoral nations (Canada, Denmark/Greenland, Norway, Russia, and the United States). It echoes earlier Far North development plans, such as the domed cities promoted by John Diefenbaker, the prime minister of Canada (1957-1963) during the birth of manned space travel.

With a footprint of 1.2 km x 800 m, and 5,000 residents, the population density of Umka will approach Hong Kong's 6,349 inhabitants per km^2. Those first 5,000 residents of the city might be scientists, engineers, and workers for oil platforms and mining companies, but scientific researchers living in Umka will also plumb the rich Arctic depths of undiscovered knowledge for the biological and physical sciences. The Umka project is best appreciated from an extraterrestrial standpoint. At 75° N latitude, there are strong winds and temperatures sink below -30°C—the same temperature as in lunar lava tubes. These hostile conditions are an opportunity to develop and apply technologies that will be needed at the frontiers of space exploration.

The Eurasian Union

One of the sillier statements in the Valdai Club's 2014 *Toward the Great Ocean-2* report is this: "Theoretically, integration in the post-Soviet space and integration in the Asia-Pacific are not rival projects.... However, given the limited amount of financial and human resources, which may not be enough for the two projects, ... there is a risk that priority will be given to Eurasian integration." If it hadn't been published earlier, one would think it were an attempt to undercut what Presidents Putin and Xi pledged in May 2014 about seeking "opportunities to combine the perspective of the 'Silk Road Economic Belt' (SREB) with the conception of the 'Eurasian Economic Union,'" cited in Helga Zepp-LaRouche's introduction to the present EIR Special Report.

In fact, the founding members of the Eurasian Economic Union (EEU)—Belarus, Kazakhstan, and Russia—and prospective members such as Armenia and Kyrgyzstan stand to gain economically from the SREB cooperation policy, if manipulated geopolitical rivalries can be set aside, and the rotten legacy of the 1990s, described above for the case of Russia, expunged.

The EEU, announced by Putin, Belarusian President Alexander Lukashenka, and President Nursultan Nazarbayev of Kazakhstan in simultaneous articles in 2011, was preceded by several economic programs among former republics of the Soviet Union, attempting to recoup some of what was lost when the physical economic ties between them were abruptly cut in 1991. Nazarbayev first proposed a Eurasian Economic Union in 1994. The Eurasian Economic Community (EurAsEC) was formed and began regular meetings in 2000, but it was not until the second half of that decade, that serious work on a Customs Union (CU) began, among Belarus, Kazakhstan, and Russia. Academician Glazyev, first as EurAsEC deputy general secretary and then as responsible secretary of the Customs Union Commission, worked from 2008 to 2010 to organize the CU, which went into effect on January 1, 2010. The founding members went on to establish the Common Economic Space (2012) and sign the Eurasian Economic Union Treaty in May 2014. Armenia subsequently adhered to that Treaty, so that there will be at least four members of the EEU when it is inaugurated on January 1, 2015, and President Atambayev of Kyrgyzstan said in October that his country also hopes to be in the EEU from the outset.

Putin, in remarks on May 29, emphasized the coordination of energy, industrial, agriculture, and transport policy within the EEU, linking this with trans-Eurasian activity: "The geographical position allows us to create transport-logistics routes of not only regional, but also global importance, attract-

Presidents Alexander Lukashenka, Nursultan Nazarbayev, and Vladimir Putin, of Belarus, Kazakhstan, and Russia, respectively, after signing the Eurasian Economic Union treaty in Astana, Kazakhstan on May 29, 2014.

ing massive trade flows in Europe and Asia." Beginning with his 2011 *Izvestia* article, Putin said that as Russia, Belarus, Kazakhstan, and other post-Soviet countries that might join a Eurasian Union integrated their economic ties, they would seek to become a bridge between Europe and the Asia-Pacific region.

Kazakhstan in the Center

The geographical location and resources of Kazakhstan (**Figure 1**), in particular, qualified that country to be featured in LaRouche's "Vernadsky strategy." At the same time, with mineral production second only to Russia's in the region, independent Kazakhstan has attracted a large presence of the international raw materials cartels, eager to exploit those resources without concern for the country's development. Kazakhstan has also been heavily targeted in geopolitical and ideological areas, typified by the notorious war-monger and apostle of liberal imperialism Tony Blair's contract as a consultant to the country's leaders. President Nazarbayev frequently embraced dubious policies, peddled from abroad, as in the case of his Green Bridge program for a transition to a supposed "green" economy.

When it comes to energy and infrastructure, however, Kazakhstan has been in the forefront of seeking development, welcoming potential cooperation in both the EEU and the SREB. China has been involved in building two railways in Kazakhstan, so far, one running to the new capital, Astana, and thence to Russia, and the other to the more southerly former capital, Almaty, with routes proceeding onward, both northwest to Russia and west through Central Asia and Iran to Turkey. The dry port at Khorgos on the China-Kazakhstan border is receiving significant investment, as a logistics center for Chinese trade with Europe and Southwest Asia. Kazakhstan is also working with Turkmenistan and Iran on the Caspian eastern shore branch of the International North-South Transport Corridor.

While the CU brought about an overall opening of the Russian market to Kazakhstan, the current highlights of Russian-Kazakh cooperation are in the energy field. While Putin was in Astana for the May 29 EEU signing, a bilateral nuclear agreement was also signed. Russia's state-owned firm Rosatom is committed to building a new nuclear power plant in Kazakhstan, while assisting with uranium processing. One of the world's biggest producers of this nuclear fuel, Kazakhstan has a 10% stake the Rosatom-run International Uranium Enrichment Center at Angarsk in Siberia.

A Mission for Ukraine

Russia tried to attract Ukraine into the Eurasian integration process, with the Ukraine-born Glazyev working intensely in 2011-2013 to organize its greater cooperation with the CU and the future EEU. Closer Eurasian integration was resisted by Ukrainian governments, under pressure from the West to opt for free-trade relations with the European Union, for more than a decade. Nonetheless, because of the historical interface of Ukraine's economy with Russia's, the issue of Ukraine's Eurasian relations did not disappear.

In November 2013, when Ukraine suspended work on an Association Agreement with the EU, the act that served as a pretext for the insurgency against Ukraine's elected government, then-Prime Minister Mykola Azarov's cabinet invoked "national security interests," announcing that it would study ways "that restore its lost production capacities and areas of trade and economic cooperation with the Russian Federation and other CIS members," and to revive negotiations with the (Eurasian) Customs Union, in particular. Detailed studies by the Ukrainian Academy of Sciences quantified, in monetary terms, the advantages to Ukraine of closer involvement with the EEU, as against its dim prospects as a source of cheap land and cheap labor in an EU framework. Ukrainian political figure and economist Natalia Vitrenko reported on those studies in 2013, saying: "If Ukraine joins the Eurasian Customs Union, its GDP will increase by an amount in the range of 1.5 to 6%.... The Russian Academy's Institute for National Economic Forecasting has projected that if Ukraine joins the Customs Union, the Ukrainian economy will gain $7 billion annually, and its exports will increase by 60%, or $9 billion annually."

The real measure of Ukraine's potential within Eurasia is not found in monetary or GDP terms. Southeast Ukraine was a leading machine-tool and engine production center in the Soviet Union. Its steel industry produced top-quality alloys. These capacities have been savaged during privatization, even before the current civil war, but they still exist and could be geared up. Likewise, the education levels of the population in late Soviet-era Ukraine were recognized in UN reports as among the highest in the world. Those skilled people are older now, but Vitrenko estimates that a portion of them could still be a factor, for economic recovery, during the next ten years.

It would be tragic, if Ukraine could not take up its natural role in Eurasian economic development. Apart from the deadly civil strife in the wake of its February 2014 coup, Ukraine's situation is similar to that of other once-advanced industrial nations, especially those with a highly skilled work force and machine-making industries, like the United States, Europe, or Australia. Their contribution, under sane policies, is indispensable.

Tumen River Initiative: A Step for Peace in Northeast Asia

By Michael Billington

September 2014

On July 18, 2014, leaders from Russia, North Korea, and, notably, South Korea attended a ceremony officially opening a state-of-the-art port facility, built by Russia, at the North Korean port of Rajin, part of the Rason Special Economic Zone (earlier called Rajin-Sonbong). Rajin is located at the mouth of the Tumen River, which empties into the Sea of Japan, and is also the end point of the recently completed rail line from Russia. China is also building a port at Rajin, and recently completed a road to the port from near the tri-border region between Russia, China, and North Korea.

The July Rajin event marks significant progress toward completion of the Greater Tumen Initiative (GTI), a development project based on building up this border region especially through rail transportation corridors. In 1991, the UN Development Programme declared its support for a collaborative effort between China, Russia, Mongolia, North Korea, and South Korea to develop the region surrounding the Tumen River, which forms the border between China and North Korea (**Figure 1**). Japan has also been involved tangentially. Despite several false starts, the project has begun to take off in the past 2-3 years, although North Korea has not been officially part of the project since 1999.

Besides the dramatic economic benefits for every country in the region (**Figure 2**), this concept is also the crucial, core development project required to end the last remaining legacy of the Cold War in Asia—the so-called North Korea problem. As Lyndon LaRouche has emphasized for many years, the solution to every crisis created by the imperial "divide and conquer" policies, is located in the common development interests of the parties involved—and, ultimately, in the common interests of mankind. If the Tumen River development project is located within the broader interest of developing the entirety of East Asia, and especially the diffi-

FIGURE 1
The Tumen River: Boundary of Russia, China, and North Korea

UNESCAP

FIGURE 2
The Greater Tumen Region

Greater Tumen Initiative

cult (but resource rich) areas of the Russian Far East, and in the even broader interest of the Pacific basin as a whole, from the Mississippi River to the western borders of China and Southeast Asia, then the project defines a basis for long-term cooperation between nations, and the uplifting of the life and livelihood of the population of the region.

Over the past year, Russia's President Vladimir Putin, China's new President Xi Jinping, and South Korea's new President Park Geun-hye have held several bilateral meetings, with a major subject of those meetings being the development of the Russian Far East and the completion of the Eurasian Land-Bridge to its original goal—from Pusan to Rotterdam. The gaping hole in that extensive development corridor is the necessary passage through North Korea.

North Korea dropped out of the original Tumen Development group in 1993, as a confrontation with the United States nearly led to war. Under President Clinton, the war was avoided through an agreement called the General Framework, with North Korea giving up those aspects of its nuclear program that could have been used for a weapons program, in exchange for food and energy support, and a U.S./South Korea project to build a weapons-free nuclear power plant for the North. This process lasted through the end of the 1990s, only to be scrapped entirely when Dick Cheney came to power in 2001. Cheney's commitment to confrontation rather than cooperation led to North Korea building a nuclear weapon.

The Greater Tumen Initiative wisely chose to proceed with its planning on the basis that the North Korean problem would eventually be resolved. A set of priority corridors for the Greater Tumen Region (GTR) has been delineated (**Figure 3**). A February 2013 GTI report, "Integrated Transport Infrastructure and Cross-Border Facilitation Study for the Trans-GTR Transport Corridors," states: "The Democratic People's Republic of Korea (DPRK – North Korea) is no longer a member of GTI after withdrawal in 2009. Therefore corridors 5 and 6 [road and rail corridors along the west and east borders of North Korea, connecting South Korea with China and Russia] originating from the Republic of Korea (ROK – South Korea) cannot reach the rest of GTI countries (except by air and sea). This poses a serious limitation to the study. However, it was decided to consider in an optimistic scenario further liberalization and opening up of DPRK with re-establishment of connections with ROK and proper functioning of the Korean Peninsula corridors."

FIGURE 3
Trans-Greater Tumen Region Transport Corridors

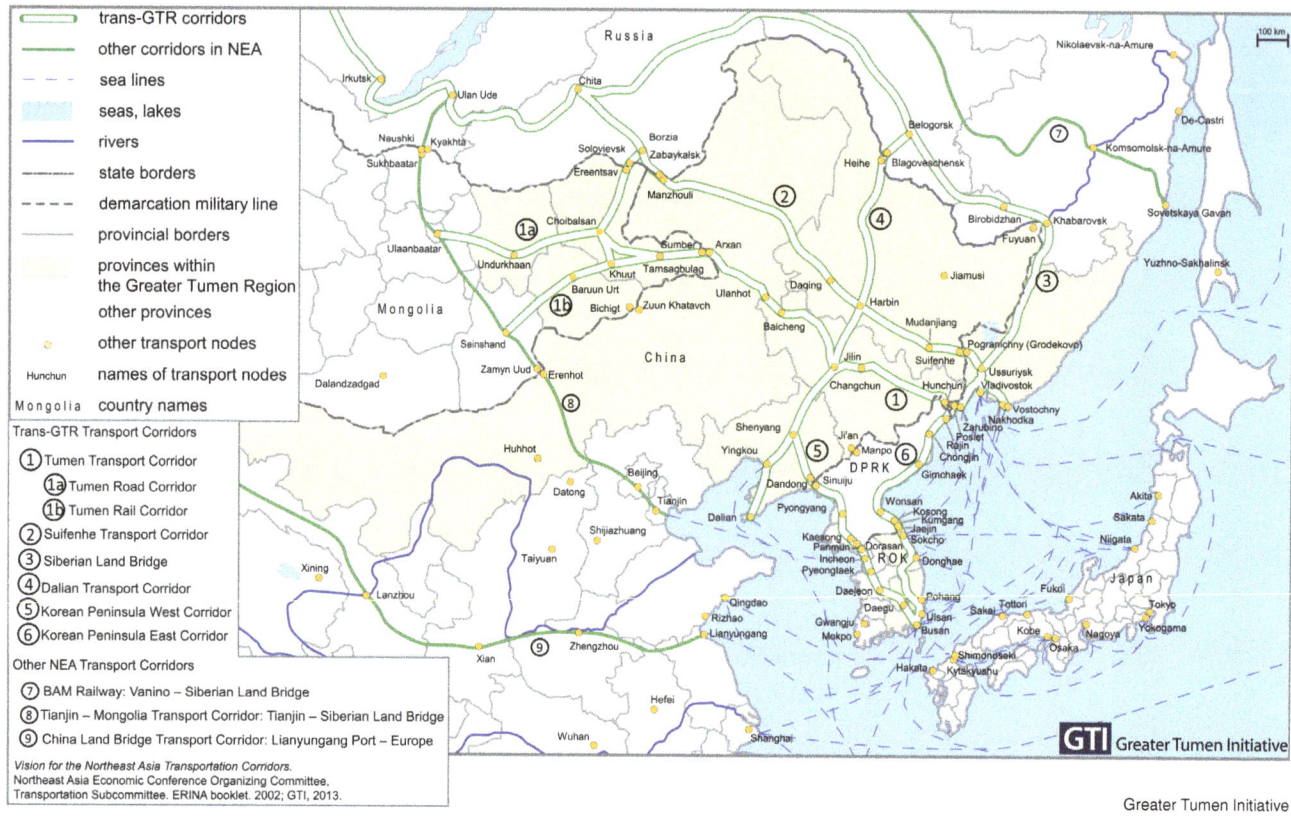

Greater Tumen Initiative

In fact, the success of Russian plans for the development of the Far East of Russia and the Arctic region depends to a great extent on the successful resolution of the Korean issue. Both Japan and South Korea have technological and construction capabilities that are essential for the development of the vast and difficult terrain of the Russian Far East and the Arctic, while North Korea has a highly skilled work force which will be an invaluable input for such projects, while also further integrating North Korea, through development of mutually beneficial projects, into the East Asian community of nations.

The Projects

The GTI region encompasses the Chinese provinces of Liaoning, Jilin, Heilongjiang, and the eastern portion of Inner Mongolia; North and South Korea; Russia's Primorsky Territory, Khabarovsk Territory, Amur Region, the Jewish Autonomous Region (all in the Far East Federal District), and Transbaikal Territory; Mongolia; and to a certain extent, Japan. The Yanbian Korean Autonomous Prefecture of China is a crucial link between Korea and Jilin Province.

The two primary east-west corridors of the GTI, labeled Corridors 1 and 2 on the map (the Tumen Transport Corridor and the Suifenhe Transport Corridor, respectively) connect the coastal zones with the interior, and to the Trans-Siberian Railway at Chita. Northeast China was the industrial heartland of China and, despite the rapid growth in the south, remains the core region for heavy industry. The region is cut off from the Sea of Japan by the Primorsky Territory of Russia, which runs down the coast to the Korean border, and by North Korea, resulting in much of the industrial inputs and outputs of China's northeast being transported through the port at Dalian in Liaoning Province far to the south. Although in need of upgrading, Corridor 2, the road and rail connections from Harbin to Vladivostok and the nearby ports of Vostochny and Nakhodka, and, in the other direction, to Manzhouli in Inner Mongolia and across to Russia, accounts for 60% of China's trade with Russia. Corridor 1, from the Russian port of Zarubino on the coast to Hunchun, Changchun, and on to Mongolia and Russia, is both road and rail on the Chinese side, but in Mongolia it is gravel road. The potential for bringing Mongolian coal and other resources into China, Russia, and to the ports is one of the primary bottlenecks of the GTI.

While these two corridors can be greatly improved, it is the connection of these corridors, as well as the Trans-Siberian Railway at Vladivostok, to the corridors through North Korea and through to South Korea, which remains as the most critical bottleneck to development and to peace in the region.

Politically, the key regional players required for solving the North Korean quagmire—China, Russia, and South Korea—are fully engaged in efforts to locate a peaceful settlement within a process of large-scale, inter-regional development projects. President Xi Jinping met privately with President Putin five times in 2013—his first year in office—and again at the Sochi Olympics this year. Their agenda in these meetings always includes the necessary cooperation in the urgent development of Central Asia, the Arctic, the Russian Far East, and in that context, the Korean Peninsula.

Putin also travelled to Seoul in November 2013, where he and President Park Geun-hye signed an historic set of agreements, including for several development projects that will necessarily engage North Korea. While the issue of North Korea's cooperation in these projects was not discussed publicly, it is certainly the case that Putin had coordinated the projects with Pyongyang ahead of time.

President Park described the set of agreements coming out of her meeting with President Putin: "We, the two leaders, agreed to combine South Korea's policy of strengthening Eurasian cooperation and Russia's policy of highly regarding the Asia-Pacific region to realize our mutual potential at the maximum level, and move relations between the two countries forward.... South Korea and Russia will join hands to build a new Eurasian era for the future."

The summit produced 17 cooperation agreements, most having to do with joint economic development, and many of them implying some level of North Korean involvement, the most important being a memorandum of understanding (MOU) on South Korean participation in the Russian-led Rajin-Sonbong (called Rason) development project in North Korea. It calls for POSCO (South Korea's steel giant), Hyundai Merchant Marine Co., and Korea Railroad Corp. to participate in the Rason development project—the first such South Korean industrial investment proposal in North Korea, other than the joint industrial park at Kaesong on the North Korea-South Korea border. The Korean consortium plans to buy a stake in RasonKonTrans, the Russian-North Korean joint venture carrying out the rail and port renovation project, including the now-concluded reconstruction of the rail link from Rason to Khasan in Russia, and on to Vladivostok. The state company Russian Railways has a 70% stake in the joint venture, with North Korea holding the remaining 30%. News reports have said that the South Korean consortium plans to buy about half the Russian stake.

The three firms did in fact visit Rason twice during 2014. Although the decision on the planned purchase of the stake in RasonKonTrans had not been finalized as of October 2014, the three south Korea mega-companies did announce in October that they are preparing to import Russian coal through the port of Rajin to the southern Korean port of Pohang in November, a 35,000-ton test run. Hyundai Merchant Marine will provide the ship to bring the coal to South Korea's state steel company, POSCO. This would be the first such Russia-North Korea-South Korea cooperative industrial project.

The project fits into Park's "Eurasian Initiative," which calls for binding Eurasian nations closely together by linking roads and railways to realize what she called the "Silk Road Express" running from South Korea to Europe via North Korea, China, and Russia. The President early in 2014 declared that a "Korean Bonanza" awaits the region and the world if reunification between North and South Korea can be achieved peacefully. "Unification will allow the Korean economy to take a fresh leap forward and inject great vitality and energy," she said.

After the project to modernize the port of Rajin is completed, the rail-connected port can be used as a hub for sending cargo by rail from East Asia to as far away as Europe. South Korean firms will be able to ship exports first to Rajin, and transport them elsewhere via Russian Railways.

The long-discussed project to link the railways of South Korea with Russia's Trans-Siberian Railway, via North Korea, and through to Europe, "from Pusan to Rotterdam," is also back on the table—the two sides signed an MOU on rail cooperation and agreed to study the project as a long-term venture—establishing "Peace through Development." Together with the construction of the Bering Strait tunnel, and nearly 5000 km (3107 mi) connecting to it, the completion of the Korean Peninsular rail project will make possible a train ride from Pusan to New York City, as well as expanded trade between Korea and western North America.

Other projects in which South Korea and Russia agreed to cooperate as long-term ventures included building a natural gas pipeline linking Russia and South Korea via the North, and developing Arctic shipping routes to reduce shipping distances and time between Asia and Europe.

PART 5
South and Central Asia: From Arc of Crisis to Corridors of Development

India Is Ready to Fulfill Its Legacy of Leadership

By Ramtanu Maitra

September 2014

India, the world's second most populous nation, with more than 1.2 billion people, is well situated to play a crucial role, in combination with fellow-BRICS giants China and Russia, in carrying out the global transformation of the world economy, toward one of cooperation for dramatic scientific and industrial progress. While in urgent need of a major modernization of its physical infrastructure—transport, power, and water, in particular—India, in the years since its independence in 1947, has developed in-depth capabilities in science, engineering, and agriculture that will allow it to undergo a qualitative leap, once the political decision is made.

Once a proud leader of the Non-Aligned Movement, India has had the benefit of two extraordinary prime ministers, who had global vision and paved the way for the current opportunity. The first was Jawaharlal Nehru, the newly liberated nation's first prime minister, who established the foundation for India's strong scientific and industrial base. The second was his daughter Indira Gandhi, who led a global battle by developing nations for a new just world economic order in the 1970s and early 1980s, demanding technology transfer and justice for the poorer nations of the planet. Mrs. Gandhi also made the indispensable contribution of promoting the agricultural revolution that made India self-sufficient in food.

Since Mrs. Gandhi's assassination in 1984, India has generally played a lesser global role, and found itself faced with a dilapidated infrastructure now crippling its progress. But now, after 10 years of insipid and self-confining economic policies led by IMF- and World Bank-trained economists, India has a new leader, Narendra Modi. Modi, unlike his predecessors, drew the electorate's attention during the 2014 parliamentary elections by promising large-scale infrastructure development in the coming years to set the stage for future rapid progress in industrial, manufacturing, and agricultural sectors, and thus create in the process millions of productive jobs for India's youth. He adjusted the pitch of his campaign to meet the expressed aspirations

creative commons/Rangilo Gujarati

India's new prime minister, Narendra Modi, before his election.

of hundreds of millions of youth who want to be part of the nation-building process to unfold a brighter future for themselves and future generations of Indians. Having succeeded in conveying to the youth what he wants to undertake, Modi won the national elections in May 2014 with a large plurality, and has thus been entrusted with the task of delivering on his promises.

Modi, as well as many other Indians, realizes that the task is not going to be an easy one. The damage to the economy wrought by earlier administrations has created a deep rot within Indian institutions through which these tasks need to be carried out. In addition, the global economic downturn since 2008 has bankrupted potential investors in Europe and Japan. What is in Modi's favor is India's solid economic foundation, which was laid soon after India broke away from the British colonial grip and became an independent nation in 1947, and the solidarity Modi shares with the BRICS nations, which are fighting to break free of the monetarist stranglehold.

Nehru's Contribution

Soon after India became independent, India's first Prime Minister Jawaharlal Nehru did a few things right. One of those was laying a strong foundation geared toward adopting frontier science and technology to build up the country's industrial and manufacturing base. Nehru understood the need to build India's physical economy and how important a role infrastructure plays in building up a high-productivity economy. At the outset, Nehru realized that to provide a future for the multitude of present and future Indians, India must move away from the British-organized coolie-labor-based economy to master the frontiers of technologies.

Homi Bhabha (right), father of India's nuclear industry, and India's first prime minister, Jawaharlal Nehru. Nehru was determined to build up his nation's scientific capability.

With the help of a few brilliant, nationalist scientists, among whom Dr. Homi Bhabha and Professor Shanti Swarup Bhatnagar stand out as the two most well-recognized, Nehru began to lay the foundation for Indian science in general, and nuclear in particular. While the fundamental research on nuclear science had already been started by Dr. Bhabha in 1945, India set up the Atomic Energy Commission, Trombay—renamed the Bhabha Atomic Research Center (BARC) in 1967 following Dr. Bhabha's mysterious death in 1966 on his way to an International Atomic Energy Agency (IAEA) meeting in Vienna—in 1954, with the intent to evolve a self-sufficient atomic energy program and to make the nation, eventually, power-independent. In the late-1950s Dr. Bhabha had laid out a three-stage atomic program, which has been pursued diligently since. India's atomic energy establishment is now on the verge of entering the second stage, which features the use of breeder reactors. The third stage of India's nuclear program will be based entirely on using India's vast reserves of thorium as fuel. An experimental and indigenously-developed 300-MW thorium-fueled nuclear power reactor is scheduled for commissioning in March 2015 at the Kalpakkam nuclear plant complex in Southern India.

For a number of reasons, which unfortunately include a concerted decades-long effort by the developed countries to prevent India from importing nuclear-related equipment or material, and the inability of the Indian leaders, particularly those who were at the helm during the past three decades, to grasp the importance of Dr. Bhabha's vision of how to make India energy independent, India's atomic energy establishment has so far contributed only minimally to reducing India's huge power shortfalls. But decades of extensive research and development work that India's nuclear scientists have carried out have put India at the frontline of many aspects of atomic science, with manpower to match, and has now placed India on the threshold of utilizing this infinite power source. It is the success of India's atomic sector in the coming years that will be the base on which India's future vast industrial and manufacturing sector is established. It is now up to Prime Minister Modi to seize the hour.

India's promise to build an agro-industrial nation based on science and technology began with the Scientific Policy Resolution of 1958, which has come to be considered as a sort of Magna Carta for science and technology in India. The 1958 Resolution laid out briefly the commitment of New Delhi to the advancement of science. It said the key to national prosperity lies in industrialization, involving the roles of technology, raw materials, and capital, pointing out that "technology can only grow out of the study of science and its application." In addition, the Resolution, making note of the accelerated pace of the development of science in the 20th Century, said that "it is an inherent obligation of a great country like India, with its traditions of scholarship and original thinking and its great cultural heritage, to participate fully in the march of science, which is probably mankind's greatest enterprise today."

The next step for India in developing its science sector was its establishment of a space sector in 1969, the Indian Space Research Organization (ISRO). Despite the country's financial and infrastructural weaknesses, India's space program is an unmitigated success, ushering in a whole new set of technologies, materials for industrial use, a gamut of spin-offs that range from artificial polyurethane feet to automatic weather stations, pressure transducers, and hundreds of other products. It has also created a brigade of scientists and technicians who are at the frontline of space technology.

ISRO's first satellite, Aryabhata, was launched by the erstwhile Soviet Union in 1975. Rohini, the first satellite to be placed in orbit by an Indian-made launch vehicle (the Satellite Launch Vehicle 3), was launched in 1980. ISRO subsequently has developed two other rockets—the Polar Satellite Launch Vehicle (PSLV) for putting satellites into polar orbit, and the Geostationary Space Launch Vehicle (GSLV) for placing satellites into geostationary orbit. These rockets have launched communications satellites, Earth-observation satellites, and, in 2008, Chandrayaan-1, India's first mission to the Moon. On November 14, 2008, the Moon Impact Probe separated from the Chandrayaan-1 orbiter and struck the South pole in a controlled manner, making India the fourth country to place its flag on the Moon. The probe impacted near the crater Shackleton, ejecting underground soil that could be analyzed for the presence of lunar water ice.

India's first interplanetary probe has completed 300 days of its mission. The Mars Orbiter Spacecraft, which was launched by PSLV-C25 on November 5, 2013, is designed to orbit Mars in an elliptical orbit. ISRO plans to put two astronauts into orbit in 2015.

It is expected India's space program will receive strong patronage from the current administration because Prime Minister Modi is a fervent backer of India's space program and has suggested putting up satellites for the use of India's South Asian neighbors. Moreover, India's space program-generated rockets have become the backbone of the Indian military's defensive and offensive weapons systems.

India also has a significant nuclear fusion program. It began in 1989 at the Institute for Plasma Research in Gandhinagar, when India's first tokamak, called ADITYA, was commissioned. It was designed and mainly fabricated in India, and has been upgraded several times. Currently, Indian scientists are moving to the next phase, which is to build and operate the Steady State Tokamak, which will use superconducting magnets to produce a 3 Tesla magnetic field, and set the stage for commercial development of fusion.

NASA/Wikimedia Commons

The INSAT 1B is part of the Indian National Satellite System, a series of geostationary satellites commissioned in 1983. INSAT is now the largest domestic communications system in the Asia-Pacific region.

Meanwhile, India is also producing nine large components, amounting to almost one-tenth of the project, for the International Thermonuclear Experimental Reactor (ITER). The biggest of these is the cryostat, a 3,800-ton pressure chamber the size of a 10-story building, which will have to be shipped to ITER's site in France in pieces.

Mrs. Gandhi Defeats Famine

These are credible successes, made despite India's failure to build an adequate infrastructure by keeping its focus on the maximization of energy flux-density. As a result of that failure, India had a massive shortage in power generation, and the country was unable to manage the vast amount of water during the 13 weeks of monsoon that visits India every year, inundating rivers and flooding cities and plains on its way to the sea. This failure gave rise to a famine-like situation in the mid-1960s.

Under the tutelage of then-Indian Prime Minister Indira Gandhi and two plant breeders, Norman Borlaug with wheat and M.S. Swaminathan with rice, the national government set up the logistics to encourage the planting of higher yielding varieties. Within a span of two decades, India, adopting high-yield variety seeds and fertilizers and using some irrigation and a large amount of pumped-out groundwater, turned India from a food-short to a food-surplus nation. This was enormously important for the country's political independence, and that commitment continues to be shown today by India's refusal to kowtow to the World Trade Organization on the question of its Food Security program. That single developmental surge focused on the agricultural sector allowed India to get set for speedy progress.

That, however, did not happen. India's agricultural sector, neglected for almost two decades now, is still far from achieving its potential. Failing to divert surplus river waters to water-short rivers during the past four decades has kept India's agricultural productivity way below that of Japan, China, and South Korea, and its agricultural production highly dependent on the strength and timely arrival of the annual monsoon rains.

Heavy Engineering Powerhouse

Notwithstanding the developmental shortfalls, anchored on its weak and dilapidated infrastructure, India has been positioned for more than decade to advance rapidly in the agro-industrial sector. Its heavy engineering capabilities, the foundation of which was laid in the 1950s; its technological capabilities, exhibited by its atomic energy and space programs; and its food self-sufficiency make India ready to move fast forward.

For instance, India's heavy engineering capabilities were exhibited in its ability to manufacture its coal-fired power plant equipment. The country has built almost 3,200 major and medium dams since independence and manufactured the nuclear reactor core vessel of the Pressurized Heavy Water and Fast Breeder reactors, made from stainless steel. It is also prepared to produce component modules for the Westinghouse AP1000 reactors and to cooperate with Russia's Atomstroyexport in building components for the next four VVER reactors at Kudankulam in Tamil Nadu. India has also signed an agreement with GE Hitachi to carry out complete construction of nuclear power plants, including the supply of reactor equipment and systems, valves, and electrical and instrumentation products for the Advanced Boiling Water Reactors (ABWR)—a third-generation reactor.

India's Bharat Forge Ltd (BFL), a multinational company that claims to be among the largest and technologically most advanced manufacturers of forged and machined components and is said to be the world's second-largest forging company, is extending its activities into the power sector. In 2008, BFL formed a joint venture with Alstom, a French multinational company, primarily for manufacturing state-of-the-art supercritical power plant equipment in India, although the enterprise may extend to nuclear applications. In January 2009, BFL signed a memorandum of understanding with France's nuclear company Areva to set up a joint venture in casting and forging nuclear components for export and the domestic market.

These capabilities did not emerge in a vacuum. India had long ago begun to develop its machine tool industry. There are about 450 firms manufacturing complete machines or their components. There are 150 units of production in the public sector. Almost 73% of the total machine tool production in India is contributed by 10 major companies. The industry employs a workforce directly or indirectly totaling 65,000 skilled and unskilled persons. Similar insufficient, but growing strength is also reflected in India's shipbuilding and defense sectors. India has 37 shipbuilding yards, four of which are major ones in the public sector domain with the capability to build ships of 50,000 Dead Weight Tonnage (DWT).

In the defense sector, the Indian Navy plans to expand to a fleet of 150 ships in the next 10 to 15 years, with 50 warships now under construction and 100 new vessels in the acquisition pipeline.

Infrastructure Needs

As we have repeatedly indicated, the blockage on full development based on this solid foundation is the lack of a modern infrastructure in water, power, and transportation infrastructure that would not only enhance the national economy, but permit an efficient linkup with the New Silk Road and the Maritime Silk Road. Indeed, *EIR* pointed out more than three decades ago, in its report "India: An Agro-Industrial Superpower in 2020," that modernizing infrastructure was the single most critical ingredient for allowing India to fill the gaps in its economy.

The following are the major development corridors *EIR* has identified as crucial for India's rapid development, and integration with the region's development perspective.

1. A northeast economic corridor, which will consist of a high-speed rail corridor that will link Myanmar borders through Dangari (India) and Tamu (India) in the central part of India's border, with Myanmar to the port of Kolkata in the southwest and Patna in the east. India is in the process of building the Jiribam-Imphal-Moreh line in the east Indian state of Manipur and the Tamu-Kalay-Segyi line in Myanmar. The northeast transport corridor needs to be linked to the Jiribam-Imphal-Moreh.

The economic corridor, which will also be supplanted with the Kunming-Kolkata Highway, will broaden a segment of the Silk Road, linking Yunnan to Kolkata, an essential element in development of the area.

The northeast corridor will need power. Water will be available aplenty from the Brahmaputra River Basin, but what the area needs beside a transport corridor is nuclear power. A number of nuclear power plants in clusters need to be set up to facilitate the agro-industrial complexes; primary, secondary, and higher education centers; healthcare facilities with research capabilities; and heavy engineering manufacturing centers.

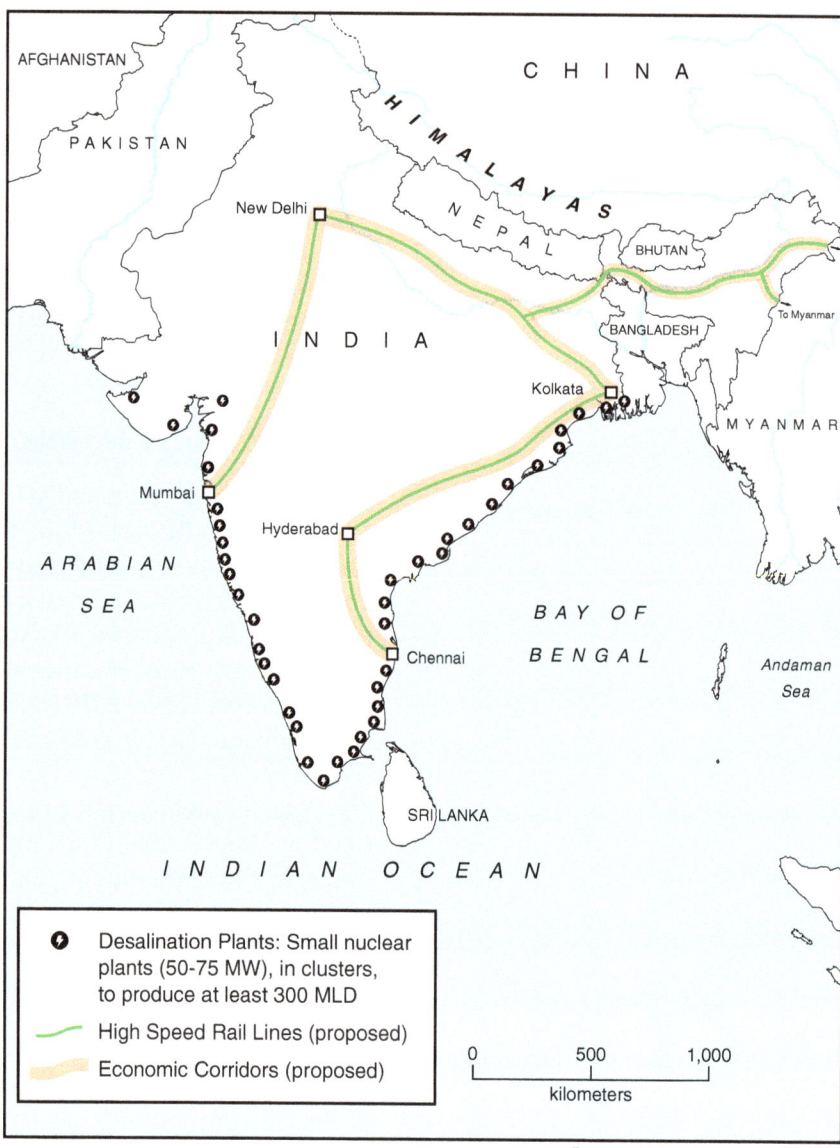

2. A Kolkata-Delhi high-speed transport corridor and an economic corridor. This is under discussion with Japan. Japan has already begun work in the Delhi-Mumbai (Bombay) high-speed (bullet train) transport corridor, and New Delhi has begun acquiring land (50 miles wide on either side of the corridor) to launch the Delhi-Mumbai economic corridor. A similar width of land will be required for the 900-km-long Kolkata-Delhi economic corridor.

3. India will need two more high-speed transport and economic corridors. These will link Kolkata with Hyderabad and Hyderabad to Chennai (Madras). Along these corridors, plants supplying 15-20,000 MW of nuclear power have to be set up. Both these corridors will en-

counter water shortages, and they require reviving the old Peninsular River linking plan.

4. India has two sets of rivers. The Himalayan rivers, e.g., the Ganga, Yamuna, and their tributaries, are those that originate in the Himalaya mountains and flow through the Northern Plains. The second category of rivers is known as Peninsular Rivers, which originate in the Western Ghats mountains. They have a large seasonal fluctuation in volume because they are solely fed from rainfall. These rivers flow in valleys with steep gradients. Major rivers of the Peninsula, such as the Mahanadi, Godavari, Krishna, and Cauvery, flow eastwards on the plateau and drain into the Bay of Bengal. The Narmada and Tapti flow eastwards. Narmada rises in Amarkantak plateau in Madhya Pradesh and enters the Gulf of Cambay in the Arabian Sea.

To make these economic corridors fruitful, long-distance transfer of water from water-surplus basins to water-short basins will be needed. In this particular situation, the interlinking of Mahanadi-Godavari-Krishna-Pennar-Cauvery is one of the four parts of the Peninsular River Development. Among the Peninsular Rivers, the Mahanadi and Godavari have sizable surpluses after meeting the immediate and known future requirements. It is therefore necessary to transfer surplus Mahanadi-Godavari water to the water-short Krishna, Pennar, and Cauvery. The canal system that will carry the water from one basin to another may not meet the overall requirements of these two economic corridors.

Because these economic corridors are not at a great distance from India's east coast, desalination will have to fill the deficit. Desalination could produce a perennial supply of freshwater. India has begun looking at the use of nuclear power for desalination of seawater. A desalination demonstration plant at Kalpakkam, using nuclear waste heat for the multi-stage flash (MSF) process that produces 4,500 m^3 per day, has already been set up. These plants can be scaled up 10 times from the current configuration without any difficulty, according to one expert. Already two methods of desalination—reverse osmosis and MSF—have been demonstrated at the Bhabha Atomic Research Center.

Bringing High-Technology Development to South Asia

By Ramtanu Maitra

September 2014

The South Asian nations under discussion—Bangladesh, Nepal, and Pakistan—constitute a grouping of countries that have close connections to each other and also are, in effect, connected to China. All of these nations are highly underdeveloped and, for reasons not under discussion here, have not developed the wherewithal to develop by their own efforts. The incremental progress that these countries have achieved over the years has not made a significant number of people in those countries more productive. The weak development in fact could not keep up with natural population growth, and as a result, poverty has engulfed more and more people in recent times, leaving them and their families with little hope for the future.

At the same time, major economic powers in the Asia-Pacific region—China, Japan, South Korea, and partially India—have developed and mastered technologies that could help these nations to overcome their economic shortcomings in the next 30 years. But that would happen only if these South Asian nations become integrated through railroads, and develop high energy flux-density-based power to provide food, water, power, and education to the hundreds of millions who have long been waiting. This paper deals with what is minimally needed to get the process going.

We present the picture of the region from East to West, with emphasis on the prospective linkups with the southern branch of the Eurasian Land-Bridge.

Bangladesh

Bangladesh, with about 165 million people, is the most densely populated nation in the world. The southern part of the country, inundated with Bay of Bengal estuaries, is where population density is the greatest. Occasional hurricanes and cyclones originating in the Bay of Bengal devastate this densely-populated southern part of Bangladesh. Bangladesh is also connected by land with India on three sides. It is also close to Myanmar in the east and Nepal in the North.

Linkage with the Asia-Pacific is the key to Bangladesh's future.

Bringing High-Technology Development to South Asia

The following projects are necessary for Bangladesh to develop its economy and become part of overall Asia-Pacific development:

1. Two agro-industry-based economic corridors feeding the east-west transport corridors. One will run from Thakurgaon to Mymensingh, and the other from Mymensingh to Narayanganj in the southwest.

2. Bangladesh requires three high-speed railroads and two linkages—one to Nepal (Saidpur to Biratnagar) and the other to India, (Jessore, a Bangladesh city in the east, to Kolkata). The three high-speed railroads will connect Thakurgaon in the northwest to the capital, Dhaka; another from Dhaka to the southeast port-city Chittagong; and the third will link Dhaka to Jessore.

3. Power: Bangladesh is in the process of setting up two 1000 MW nuclear power plants, one in Rooppur by the Padma River, and the other close to where the Jamuna River meets the Padma. These two reactors have been contracted to Russia's Rosatom, and the first 100 MW reactor is scheduled to become operational in 2018.

Bangladesh, with installed power capacity of about 8,500 MW, has a dire shortage of electrical power. It has little known reserves of oil and coal. Bangladesh's recoverable gas reserves are widely estimated as 16.36 trillion cubic feet. It is likely that more reserves will be found, but even if gas consumption goes up to 4-5 billion cubic feet per day, which is quite likely in the coming years, the known recoverable reserves may not last more than 8 to 10 years.

Bangladesh needs a large number of nuclear power plants. It had set up a research reactor decades ago and has a significant number of trained personnel. Nonetheless, the nuclear engineering and technology needs to be a given a big push in order to facilitate future nuclear power growth. What is needed immediately is to fulfill manufacturing requirements for the agro-industries. This will require education: setting up more engineering colleges, agricultural universities and overall medium-to-heavy engineering capabilities. Nuclear power plants, again in clusters, need to be set up not only to facilitate the two agro-industrial corridors, but also to provide electricity, healthcare, education, small and medium-scale industries, and bolster agricultural production.

Bangladesh, with its abundant water and highly arable soil, could feed not only its own burgeoning population, but also become a major source of rice for the world. However, it requires organizing its agricultural sector using high energy-flux density nuclear power to manufacture fertilizers, establishing high-yield-variety (HYV) seed research centers, promoting mechanization of agriculture to enhance productivity, and protecting harvested agro-products. Because of the climate, Bangladesh needs large-scale refrigeration capabilities.

4. In addition, Bangladesh needs to link up with the proposed

Kunming-Kolkata highway (linking China to India) in the north, as a feeder to the proposed major Eurasian economic corridors.

The Kunming initiative, which was later called the Bangladesh-China-India-Myanmar Forum (BCIM Forum), is to link Kunming in China's Yunnan province with Kolkata in India. There also is a plan to extend this road to Moreh-Tamu (India) in the India-Myanmar northeastern border state of Manipur, and eastward to the Myanmar town of Kalewa, across the Chindwin River in Myanmar. It would be a north-south road linking China, Myanmar, Bangladesh, and India— running a distance of about 2100 miles starting in Kolkata, and weaving through the Phangsu Pass to Ledo in Assam and Mytkina in Myanmar, until it reaches Kunming in Yunnan, China. The Ledo to Mytkina stretch is very difficult due to the mountainous terrain.

Nepal

One of the most troubled nations in South Asia, Nepal, years ago, was a rice-surplus nation. Over the years, lack of development, population growth, and political uncertainties have turned Nepal into a perennial rice-importing nation. In order to reverse the situation and stabilize the country, which is located between China and India, two major projects are necessary. These projects will connect the more populated eastern part with western Nepal, where more than 3 million of Nepal's approximately 29 million people, regularly suffer from starvation. To make Nepal a part of Asia-Pacific development plan, the following is required:

1. A high-speed railroad that will connect Biratnagar in the east to Kathmandu and another that will link Kathmandu to Dhangadhi via Nepalganj.

Nepal is a land-locked nation with a scanty rail transportation system. Two lines, the Raxaul-Amlekhaganj (NGR) and Jayanagar-Janakpur-Bijalpura (NJJR), two short distance 762 mm (2 ft 6 in) narrow gauge railways, were built in 1927 by the Ranas who ruled Nepal then. The Amlekhaganj line closed in the 1960s. About 6km of it is used for a broad gauge (5ft 6in) line from Raxaul (India) to a Dry Port at Birganj.

Nepal also needs a North-South Railway line to connect the railway system with Bangladesh (near Saidpur) and two links to India: one connecting Nepalganj (Nepal) to Lucknow (India), and the other connecting Kathmandu and Birganj (both Nepali cities) to Kolkata via Patna (both of the latter cities being in India).

2. Water needs to be harnessed from the Karnali, West Rapti, and southern river basin in the west; and Kamala, Bagmati, and southern river basin in the east. This would call for an intricate canal system to feed the arable landmass.

3. Two agro-industrial complexes, both in southern Nepal, running east-west and about 25-50 miles wide.

4. Power: Nepal has hydropower potential close to 45,000 MW, of which less than 900 MW has been exploited. Opponents claim that such power development will only facilitate India buying and using power generated in Nepal. In addition to developing some of its hydropower capabilities, Nepal must immediately set about to develop its nuclear power program, which would enable it to set up 4,000 MW clusters to prosper the agro-industrial complexes, open up the door for educating manpower for participation in small and medium-sized industries, and provide its present population with health care.

This use of high energy-flux density power would also entail educating and training individuals in various aspects of agro-industry, such as manufacture of fertilizers, use of equipment necessary to enhance agricultural productivity, and establishment of HYV seed production in order to bring about a Green Revolution in Nepal.

Pakistan

Pakistan was once a less economically-unstable nation than it is now. It has now reached crisis conditions that require immediate action. The following are some of the projects needed.

1. Power Plants: Pakistan needs at least an additional 30,000 MW of power within the next 10 years. At present, it is being starved for electricity, with only 14,000 MW of installed power for a population of 185 million. Over a longer period of time, say 30 years, which is our time-span in evaluating what the South Asian countries require, it needs a lot more. That amount of power can be generated only through setting up nuclear power plants. The location of these power plants should be selected in order to meet the requirements of agro-industrial and transport corridors, as well as population centers, to enhance its small and medium-scale industries, education, and healthcare.

2. Setting up large-scale nuclear waste-heat based or cogeneration plants to desalinate seawater along the coast in the provinces of Sindh and Balochistan, and brackish water in the provinces of Punjab and Khyber Pakhtunkhwa.

3. Railroads: While a com-

plete overhaul of the railroads is necessary, in addition, Pakistan must have two east-west railroads, connecting it to India in the east, and to Iran in the west. Meanwhile, China has proposed a north-south railroad. A consortium of two leading Chinese construction companies, CREC and Sinotec, has offered to construct the strategic Gwadar-Khunjerab rail link, which would run from Xinjiang Province down to the Pakistani port of Gwadar. The Xinjiang Regional Development and Reform Commission has described this line as the Southern Route of the New Silk Road Economic Belt.

The Southern Route of the Silk Road Economic Belt, when completed, would accrue benefit for both China and Pakistan. The rail corridor would stretch from Kashgar in China›s Xinjiang province to Pakistan's port of Gwadar, which is situated on the Arabian Sea, a stone›s throw from the Strait of Hormuz, and next door to Iran. The rail corridor will run through the entirety of Pakistan, from its northern-most point to the southern-most point, traversing the country›s heartland. The length of this rail corridor will be close to 1,800 km or 1125 miles. The building of pipelines and highways hugging this rail corridor will further enhance its economic importance. The project, however, could be an engineering challenge. The rail line will have to run through the high Pamir Plateau and Karakoram Mountains, and besides the construction and maintenance challenge that it poses, the cost of construction is expected to be very high. But it is also full of economic promise.

4. Economic Corridors: While the railroads will allow Pakistan to interlink with China and Central Asia in the north, and with Afghanistan, Iran, southwest Asia, India, and southeast Asia in the east, these links will benefit others at least as much as they will Pakistan. In order to optimize benefits from the transport corridors, Pakistan needs to develop at least two economic corridors along the railways—each about 150-200 miles long and about 50 miles wide. These economic corridors will connect the provinces of Punjab, Khyber Pakhtunkhwa, and some of the northern districts in one, and Sindh and Balochistan in the other. These economic corridors will be developed with the transport corridor and power plants as their backbones.

FOR FURTHER READING

"Pakistan's Neglected Infrastructure, a Barrier to Rapid Development," *EIR*, April 27, 1990.
"Pakistan Must Update Its Water Policy: It's Time to End Decades of Complacency," *EIR*, March 19, 1999.

Central Asia: Ending Geopolitics

by Ramtanu Maitra

October 2014

Central Asia, the geographic midway of the Old and New Silk Roads, has also been dead center in the Arc of Crisis—the target zone for a geo-strategy of strife in Eurasia, conducted by the decayed British Empire. The very name, "Arc of Crisis," was coined in the 1970s by Zbigniew Brzezinski, taking from the concepts of British intelligence operative Bernard Lewis, to refer to the region stretching from Egypt, across Central Asia, to the Indian Subcontinent, for warfare power-plays in a renewal of the Great Game of 19th-Century British Imperialism. For this and related reasons, Central Asia today ranks at the top, with Southwest Asia and Africa, of world priorities for deliberate, collaborative intervention to end the destruction, rebuild, and foster peace through development.

The physical geography itself is a challenge, including the world's highest mountain ranges, advancing deserts, and recession of the Aral Sea. But the biggest challenge is the legacy of deadly geopolitics.

If geopolitics can be defeated in Central Asia, it can be defeated anywhere.

We are now at a breakpoint. By the beginning of 2015, the pullout of U.S. troops from Afghanistan is to occur, after a 12-years-long NATO operation, whose character was textbook "Great Game" devastation. The nation of 31 million people is in shambles. Because of the modern continuation of the British East India Company's 19th-Century opium policy, Afghanistan today accounts for 90% of the world's output of opium poppy, and much hashish. The area of poppy cultivation has risen in Afghanistan from 8,000 hectares (ha) (roughly 18,760 acres) in 2001 up to 209,000 ha (516,230 acres) in 2013.

Nor has the problem remained within Afghanistan. Transit routes for the drugs into Russia and Europe, and terrorist gangs that thrive on the drug and arms trade, have permeated Central Asia as a whole, exacerbating the tensions created by underdevelopment and poverty.

Russia's Federal Drug Control Service director Victor Ivanov has called for international cooperation on an all-out "alternative development" program in Afghanistan and Central Asia as a whole, to completely end the "planetary crisis" of drug production. A Russian

FIGURE 1
Central Asia—Political Boundaries, Topography

ian.macky.net

Bodies of water other than the Caspian Sea are colored dark blue on the map. In Kazakhstan and northwest Uzbekistan are the remains of the Aral Sea.

program to accomplish this through "crash industrialization" was prepared for the June 2014 Group of Eight meeting in Sochi, Russia (see Appendix to this Chapter). But the meeting was scuttled by London and the Obama Administration. Now the Eurasian BRICS nations—Russia, China, and India—are themselves taking the lead to back development in Central Asia.

The BRICS members' commitment to develop this region (**Figure 1**) is embodied in the venue chosen for the inaugural announcement of the Silk Road Economic Belt (SREB), which Chinese President Xi Jinping made Sept. 7, 2013 at Nazarbayev University in Astana, Kazakhstan. In the latest expression of the expanding BRICS involvement with Central Asia's future, Afghanistan's new President Ashraf Ghani Ahmadzai spent four days in Beijing, Oct. 28-31, 2014, where he committed Afghanistan to participate actively in the new SREB; and, in turn, Chinese leaders, including Premier Li Keqiang, announced concrete ways China will help rebuild Afghanistan and the region. "In the past 13 years, the Chinese government has rendered enormous help to Afghanistan, to facilitate its peace and reconstruction process. That assistance has focussed on projects promoting people's livelihoods, like education, medical services, and water resource exploration," and much more is to come, said the Chinese Ambassador to Afghanistan, Deng Xijun, in an Oct. 28 CCTV interview.

Support for Central Asia security through economic growth also comes from the Shanghai Cooperation Organization (SCO), a regional grouping active since 2001 (members: China, Kazakhstan, Kyrgyzstan, Russia, Tajikistan, and Uzbekistan). At the 14th SCO heads of state meeting, Sept. 12, 2014, Chinese President Xi called for full membership for the SCO observer nations: Afghanistan, India, Iran, Mongolia, and Pakistan. SCO also has official dialogue partners: Belarus, Sri Lanka, and Turkey. It is from this activist vantage point for development that key initiatives and impediments can be reviewed in the Central Asia region at large.

Economic Geography

The current regional population numbers 97.6 millions, with 47.9 millions in the four inner nations of Central Asia—Kyrgyzstan (5.6 million), Tajikistan (8.2 million), Turkmenistan (5.2 million), and Uzbekistan (28.9 million)—plus 17.9 million in Kazakhstan and 31.8 million in Afghanistan. Minus Afghanistan, these countries were formerly republics in the south central part of the Soviet Union; the Soviet administrative boundaries became national ones in the early 1990s, upon independence.

The physical landscape, depicted in **Figure 1**, includes the dramatic mountain ranges in the eastern areas, of the Alai Mountains, Tien Shan, and Pamirs, sloping down westward to the extensive desert plains of the inner Aral Sea Basin and the Caspian Sea. Some of the world's highest peaks are here, including Ismoil Somoni Peak, 7,595 meters (24,590 ft) above sea level, in Tajikistan.

The two major river systems, the Amu Darya and Syr Darya, arise in these mountainous regions, flowing westward to the Aral Sea. Their flow comes from the highlands rainfall, snow, and glacial melt; but the highest mountain ranges are so tall, they block potential precipitation from the monsoonal currents coming from the southeast. Water scarcity is present in many places, except Kyrgyzstan. The shortages have been aggravated by the prevalence of cotton monoculture. Central Asia's cotton industry dates from soaring prices during the U.S. Civil War (when King Cotton was kept off the market), but monoculture intensified in the Soviet period and beyond. The cultivation of cotton—a very thirsty crop—reduced the river flow reaching the Aral Sea so much, that its volume has declined 75% since 1975.

Wikimedia Commons/Arian Zwegers

The ship graveyard at Moynaq, Uzbekistan, formerly a seaport on the Aral Sea.

The mineral and fossil fuel wealth in the region is significant. The underlying sedimentary formations of the western areas have sizable oil and gas deposits. Coal reserves are present in Kazakhstan and elsewhere in the intermontane. Deposits of iron and other ores, and minerals, gold, and uranium are identified, and many are being mined.

Central Asia has been densely settled for more than 2,500 years, with population concentrations in the piedmont of the southeast, and along the river valleys. Agriculture and mining are major areas of economic activity, but in most of the nations, except Kazakhstan, the poverty level is extremely high. Now, the prospect of new corridors of development, to allow the creation of new, man-made natural resources of water, fertile land, and power sources, opens up a new era for the region.

Rail Corridors and Connectivity

The Central Asia region has a unique role, along with Xinjiang Province in China, as the "traffic hub" of the Eurasian Land-Bridge. The old Silk Road Asian stopovers here are now legendary names— e.g., Samarkand and Tashkent in Uzbekistan and Xi'an in China. But as the new Eurasia-spanning rail corridors come into being, they give rise to potentially new growth-point cities. The challenge is to build nation-serving webs of rail networks and new settlements, to foster overall regional development, not just "stopover" towns, serving through traffic and out-of-region trade.

The concept map in **Figure 2** presents this idea schematically. The black lines indicate the main Eurasian Land-Bridge lines. The Trans-Siberian Railway (TSR) runs across Russia, at the top of the map, through Chelyabinsk and Omsk. Running off the TSR, northwest to southeast, is the line going through Kazakhstan, via Aktobe, Saksaulsky,

FIGURE 2
Central Asia—Silk Road Rail Lines and Proposed Regional Development Corridors

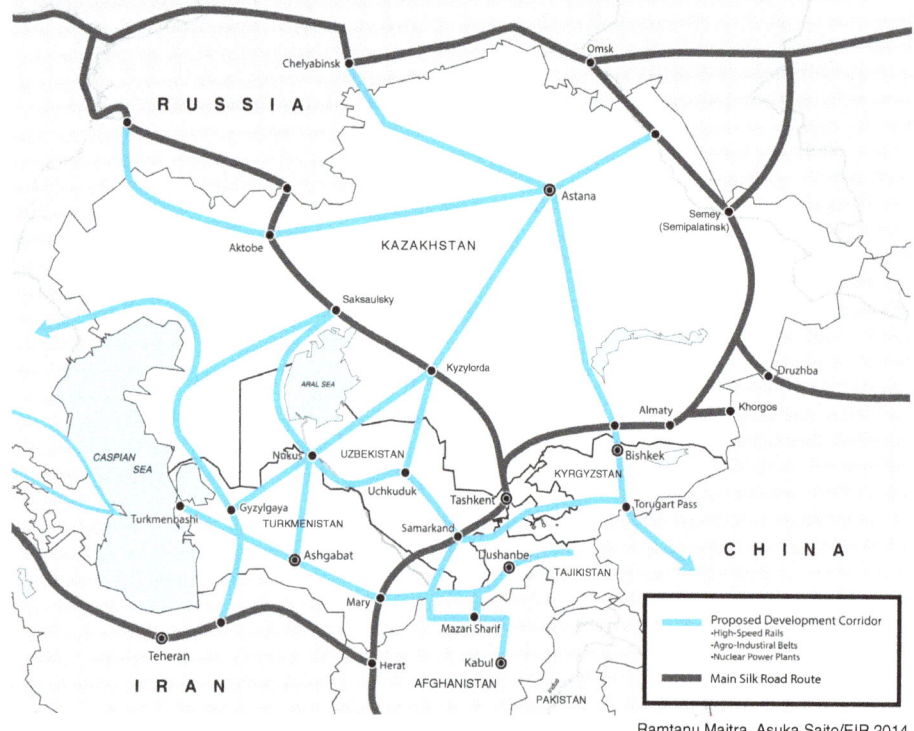

Ramtanu Maitra, Asuka Saito/EIR 2014

In addition to the Trans-Eurasia rail routes of the Silk Road crossing Central Asia, these nations require the development of their own regional rail network, for which key segments are shown here schematically. Some are under way, planned, or intended.

and Kyzylorda, into Tashkent, thence either southward through Mary in Turkmenistan, to Herat, Afghanistan, and onward to the Arabian Sea coast; or from Tashkent to Almaty, the former capital of Kazakhstan, and thence to China through Xinjiang.

From Omsk runs another trunk line off the TSR, through Semey (formerly Semipalatinsk) in far eastern Kazakhstan, thence to China, via the border crossing at Druzhba; or southward, via Almaty, through the core central Asian region and to South Asia.

The blue lines show several of the other proposed, planned, or in some cases, under-construction, rail routes that constitute the potential for development corridors for high-speed service, agro-industrial belts, and siting of nuclear power plants for plentiful electricity and desalinating and purifying water.

Several new rail projects stand out as transportation initiatives; what is required is collaboration for an integrated system in this heartland of Eurasia. Even the track gauges pose a problem, but it can be dealt with. Because Central Asia's railroads were built when these nations were part of the Russian Empire and then the Soviet Union, they have the broad gauge of 1,520 mm, while most neighboring countries use the standard gauge of 1,435 mm. In Pakistan, the width is the wide Indian gauge of 1,676 mm. As a result, trains going, for example, between Iran and Turkmenistan, or China and Kazakhstan, or into Pakistan, must change bogies (wheel trucks), or the passengers and goods must transfer to new rail cars. Among the notable new or anticipated Central Asia rail links:

North-South

• Turkmenistan and Kazakhstan have launched a direct railway linking their oil-and-gas-rich Caspian Sea regions, bypassing the former circuitous route through Uzbekistan, and going directly on to Iran. This forms the first phase, scheduled for completion in Winter 2014, of a north-south 1,520 km (945 mi) rail corridor linking Kazakhstan, Turkmenistan, and Iran, which is the spine of the new International North-South Transport Corridor from India to Russia, by sea and rail. Within Iran, construction is under way on key links from the Turkmenistan border to the existing Iranian rail grid entry point at Gorgon, and thence to the coast.

• A short, but critical new rail track into Afghanistan from Uzbekistan, was completed in 2011 by UTY, the state-owned Uzbekistan Railways. The 75 km (47 mi) line connects Hairatan on the Uzbekistan-Afghanistan border to Gur-e Mar, outside of Mazar-i-Sharif in northern Afghanistan, that nation's second largest city. The new line has been important in transport of humanitarian aid as well as commercial merchandise. Formerly, these goods had to be offloaded onto trucks at the border.

East-West

• Kazakhstan, in late 2013, completed a 293 km (183 mi) stretch of rail from Zhetygen (just north of Almaty) to Khorgos at the Chinese border, looping it into its existing national railway, thus opening the second China-Europe link across its territory. As a result, it takes just 15 days for trains carrying all kinds of cargo, to cover the 10,800 km (6,750 mi) route from Chongqing in southern China to Duisburg in Germany's industrial Ruhr region. Khorgos, straddling the Kazakhstan/China border, is now a rapidly expanding land port.

• China is planning a rail link west from the far western Xinjiang city of Kashgar, through the mountains into Kyrgyzstan (Irkeshtam Pass into the Alia Valley), through Tajikistan (Rasht Valley), into Afghanistan (across to its western city Herat), thence to Iran, and westward. This would create yet another band of the Silk Road Economic Belt.

• Tajikistan, Afghanistan, and Turkmenistan agreed in March 2013 to build a 160-km (99.4 mi) rail section across far northern Afghanistan, which will be Tajikistan's first line to several major Eurasian rail routes. (**Figure 6**, Appendix to this Chapter.)

National

• Uzbekistan is expanding its internal rail routes and service, in order to forego using parts of the old Soviet-era lines through neighboring countries. New Uzbek rail projects link cities in its far west directly to the capital, Tashkent, in the east. Previously, rail service was possible only via Turkmenistan. Tashkent is also seeking to link its densely populated part of the Fergana Valley, with the rest of the country, via the Kamchik Pass, thus avoiding transit through Tajikistan.

• Tajikistan and Kyrgyzstan have very few internal rail lines and little service. Tajikistan has only 680 km (420 mi) of track, all of it 1,520 mm broad gauge. The system connects the main urban centers of western Tajikistan with points in neighboring Uzbekistan and Turkmenistan, and soon, Afghanistan. Kyrgyzstan is virtually "rail free." The small bits of rail lines within the nation add up to only about 370 km (230 mi) of 1,520 mm broad gauge track.

The history of this limitation is that, during the Soviet Union period, Kyrgyzstan's Chuy Valley in the north and the Fergana Valley in the south, were end-points of the rail system in Central Asia. At independence, rail lines that had been built without regard for administrative boundaries, were suddenly in different countries. The dysfunctional railways of these countries are a priority to remedy; one set of ideas from Russia, for addressing this problem, is documented in the Appendix to this Chapter.

Nuclear Power

A top priority for the region is the provision of plentiful power through nuclear fission, including for large-scale desalination along the Caspian Sea littoral, and at other sites, for recycling and upgrading wastewater, as well as for electrified rail.

None of the four core Central Asian nations operates, or is in the process of building, a nuclear power plant. To their north, Kazakhstan has an active intention to do so. In the past, Kazakhstan had a long-running nuclear facility, including producing desalinated seawater. In 1973, the Soviet Union opened an experimental fast breeder reactor on the Caspian Sea near Aktau, which ran until 1999, when it was decommissioned and taken down. On May 29, 2014, the day on which the Eurasian Economic Union treaty was concluded in Astana, the country's Kazatomprom nuclear corporation signed a memorandum of understanding with Russia's Rosatom, to build a 300 to 1,200 MW nuclear power plant with the water-cooled, water-moderated Russian VVER reactor model, near Kurchatov in far eastern Kazakhstan. A town with a long nuclear history, named after the Russian physicist Igor Kurchatov, this was a Soviet "closed city," housing research facilities for the nuclear weapons test site at nearby Semipalatinsk, modern Semey.

Uzbekistan possesses two operational nuclear research reactors. One is outside Tashkent, at the Institute of Nuclear Physics in Ulugbek. In addition to the reactor (a 10 MW VVR-SM), the institute has two cyclotrons, a gamma source facility, a neutron generator, and a radiochemical complex. The other research reactor, a 20 KWt (static) pulse reactor, is operated by owner JSV Foton.

Within the Soviet Union, Uzbekistan provided much of the country's uranium. The Navoi Mining and Metallurgy Combine operates six in-situ leaching mines, with nine additional mines under development, and five other commercially viable deposits identified. The processed yellowcake is shipped to various countries, including the United States and South Korea. All these installations and activities are useful precursors to building out a full-scale nuclear power platform in the region.

The current profile of energy supply throughout the region, is that certain areas rely on their fossil fuel for electricity, while several areas rely on hydro-power, a legacy of Soviet dam-building for both the energy grid and agricultural water management.

Kazakhstan is the leading oil producer of the region, with output of roughly 1.6 million barrels per day (bbl/day), of which approximately 90% is exported. The first pipeline connecting the Caspian Sea shore oil patch with Xinjiang, China's westernmost province, is one of the longest in the world, at nearly 2,300 km (1,429 mi).

Turkmenistan possesses the largest known natural gas deposits in central Asia, and among the largest anywhere in the world. It is the main gas exporter in the region. Turkmenistan itself gets almost all its electrical power generation of 4 GW from natural gas. Adjacent Uzbekistan currently produces even more gas (60 billion cubic meters annually) than Turkmenistan, but uses 85% of its output for domestic electricity production, leaving little for export. In fact, Uzbekistan serves as a transit country for gas conveyed from Turkmenistan, on its way to China and Russia. In 2007-2010, the first two lines of the Turkmenistan-China (or Central Asia-China) natural gas pipeline, largely financed by China, were built from the Bagtyyarlyk gas-field area in southeastern Turkmenistan, through Uzbekistan and Kazakhstan, crossing into China at Khorgos and connecting to China's second West-East Gas Pipeline. A third pipe on the nearly 2,000 km (1,243 mi) export pipeline started operating in June 2014, and a fourth is planned.

Tajikistan and Kyrgyzstan rely on hydro-power, because they have no substantial developed hydrocarbon deposits, although it is believed that Tajikistan's Bokhtar field has sizable oil and gas reserves. A regional plan exists on paper to construct the "Central Asia-South Asia" (CASA) electrical transmission line, the CASA-1000, to run Summer-surplus electricity from the dams in Tajikistan and Kyrgyzstan, south to Pakistan and Afghanistan, a distance of 1,173 km (759 miles). There are objections and threats of violence against it along the way, but the technology is realistic.

The Power-Water Nexus

Central Asia is an arid region that grapples with serious constraints related to the water resources necessary for advanced agro-industrial development. However, action on multiple fronts, in a growing economy, can provide for "new" water resources, despite the stark example of the eco-disaster in the Aral Sea Basin. This region is a planetary priority for the earliest application of yet-to-come breakthroughs in knowledge and methods of "rain-making" (see Part 2 of this report, Water).

Most of the region's water comes from the mountain ranges of the upstream nations of Kyrgyzstan and Tajikistan (and to a lesser extent from Afghanistan) channeled to the downstream nations of Kazakhstan, Turkmenistan, and Uzbekistan, mainly through the Amu Darya and Syr Darya rivers, and some lesser rivers. In short, roughly two-thirds of the water resources is generated in the mountains, and two-thirds of that is consumed downstream.

Concern over future patterns of water supply, comes from various weather shifts. For example, the glaciers are shrinking. Between 1957 and 1980, the Central Asian glaciers diminished by about 19%, according to most estimates. The glaciers surrounding Lake Issyk-Kul in Kyrgyzstan shrank by about 8% over this time period.

Technically, however, much of Central Asia is not among the world's most water-short regions. Uzbekistan, for instance, has almost double the amount of water available per capita as Spain, one of the major agricultural producers in Europe. Thus part of the interim solution to the problem of water supply lies on the demand side, if the most effective uses of water are introduced, and the water-consuming production systems modernized (see Part 6 of this report).

Modern water-storage methods are being implemented in Turkmenistan and can be expanded widely. In 2013, Turkmenistan began building Turkmenkol, an artificial lake at a natural depression in Garashor. The lake will collect drainage water and sewage from the provinces, for purification and reuse. Two more such reservoirs are planned for construction in 2014.

One aspect of improving infrastructure to reduce water losses, is to rehabilitate the sections of Soviet-era irrigation canal systems that are in disrepair. The big contributing factor to the problem is that national boundaries now cut through what were formerly Central Asia-wide systems on the mid-level Amu Darya and Syr Darya rivers.

The challenge in the coming decades will be to accommodate large-scale industrial development and natural growth of population, which would put a great deal of stress on regional water availability. Because Central Asia is land-locked, desalination of sea water is not an option. The western parts of Turkmenistan and Kazakhstan, however, will be able to generate significant amounts of potable water by nuclear desalination on the Caspian Sea. In the short term, this does not relieve the environmental aridity from the drying up of the Aral Sea. The exposed beach on the receding shoreline, for example, has resulted in toxic sand storms, which must be dealt with. But nuclear power is essential for water for the future.

Another legacy of the Soviet era in Central Asia was the creation of an electricity-water nexus, whereby the generation of electricity from hydro-power dams in upstream countries was linked to supplying the power and water needs of those downstream. This system operated relatively smoothly under a common management system, and shared energy arrangements through regional electricity grids. This mode came to an abrupt end with the collapse of the Soviet Union, and overnight emergence of international borders. Now a stance of opposition is in effect between the "upstream" dam operators—Kyrgyzstan and Tajikistan—and the water- and electricity-dependent downstream nations, Uzbekistan and Kazakhstan. Although the 1992 Almaty Agreement reduced water use by "upstream" countries, and Kazakhstan is currently promoting formation of a Central Asia regional water committee, what is needed is more water.

The advent of nuclear power throughout the region can remove the apparent necessity for competition and no-win trade-offs for power and water. It can also open the way to other beneficial uses of natural gas, instead of electricity-generation, thus lifting up the entire economic platform for each nation in the region.

The story of one large dam project underscores the principle involved. Since 1960, the Rogun Dam, on the Vakhsh River, a major tributary of the Amu Darya in Tajikistan, has been proposed, designed, and partially built, but never completed. Downstream nations now

Central Asia: Ending Geopolitics

The Nurek Dam reservoir on the Vakhsh River, Tajikistan.

oppose ever going ahead with this dam project, for fear of losing river flow, while Tajikistan is trying to finally get it under way. As originally envisioned, Rogun (at 335 meters, the highest dam in the world) was to have been one of a trio of dams contributing to water-for-power swaps with the Uzbek and Turkmen Republics. They would provide power for the Tajik Republic and these downstream neighbors, and water for the neighbors. When power-generation was suspended, seasonally, Uzbekistan and Turkmenistan would compensate Tajikistan with power generated at fossil fuel-burning plants.[1] One of the other two dams in the plan, the Nurek Dam, was built in the Soviet period, while a second, Sangtuda-1, went operational in 2008.

Agro-Industrial Development

There is great potential for expansion of agriculture and industrial activity in the pre-existing zones of light and heavy manufacturing concentrations across Central Asia, as well as in new urban centers and potential croplands.

The region is rich in industrial raw materials, as well as oil and gas, all of which are disparately distributed. In the piedmont and intermontane basins in the east, are deposits of iron ore and coal, copper, lead, zinc, antimony, gold, and others. The large natural gas fields are in the dry western lands.

The task ahead is to foster expansion of the industrial base. Heavy industry in recent times has been concentrated in northeastern Kazakhstan, also in the foothills of the Tien Shan range near Tashkent, and in places in Kyrgyzstan and elsewhere. This is connected to steel-making, processing minerals, ore-smelting and refining, and certain kinds of manufacturing—for example, agriculture implements for high-slope field work, and other specialties—as well as food processing and other light industry.

There is a vast agricultural land area of 306 million ha (756 million acres) currently in use in the four core Central Asian nations and Kazakhstan combined, and still more potential area in the future, with plentiful power and water. Kazakhstan, according to its Ministry of Agriculture, has 222 million ha (549 million acres) of farmland, a majority of which (189 million ha, or 85%) is being used as pasture land, and 24 million ha (10%) as cultivated land. Nearly two-thirds of the latter is devoted to growing cereals and one-third to fodder crops. Kazakhstan is an exporter of winter wheat.

A priority in the existing irrigation regions, is to upgrade the physical systems, not only those used for the conveyance of water, but also water-saving methods of precision irrigation and soil-drainage. Switching from cotton to less water-consuming crops is also important, and can provide an increase in fruit and vegetable output for better nutrition regionally and for export. Expanding meat production

---

1. Eli Keene, "Solving Tajikistan's Energy Crisis," Carnegie Endowment for International Peace, March 25, 2013, provides a detailed history of the planned arrangement.

to improve the diet is another priority, which includes not only cattle, goats, and sheep, but also poultry, which can increase production the quickest.

The Fergana Valley Challenge

The Fergana Valley is one of the world's leading agriculture centers, home to almost 25% of the population of Central Asia, in a land area only 5% of the region's total. The Valley is a major food supplier to all Central Asia, including rice, wheat, fruit, and vegetables, as well as cotton. The Fergana Valley is strife-ridden, however, because of water and land disputes, and for political reasons. The only solution lies in the overall Land-Bridge transformation of Eurasia.

The triangular-shaped valley is defined by the Tien Shan Mountains to the north, and the Gissar-Alai range to the south. It is a flat plain of 22,000 km^2 (8,500 square miles), with fertile soils, and water resources, due to being at the intersection of the Naryn and Kara Darya rivers, which, entering the Valley from the east, then join up near the town of Namanagan, to form the Syr Darya River.

The population density in the Fergana Valley is more than 250 people/km^2, in contrast to the average of 14 persons/km^2 in Central Asia at large.

The challenge arises from the fact that Tajikistan, Kyrgyzstan, and Uzbekistan share convoluted borders, a relic of Soviet so-called "nationalities policy" in the 1920s. Moreover, a significant portion of the national population of each country resides in the Valley: 30% of Tajikistan; 50% of Kyrgyzstan, and 27% of Uzbekistan. There have been endless cross-border conflicts, mostly related to access to and management of land, water, and other natural resources, and of physical assets (for example, canals, gates, and pumps) put in place during Soviet times. Borders cut across local roads, orchards, fields, sluices, and even private home properties.

Violence periodically erupts among the Kyrgyz, Tajiks, and Uzbeks from territorial disputes, especially in densely populated areas. As the newly independent states began to privatize state farms in the 1990s—and this program was carried out throughout the region—these private operations became the only reliable sources of family income, or, at least, subsistence farming. Agriculture in the Valley shifted from cattle breeding to subsistence farming. As a result, individual demand for irrigation water on small fields has increased sharply, along with conflict.

At the same time, there is less water available, because after the end of the Soviet Union, there have been no entities to fund maintenance of the distribution systems—mostly open concrete canals—which have fallen into disrepair. Parts of them have been broken up and sold off for scrap.

All these conditions can be superseded, and untold new productivity take place in agriculture and industry, under the impetus of the Silk Road Economic Belt, criss-crossing this region with corridors for progress.

A Future for Afghanistan

The importance of a stable Afghanistan is not to be underestimated. Its stability is not only crucial for its 30 million people, but, as one good look at the location of Afghanistan makes evident, unless Afghanistan is stabilized, Central Asia, parts of South Asia, and even western China cannot be secured. All of the core Central Asia countries have been affected by Afghanistan's wars of recent decades and its transformation into a huge narcotics producer—from their relationships with the heavily Tajik and Uzbek ethnic Northern Alliance movement within Afghanistan since 1978, to the extensive activity of drug-running networks through the Central Asian mountains.

The damage inflicted on Afghanistan by almost continuous warfare over the past 35 years has been devastating. Two million Afghans have died in those conflicts.

A reconstruction plan must start by looking at the state of Afghanistan's infrastructure, including the fundamentals such as water, power, food security, transportation, and skilled manpower—and the lack of them. It is fair to say that the priority must be adoption of a comprehensive development plan, which would include agricultural, industrial, infrastructure, and energy projects. The requirements include those listed below.

Agriculture. Afghanistan's agricultural sector needs development of bulk transportation capability, preferably a railroad network. The rough Afghan terrain means that the initial transportation network will have to be based on roads. However, in the southern part of the country, in the fertile lands of Dasht-e-Khas, Dasht-e-Margow, and the Rigestan plains abutting Iran, an extensive railroad network can be developed to facilitate interaction between agricultural lands and urban centers. The agricultural sector will also require agro-machinery, such as tractors, harvesters, and cultivating machines. The manufacture and maintenance of such machinery will introduce industries that will help train skilled workers and technicians.

In 2010, the Afghan government announced a 25-year plan to develop its railroads. The study on railway development for Afghanistan was completed for the following routes: (1) From Hairatan at the border with Uzbekistan to Herat in the west, via Mazar-e-Sharif; (2) from Sher Khan Bandar at the border with Tajikistan, via Kunduz to Naibabad (which is on the line under construction from Hairatan), joining Mazar-e-Sharif to Herat; (3) from Torkham at the border with Pakistan to Jalalabad; and (4) from Spin Boldak at the border with Pakistan to Kandahar.

The first Afghanistan or Central Asia program issued by the Russian Institute for Demography, Migration, and Regional Development, in 2008, concentrated on a plan for reviving agriculture in Afghanistan, to address a situation where, at that time, more than half the GDP was based on cultivation, production, and illegal trafficking of drugs. The program pointed to the fact that the southern provinces of Afghanistan are famous for their fruit, and proposed creation of a nationwide network of agricultural educational institutions, and delivery of agricultural machinery and the prerequisites for a processing industry. The researchers proposed Nangarhar and Helmand as model provinces for establishing agro-industries in sugar, olive, citrus, sunflower, pomegranate, and vegetable production (see *EIR*, Feb. 27, 2009).

Minerals. Afghanistan sits on a treasure trove of mineral resources, but their exploitation begs the question of developing the necessary infrastructure.

The Hajigak iron ore project, located 180 km from Kabul, has been awarded to a consortium of Indian companies, but has been delayed beyond the six months stipulated time because of some conditions that Kabul has set which the consortium has not yet agreed to meet. When Chinese investors won a bid in 2006 to mine copper at Mes Aynak, 50 km south of Kabul and the site of one of the world's biggest copper deposits, they had pledged to lay a stretch of rail, according to Afghanistan's Finance Minister Omar Zakhilwal. The copper mine project, under a 30-year lease to the China Metallurgical Group Corporation (MCC) and Jiangxi Copper Company, is hanging fire because of security concerns. The $3 billion contract includes a railroad to carry coal to the mine area, a smelter, and a 400 MW power plant. It could provide Kabul with as much as $500 million in royalties. However, MCC now wants to renege on building the railway, power plant, and processing factory, as stipulated in its deal to mine at Mes Aynak.

In addition to copper and iron ore, according to Paul A. Brinkley, U.S. Deputy Undersecretary of Defense and director of Task Force Business and Stability Operations, Afghanistan has significant deposits of niobium, cobalt, gold, molybdenum, silver, and aluminum, as well as sources of fluorspar, beryllium, lithium, and other resources. What, however, should be key in developing Afghanistan's mining industry is to build the basic infrastructure—power; bulk transportation; water for industrial, commercial, and domestic use; and communications—to make it a success. Because these mineral reserves are dispersed, that also requires setting up institutions to train people throughout the country, cutting across ethnic backgrounds.

Electricity. Afghanistan is almost without electric power. Currently, the country produces about 500 MW of electricity—less than some Caribbean islands. It imports another 500 MW from neighboring countries. Afghanistan ranks among the countries with the lowest electricity production per capita in the world. Despite billions of dollars in projects over the past decade, at best one-third of the population has access to regular power.

Like the rest of the world, Afghanistan has no alternative but to develop nuclear fission power to stabilize the country, exploit its mineral wealth, set up agro-industrial corridors, and educate and provide people with water, food, education, healthcare, and a future. Nuclear power plants in clusters will be necessary to provide the power necessary to meet those demands.

APPENDIX

The Industrial Development of Afghanistan and Central Asia: A Russian Vision

Reports published by the Institute for Demography, Migration, and Regional Development (IDMRD), a Moscow non-governmental organization (NGO), represent what Russia was bringing to the table at the June 2014 Group of Eight (G8) summit in Sochi, Russia. President Vladimir Putin's longtime colleague, Federal Drug Control Service (FDCS) head Victor Ivanov, had announced a campaign to eliminate the "planetary narcotics production center" in Afghanistan, as a focus of Russia's G8 chairmanship.[1] But the summit was cancelled, when the G8 expelled Russia over the Ukraine crisis.

The first and third of the reports excerpted here were published in Russian, and the second one in Russian and English. Italicized notes have been supplied by EIR. Graphics are those of IDMRD unless otherwise noted.

How the South of Western Siberia Will Become an Economic Center of the Planet

This 2012 report was mentioned in Part 4, in connection with Russian concepts of Eurasian development corridors and new technologies for freight transport in the far North. It proposes that the industrial and science cities of Siberia be revived to power the development of Central Asia and Afghanistan. The report is relevant for all once-industrialized countries, such as the United States, European nations, and Australia, which are threatened with becoming unproductive, post-industrial wastelands. The excerpts were translated by EIR.

The New Central East

Western Siberia has a special opportunity, linked with a promising macroregion now taking shape: the New Central East or Central Eurasia, including the classic [in Russian terminology] Middle East (Iran, Afghanistan, Iraq, and Pakistan), Central Asia, and western Siberia itself (**Figure 1**).

If Russia pursues the right strategy, this macroregion will become a new market, with nearly 400 million inhabitants, by 2025. Siberia's unique role can be to organize a planetary center of third-stage indus-

---

1. Rachel Douglas, "After Ukraine's EU Refusal: Eurasian Development vs. Collapse and Chaos," *EIR*, Dec. 6, 2013.

FIGURE 1
The New Central East, from the Science Cities of Siberia to the Persian Gulf

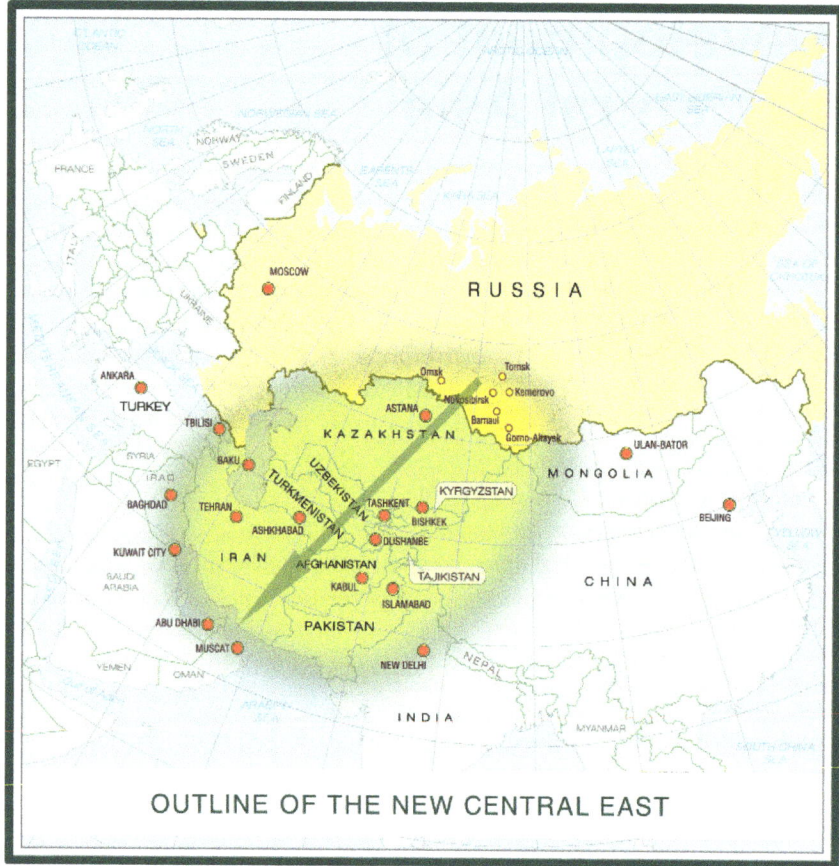

OUTLINE OF THE NEW CENTRAL EAST

trialization, which would not only pioneer this type of industrialization within Russia, but also serve as an organizing capability for the primary industrialization of Afghanistan and the second-stage industrialization of former Soviet Central Asia, as well as Iran and Pakistan, which will all become priority markets for Russian capital goods and advanced technology exports.

Russia's Caspian Sea port of Astrakhan will play an important role in organizing the New Central East.

Third-Generation Industrialization vs. 'Assembly' Industrialization

A new industrialization is an imperative not only for Russia, but for all mankind, both the least developed countries of the "third" and "fourth" worlds, and the leading world economic power, the USA, whose government debt is $17 trillion greater than its GDP.

Extremely dangerous for Russia, however, is the current tendency at all levels of government, to reduce our own new industrialization to a semi-colonial model of "assembly" or semi-knock-down industrialization, under which strategic planning and advanced technology development take place outside of Russia, while our country is flooded with parts and trimmings of industrial machinery from elsewhere, to be assembled. The mission of the southern part of western Siberia should be to prevent this substitution, while creating in Russia a leading planetary center of industrialization and production of public wealth.

The formation of a planetary center of vanguard third-generation industrialization will make it possible to organize a cascade of industrializations: from third-stage down to primary (**Figure 2**). The basis of the tertiary industrialization will be robotization of production on a large scale, advanced machine-building, and third-generation infrastructure, especially transport and multimodal systems.

Creating a planetary center of new industrialization in this region is the only feasible way to destroy the planetary center of narcotics production in Afghanistan, which kills 100,000 people annually, at least 50,000 of them young people in Russia.

FIGURE 2
A "Waterfall" of Industrialization

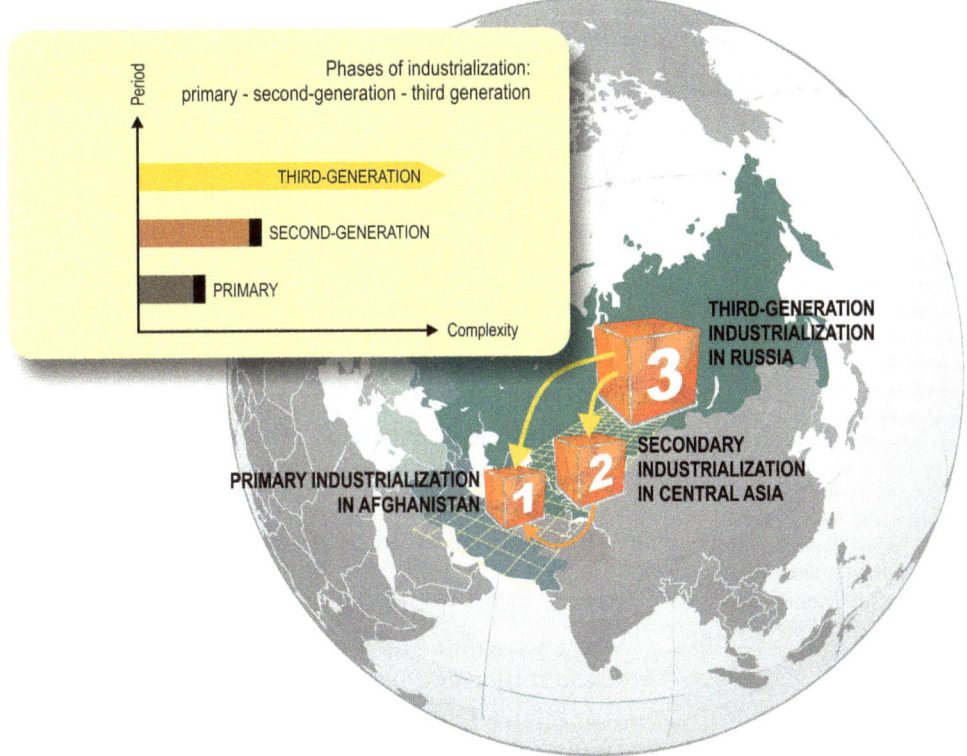

With modernization of their productive capacities, the science and industry cities of western and central Siberia could export capital goods for the second-stage industrialization of the Central Asian countries and the primary industrialization of Afghanistan.

A New Generation of Alternative Development Programs for the Elimination of Drug Production in Afghanistan

This report was prepared for a Spring 2014 G8 pre-meeting, by IDMRD and the Belarus-based Center for Strategic and Foreign Policy Studies. Yuri Krupnov, Supervisory Board chairman of IDMRD, presented it at that March 25 Moscow session, which Russian FDCS head Victor Ivanov chaired.[2] The full report is available in English at www.idmrr.ru.

Alternative Development Is Development

In the concept of "alternative development," the core concept is "development." "Alternative development" is about organizing accelerated development that can effectively and sustainably replace the drug industry and disrupt its social basis.

In international law, "alternative development" is directed at the implementation, in an unfavorable environment shaped by drug production in a country or region, of the fundamental human right to development. This right, as stipulated in the Declaration on Social Progress and Development (1969) and Declaration on the Right to Development (1986), is an inalienable human right.

---

2. Douglas, "U.S. Sanctions Don't Stop Russian Anti-Drug Proposal," *EIR*, April 4, 2014.

This Moscow meeting on Alternative Development for Drug-Producing Regions, held March 25 in preparation for the subsequently cancelled 2014 Group of Eight summit, was attended by over 100 experts from 27 countries. Chairing was Russian Federal Drug Control Service Director Victor Ivanov (second from right at table). Yuri Krupnov (third from right at table), chairman of the Supervisory Council of the Institute for Demography, Migration and Regional Development, presented the institute's report on alternative development in Afghanistan.

International aid for alternative development should be a special instance of a mutual development policy. The concept of mutual development (co-development), elaborated by IDMRD, proposes that international cooperation will be most effective, when aimed at creating new social wealth and value, rather than merely redistributing existing assets. Such an approach turns the fight against drug production in Afghanistan into a win-win game for all participants—from the international community, to the farmers of Helmand, Nangarhar, Kunduz, and Badakhshan Provinces.

A Russian Plan for Crash Industrialization: A New Method of Alternative Development

Afghanistan lacks development in any form, of which alternative development could be a part.

The IMDRD offers a vision of alternative development in Afghanistan, based on crash industrialization and basic economic infrastructure development, under a Comprehensive International Plan for Alternative Development in Afghanistan. Crash industrialization should be considered a new method of alternative development, involving urban as well as rural areas. Though urban areas have never been high on the alternative development agenda, it is obvious that an economic boom in the cities, coupled with a comprehensive urban planning policy, is crucial for the success of alternative development in Afghanistan. Industrialization will inevitably be accompanied by urbanization, which needs to be balanced and economically sound.

Russia has unique, and generally positive, experience in organizing the industrialization and uplifting of the Afghan economy, dating from cooperation between the USSR and Afghanistan in the 1970s and 1980s.[3] Soviet investments helped build 142 major infrastructure and industrial assets, which became the basis of the national economy. This continued even during the war there against an internationally backed insurgency. If the international community were united in providing security, stability, and economic prosperity in Afghanistan, such efforts could be many times more effective today. They represent the only viable strategy for fighting drugs there: ensuring security through development.

A Comprehensive International Plan for Alternative Development, with a crash industrialization program at its core, would allow implementation of the policy of a transformation toward self-reliance, adopted at the [July 2012] Tokyo Conference on Afghanistan. Such a Plan could become a key instrument in concentrating international

---

3. *The authors refer inclusively to Soviet cooperation with the post-1973 Mohammed Daud Khan regime, not only the period after the Soviet invasion in late 1979.*

efforts and achieving substantial improvement of the situation in Afghanistan within a feasible time frame.

A Comprehensive International Plan for Alternative Development in Afghanistan

A Comprehensive International Plan for Alternative Development in Afghanistan, based on a crash industrialization program, will focus on four areas:

1. Basic infrastructure, both for the general welfare and for economic programs to create a large number of new and steady jobs.
2. Building new and expanding existing industries, to provide mass employment and raise the income level of a substantial part of the population.
3. Enhanced access to social and cultural infrastructure, especially education and health care, and development of a skilled work force for the new Afghanistan economy through professional and vocational education, including training Afghan youth abroad, as well as new educational institutions within the country.
4. A security policy based on national reconciliation and an uncompromising fight against drug production and trafficking, corruption, and extremism.

Implementation of the Plan, and related economic projects in neighboring countries, will help to create a common market with more than 300 million consumers, in Afghanistan, Pakistan, Iran, Tajikistan, Kyrgyzstan, Uzbekistan, and Turkmenistan.

The following vision of projects and programs constitutes a part of the Plan that could be implemented with organizational help from the Russian Federation.

Infrastructure for Development

A top priority for the Afghanistan economy is the development of electric power, to drive industrial development and drastically alter the quality of life for people in Afghanistan. A major strategic investment project in this field, capable of providing enough energy to carry out primary industrialization of the country, is the construction of a chain of hydroelectric power plants (HPP) on the Panj River, which forms the border between Afghanistan and Tajikistan. This project calls for building 12 dams, with a total of up to 17.5 GW of installed generating capacity. The first stage of the project would be construction of the Dashtijum HPP and Rushan HPP, providing up to 7 GW of installed generating capacity.[4]

---

4. *Most of Tajikistan's existing HPP are on the Vakhsh River, which arises in Kyrgyzstan, flows through north-central Tajikistan, and joins the Amu Darya River. The Panj River is another tributary of the Amu Darya. In view of the serious damage to the Amu Darya and the Aral Sea, particularly from cotton monoculture in Uzbekistan in the Soviet period, IDMRD programs call for cooperative water and energy programs and planning in the region. Since 2013, Kazakhstan has been promoting the formation of a Central Asia regional water committee, under the auspices of the Shanghai Cooperation Organization, to address*

FIGURE 3
Concept of an Indo-Siberian North-South Railway and Development Corridor

- Trans-Siberian Railway
- Kazakhstan section of Indo-Siberian Mainline (1520 mm)
- Proposed section of Indo-Siberian Mainline (1520 mm)
- Pakistani Section of Indo-Siberian Mainline (1676 mm)
- Dry port and 1520-1676 gauge change
- Port of Karachi

The Indo-Siberian Railway is a proposed north-south development corridor, running south from Omsk, Russia, on the Trans-Siberian Railway, through Kazakhstan, Kyrgyzstan, Tajikistan, Afghanistan, and Pakistan to the port of Karachi on the Arabian Sea. At an inland terminal ("dry port") in Peshawar, Pakistan, there will be a gauge change from the Russian broad-gauge rails (1520 mm) to the even wider "Indian gauge" (1676 mm) used in Pakistan. The route intersects or parallels planned Chinese Silk Road Economic Belt railroads in both Central Asia and Pakistan.

FIGURE 4
Railways for Afghanistan's Development: A Russian Proposal

Details of existing (black) and proposed (orange) railroads for the Central Asia – Afghanistan – Pakistan sections of the Indo-Siberian north-south corridor.

If this power is divided equally between Afghanistan and Tajikistan, even the first-stage capacities will cover all the energy needs claimed by the Ministry of Energy and Water of Afghanistan until 2020. This will make it possible to launch primary industrialization. At the same time, the creation of water reservoirs as an HPP by-product will provide water resources for the reconstruction and expansion of irrigation infrastructure in Northern Afghanistan. The required investment for the first stage of the project is an estimated $7 billion.

A major consumer of electric power generated by this chain of HPP could be an electrified railway section, connecting the republics of former Soviet Central Asia to Pakistan, across Afghanistan (**Figures 3 and 4**). This section of a proposed new "Indo-Siberian" railroad would pass

water needs and resolve conflicts.

The Dashtijum HPP is in the advanced design stage, with organizations in both Russia and India expressing interest in the project. A 2011 summary, "Tajikistan's Hydro Power Potential," including some of the projects mentioned here, is available on the UNECE website www.unece.org.

FIGURE 5
Afghanistan: Resources and Future Rail

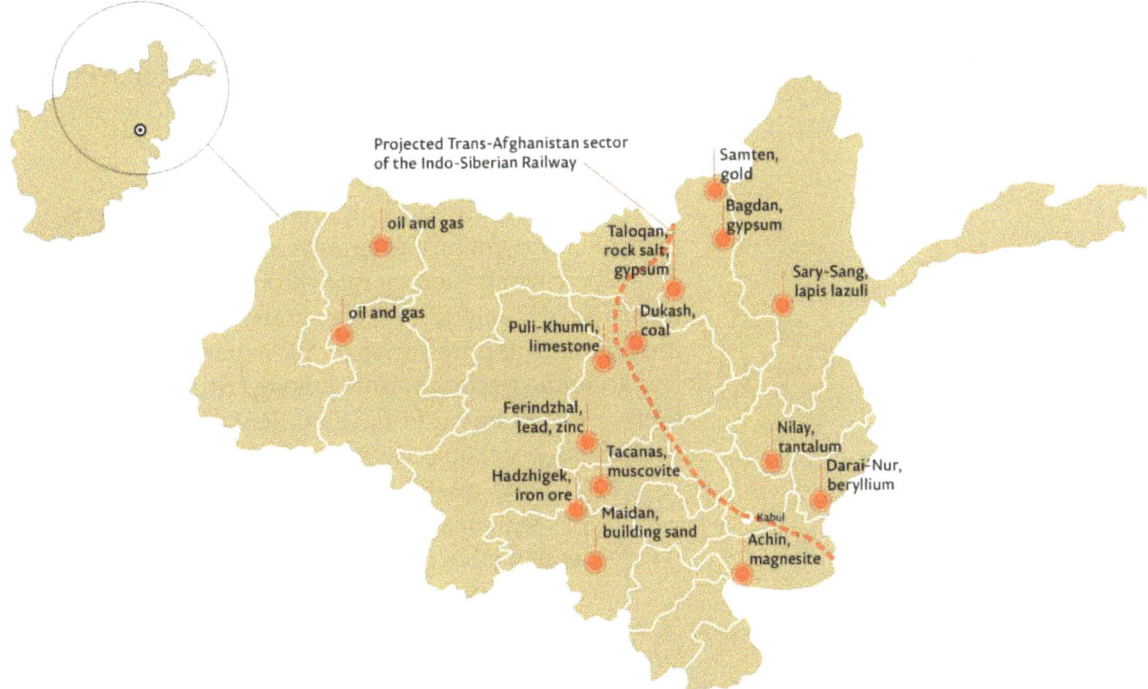

through areas of Afghanistan that are rich in mineral resources, particularly rare earth metals (**Figure 5**). The potential value of these mineral deposits is an estimated $2 trillion. Construction of the railroad will boost their availability and use value.

New Industries

Crash industrialization of Afghanistan will have two major directions. The first is to create new, large-scale industries such as mining, engineering, chemicals, and machine-building. The second is to create, expand, and support so-called network industries, based on local traditional crafts. In our vision, the principal strategic investment projects for the first direction are the following.

An energy-intensive chemical plant for hydrogen electrolysis and processing. It will produce pure hydrogen, nitrogen, and nitrogen fertilizers, partly for use in Afghanistan and partly for export to Southeast Asia and other regions.

Mining projects and new ore-processing plants will become possible with the construction of railways in the region. Attractive areas of investment include exploration and production of oil and natural gas, modernization of gold mining, and developing a construction supplies industry.

Developing the transport infrastructure, as well as energy and raw material supplies, will create the conditions for **processing facilities and assembly plants**, working for the internal as well as international markets (producing automobiles, agricultural vehicles, diesel-generator sets, mini-HPP equipment, and other machinery).

Particular attention should also be paid to agriculture and food-processing: cultivation and production of vegetables, dried fruit, cotton, and other consumer goods. This work would build on the positive experience of growing and processing saffron, which has received international support as a viable alternative to illegal crops.

Social and Cultural Infrastructure

The alternative development of Afghanistan requires basic socio-cultural infrastructure programs in health care and education.

The current literacy rate of only around 35% is a major impediment to economic growth and national development. Removing this impediment requires, first of all, expanding the network of primary and secondary schools, thus raising enrollment rates. Simultaneously, there needs to be rapid development of professional education to accomplish several objectives:

1. Provide qualified workers for the primary industrialization of Afghanistan, including large enterprises (engineers and skilled labor) and network industries (lawyers, managers, and entrepreneurs). University and academic science, oriented to national economic development goals, must be launched.
2. Ensure that educated Afghans can become a stable and self-sustaining leading layer of society (teacher training).
3. Provide temporary employment for a large number of young Afghans.

In the 1970s and 1980s, the USSR educated about 200,000 Afghan university graduates, as well as specialists with other post-secondary education. Despite the intervening years of chaos, those people still constitute an important part of the country's administrative personnel. This experience should be reproduced, now using the resources and capacities of the entire international community.

Another important effort is the expansion of medical facilities to reach the greatest possible portion of the population with health-care services. Mother and child centers are a priority, to reduce infant mortality. It is crucial to develop further the network of rehabilitation centers for the treatment and social reintegration of drug addicts, of whom there are over one million in Afghanistan.

Investment for the Alternative Development of Afghanistan

Part of the funding for implementation of the Plan will come to Afghanistan as international technical and financial aid, but most of the resources for these alternative development programs should be in the form of investments that produce a return, even if they are offered on very favorable terms. The projects mentioned above would require investments, in the first stage, totaling $17.5 billion.[5]

---

5. *IDMRD monetary estimates are included not as a blueprint for financing, but to indicate the scope of development intended.*

These projects, and further development of the new facilities, will require the construction of railroads in Kyrgyzstan and Tajikistan, costing $10.5 billion or more. The payback period for these projects, with the economies deriving from their simultaneous and integrated implementation, will be about 10 years.

A fundamental requirement for the mobilization and spending of those investments is their centralized distribution through a dedicated office, the Plan operator, under public oversight. The investment pool for implementation of the Plan could be created and managed as a World Bank special trust fund—the Fund for Alternative Development in Afghanistan.

Russia as a Major Donor of Afghanistan Development

As the successor state of the Union of Soviet Socialist Republics, the Russian Federation finds itself among the major donors to uplifting Afghanistan economically. In the 1970s and 1980s, the Soviet Union funded the construction of 142 industrial and infrastructure assets, forming the basis of the national economy. The USSR invested more than $3 billion in the geological exploration of Afghanistan's northern provinces.

In July 2010, Russia wrote off $11.5 billion of Afghanistan government debt.

Thus, Russia is already a major stakeholder in the economy of Afghanistan. At the same time, as a major market for Afghan-produced opiates, suffering the economic, political, and security impact of this drug-trafficking, Russia is eager to play a leading role in developing a Comprehensive International Plan for Alternative Development in Afghanistan, based on the accelerated industrialization of that country.

An Album of Strategic Investment Projects for Central Asia

*The Afghanistan package presented at the March 25, 2014 conference implies the creation of a network of railway-anchored development corridors in the four main Central Asian countries: Kyrgyzstan, Tajikistan, Turkmenistan, and Uzbekistan (***Figure 4***). For a 2013 conference in Dushanbe, Tajikistan, IDMRD prepared "An Album of Strategic Investment Projects," exemplifying potential economic development in these corridors. IDMRD defines Strategic Investment Projects (SIP) as 7- to 12-year projects, with "autonomous and achievable goals, which can be reached under existing economic and political conditions." They are designed to create a Central Asian economic model, "based on the principle of sovereign co-development as an alternative to the neocolonial models, promoted in Southeast Asia by Western-based global capital."*

When the international community is won over to a cooperative development policy, ending geopolitics and drug production, these projects may be superseded by the nuclear-powered transformation of Central Asia's mountainous and arid regions. In the meantime, they have the advantage of being "shovel ready," for immediate implementation.

The project descriptions have been summarized by EIR from the IDMRD presentation.

FIGURE 6
Planned Tajikistan – Afghanistan – Turkmenistan Rail Link

Tajikistan, Afghanistan, and Turkmenistan have agreed to build a 160-km railway section across far northern Afghanistan, which will be Tajikistan's first link to several major Eurasian rail routes. The "needed section" of railroad at far left is slated to open in Autumn 2014 as a Kazakhstan – Turkmenistan – Iran link along the eastern shore of the Caspian Sea (located just to the west of the region shown in this map), which is part of the International North-South Transport Corridor from India to Russia.

Tajikistan-Afghanistan-Turkmenistan Railroad

Purpose: Elimination of Tajikistan's transportation dependence
Size: $270 million

Tajikistan's only international rail link today runs through Uzbekistan. A 160-km railroad section across far northern Afghanistan to Turkmenistan (**Figure 6**) will give Tajikistan other routes to the Eurasian rail grid, as well as access to the Caspian Sea. In March 2013, Tajikistan, Afghanistan, and Turkmenistan signed an agreement on building this railroad, including two 700-meter bridges.

Development of Tajikistan's Aluminum Industry

Purpose: Local raw materials for the aluminum industry
Size: $1.72 billion

The project includes mining staurolite and muscovite (a type of mica) ores, from which aluminum oxide (alumina), the intermediate raw material for aluminum production, may be extracted using non-traditional technologies; refineries for this process; and infrastructure to bring the alumina to the Tajik Aluminum Company (Talco) plant in Tursunzade—the largest in Central Asia, currently responsible for 60% of Tajikistan's total exports. Domestic supplies will replace expensive imported raw materials for the industry. There are confirmed staurolite and muscovite deposits in western Tajikistan.

Appendix: The Industrial Development of Afghanistan and Central Asia: a Russian Vision

FIGURE 7
Dushanbe – Khujand String Rail High-Mountain Transport Route

String-Rail Transportation

Purpose: Hi-tech regional transport solution
Size: $2 billion

Innovative string-rail transport technologies, a Russian design,[6] are promising for mountainous Tajikistan and Kyrgyzstan. An initial Dushanbe-Khujand line (**Figure 7**) could be extended to Osh and Bishkek, Kyrgyzstan. Building and operating the system will create jobs.

These systems are designed to carry freight and passengers at up to 500 km/h. Special models feature dedicated tubes for transporting fresh fruit and vegetables, or oil and gas.

Poultry Plant Network in Central Asia

Purpose: Ten-year food security plan
Size: $7 billion in the first five years, self-sufficient thereafter

Chicken and egg production is key for boosting protein consumption in countries with unstable economies, but Central Asia, to date, lacks a feed base for this industry. The project calls for starting up 15 new, modern poultry plants in the region annually, together with the associated transport and marketing infrastructure. Feed for the birds will come from expanding grain and legume crops on irrigated land, with additional feed to be grown in Russia (southern Siberia) and Kazakhstan.

High-Speed Ekranoplane Services Across the Caspian Sea

Purpose: High-speed transport between Caspian coastal cities
Size: $30 million

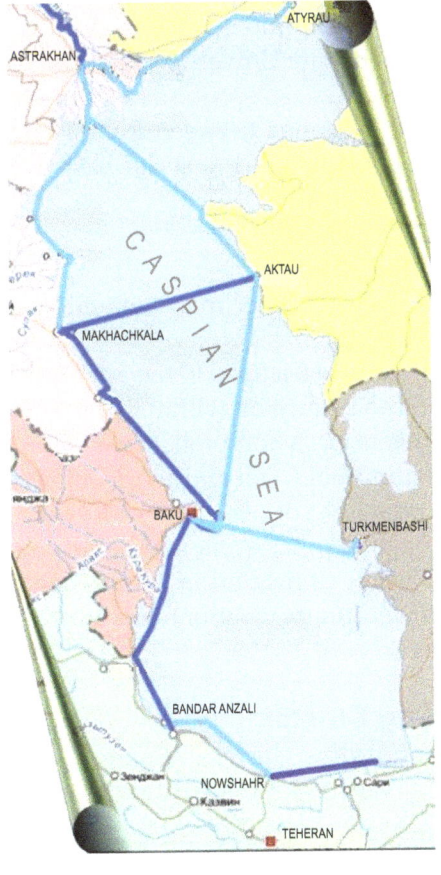

FIGURE 8
Caspian Sea Ekranoplane System

Russian engineers pioneered designs for amphibious very-low-altitude air transport, based on the ground effect—interaction between the craft's wings and the surface of the Earth. A Soviet ekranoplane tested in the 1960s was dubbed the Caspian Sea Monster. It is proposed to use modern ekranoplanes for transport around the Caspian, linking coastal cities including Astrakhan and Makhachkala (Russia), Atyrau and Aktau (Kazakhstan), Baku (Azerbaijan), Turkmenbashi (Turkmenistan), and Bandar Anzali and Nowshahr (Iran).

An ekranoplane (ground effect vehicle) production cluster in Astrakhan, Russia will provide high-speed vehicles to fly on routes between Caspian Sea coastal cities (**Figure 8**). The Burevestnik-24 model, with its engines attached to the upper wings in an innovative biplane configuration, carries 3.5 tons, including 24 passengers. It cruises at more than 200 km/h, with a range of 2,000 km.

---

6. An English overview of the string-rail design is available on the website of its developer, Anatoli Yunitsky, www.yunitskiy.com.

Social Rehabilitation System for People with Substance Dependency

Purpose: Intergovernmental cooperation on recovery and development of human potential

Drug addiction and alcoholism are shared problems of the post-Soviet countries. Russia should take the lead in addressing them throughout the new Eurasian Economic Union (EEU). Cooperation among the governments of Russia, the Central Asian nations, Afghanistan, Pakistan, and Iran, is needed to create an EEU-based Social Rehabilitation System.

A network of rehab centers, for 1,500 to 2,000 people at any one time, can be linked with economic development projects in the participating countries. The agricultural and industrial assets built for this program will eventually become full-fledged economic units, continuing to employ people that have successfully completed rehabilitation.

Multilevel Training Center

Purpose: Education and professional training

The Tomsk State Pedagogical University (TSPU) in Tomsk, western Siberia, will host an education and training program for youth from Central Asia, to deal with the lack of a skilled workforce in the region. The project includes preparatory courses for those planning to enter Russian universities, skills training for future workers on projects of the Central Asia Development Corporation, and the eventual establishment of a Russian-Central Asian State University, attached to the TSPU.

Enhanced scientific and cultural exchange between Russia and these neighboring countries will also strengthen the social basis for fighting narcotics addiction.

Water Management for Central Asia

Purpose: Development of advanced water management systems

To stabilize Central Asia and reduce frictions in the area, a regional water management system can be formed within five to seven years. The project's main elements are a comprehensive water-monitoring map, arrangements for mutual water and energy offsets among the countries of the region, oversight of new HPP in Kyrgyzstan and Tajikistan to preclude damage to other economies, and programs to improve the efficiency of water utilization in Uzbekistan's agriculture, which consumes a disproportionate amount of water,[7] as well as in Kyrgyzstan and Tajikistan.

Central Asia Silkworm-Breeding Cluster

Purpose: Organize a silkworm-breeding cluster in Central Asia, revive traditional silk-making, and create "Silk from Central Asia" as a world-famous brand

---

7. *A major cause of the drying up of the Amu Darya and Syr Darya rivers, and the Aral Sea, is the decades-long practice of cotton monoculture in Uzbekistan's economy.*

Tajikistan and Uzbekistan formerly were world leaders in silk production, a 4.5 thousand-year-old craft, with techniques passed down over the generations. Today, few master silk-makers remain, and production does not meet even domestic demand, but the Tajik-textilmash textile machinery plant and others can be modernized and geared up. The Namagan and Fergana Regions of Uzbekistan and Tajikistan's Sughd Region are ideally suited for expanded production, modernization of processing techniques, and training new specialists. Job-creation in the Fergana Valley is important for Kyrgyzstan, Tajikistan, and Uzbekistan.

Vitamin Bridge

Purpose: Central Asian fruit production to supply the Russian market

Size: $5 billion, to build 280 food-processing plants in Central Asia and Russia

Russia consumes 7 million metric tons of fruit annually, only half the recommended level. Central Asian fruit is available only seasonally, while lower-quality produce from far overseas predominates. Expanded fruit production for supplying high-vitamin produce to the Russian market year-round will create jobs in Central Asia, as well as orders for equipment and technology from Russian producers. New growing and processing technologies preserve high vitamin levels.

Eurasian Bank of Industrial Silver

Purpose: Develop silver mines in Central Asia to build up a 25,000-ton reserve of industrial silver, which may also serve as collateral for large development credits

Size: $4 billion

Dashtijum Hydroelectric Dam

Purpose: Hydroelectric power for agriculture and industry

Size: $5 billion

The Dashtijum HPP on the Panj River, which forms the Tajikistan-Afghanistan border, will provide 1.5 million cubic km of water annually for irrigation and create 6 million jobs in agriculture and industry in Afghanistan. The project depends on international cooperation and joint international financing by Russia, China, Tajikistan, Afghanistan, and Pakistan. It implies the development of northern Afghanistan, where the power generated will be consumed.

Multifunction Vehicle Assembly Plant in the Panj Free Zone

Purpose: Produce new types of farm machinery and urban maintenance vehicles

Size: $10 million

A heavy equipment assembly plant in the Panj Free Zone, Tajikistan, will produce equipment for Tajikistan and Afghanistan, using designs and parts from Russia's GAZ automotive complex.

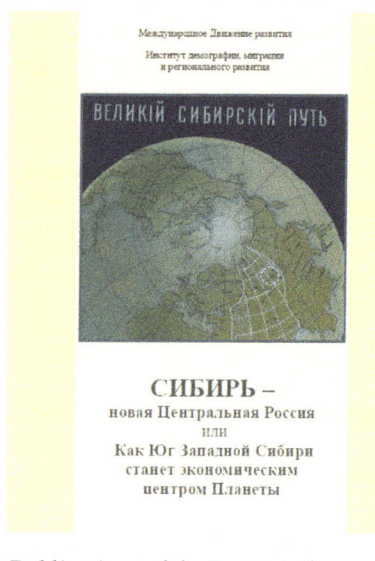

Publications of the Institute for Demography, Migration and Regional Development (IDMRD), related to the economic development of Afghanistan, include "The Path to Peace and Concord in Afghanistan Will Be Determined by the Position Russia Takes" (2008), "How the South of Western Siberia Will Become an Economic Center of the Planet" (2012, in Russian only), and "A New Generation of Alternative Development Programs for the Elimination of Drug Production in Afghanistan" (2014).

Strategic Pipeline from Russia to Central Asia

Purpose: Import water from Russia for irrigation
Size: $5 billion, 30 cubic km (7.9 trillion gallons) per year

This project is a component of a future unified Central Asian water and energy management system. A modification of the once-planned Ob-Irtysh River diversion project, it will enable reconstitution of the lower reaches of the Amu Darya and Syr Darya rivers, which flow into the currently desiccated Aral Sea, while allowing hydroelectric power development in the mountainous regions closer to their sources. Iran and Turkmenistan are candidates to supply wide-diameter pipe.

PART 6
Southwest Asia: Crossroads of the Continents

Southwest Asia and the Eurasian Land-Bridge

By Hussein Askary

August 2014

In a speech delivered at the Zayed Center in Abu Dhabi in June 2002, Lyndon LaRouche presented the economic perspective for the Gulf nations, being "on the cross-roads" of the Eurasia-Africa continents. Indeed, the geographical location of Southwest Asia nations on the most vital trade and transport routes between three continents gives them a unique advantage. If these nations were to integrate and transform their economies to facilitate the future economic development of the Eurasian landmass and Africa, they will play a key role in these developments, ensuring at the same time their long-term economic, political, and cultural survival beyond the era of fossil fuels.

This region is immediately important for economic development in other parts of the world, especially East Asia, because of its oil and gas resource base. For example, 48% of the world's crude oil exports originate from this region, and 80-90% of that oil goes to the Asian nations of China, India, Japan, and South Korea. Almost two-thirds of the oil and natural gas reserves of the world are located in this region and proximately adjacent areas in North Africa and Central Asia. Thus, this region is a choke-point for the supply of energy. Although still dominated by Anglo-American geopolitics, with an anti-imperialist policy this region could become a bridge for peace and development.

Lyndon LaRouche with Dr. Ubaid bin Masood al-Jahni between sessions of the conference on "The Role of Oil and Gas in World Politics" at the Zayed Centre in Abu Dhabi in June 2002, where he was a featured speaker.

Broader Development Potential

In Southwest Asia there is a very paradoxical situation when it comes to the correlation of living standards, culture, education, and economic and financial wealth. Since the oil crisis of 1973 in particular, these countries have been divided into two categories, the so-called rich ones and the poor cousins. The rich ones are the oil exporting countries in the Gulf, members of the Gulf Cooperation Council (GCC), which have small populations and large mineral wealth. They are also members of the British imperial club, and are coddled by the U.S. and Europe. The other ones have fewer resources and larger populations, and have been cursed by the British and the United States. The poorer ones include Iran, Iraq, Syria, Lebanon, the Palestinian people, and Egypt. Jordan has vacillated between these two camps.

The paradox here is that the population in the seemingly poorer countries has much higher levels of education, more advanced labor skills, and a deeper sense of historical identity. The richer ones are living in a strange dichotomy between material wealth and primitive traditions and religious extremism. Technological progress is welcome, but only as a pragmatic power tool, not for the improvement of the cultural and physical conditions of the citizens of the states or their future missions. An educated middle class is obviously a political threat to the ruling families. The discrepancy between the small native labor force and the foreign workers (constituting 80-90% of the labor force in the private sector in most of the richer countries) threatens to pose serious problems in the near future, as mass unemployment among the natives increases, and the lack of basic labor rights among guest workers become more tangible as their wages do not match the real increase in prices globally. Obviously it is difficult to sustain a society of half slaves.

In the poorer group of nations, a great number of the best brains and educated persons have fled these countries due to civil wars, political oppression, and wars or invasions by foreign armies such as in the case of Iraq or foreign-backed terrorist groups as in the case of Syria today. Economic sanctions against Iraq, Iran, and Syria, and International Monetary Fund (IMF)/World Bank policies imposed on Egypt, have led to the deterioration of living standards, infrastructure, and educational systems. All this has set back the development of these nations many decades.

Our program for the development of the region will shift this imbalance dramatically because the financial and mineral wealth, human resources, and skills will be directed toward one unified mission for all the countries—national and regional integrated development. Youth among the native populations would be trained to join the labor force to build their nations and green the desert, in similar fashion as the Franklin Roosevelt era (1933) New Deal and associated Civilian Conservation Corps (CCC) program during the Great Depression pulled unemployed people in America off the streets into the national reconstruction projects, helping turn the United States into the most powerful economic nation on earth during World War II. The brain drain will be stopped, and hundreds of thousands of scientists and well-educated people working in exile or as expatriates in Europe and the Americas will feel safe to come home and serve their nations. The financial and mineral wealth and whatever national credit that can be generated in the rich countries, can be matched up with the skills of the labor of the others in the short term to launch the reconstruction process.

A common credit system established through a development

bank can fill the credit gaps among the oil-poor or water-poor countries. Nations such as Yemen and Jordan will not be left at the mercy of the IMF just because they cannot pull together their credit potentials to launch an economic development process. A nation such as Jordan will be aided to build its first nuclear power plants, and will upgrade its human potential and processing of natural resources (such as phosphate and uranium) and become a rich nation within one generation, rather than waiting desperately for hand-outs from the United States, European Union (EU), or IMF and World Bank. Sharing know-how, for example in dealing with desert conditions, agricultural problems will be dealt with most effectively through establishing a unified scientific research center functioning under a common executive authority.

Right now, due to the policies of fomenting religious strife and wars throughout the region all the way to the Caucasus and western China, the entire region is threatened by 30-year religious/sectarian war from which it might never recover. This vicious cycle must be broken. There are global preconditions, of course, such as shifting the murderous geopolitical system of divide and conquer of the British Empire, that are required to assist these nations to shift focus from destruction to construction. The integration of this region into the Eurasian-African Land-Bridge will be key, and will benefit the world and these nations.

The Bridge Among Continents

Many links integrating the region into the Eurasian-African Land-Bridge are already underway (**Figure 1**). The inauguration of the Mashhad-Sarakhs (Turkmenistan) railway in 1996 by then-President Hashemi Rafsanjani connected Iran to China and further revived the ancient Silk Road. Two years later, Iran completed its connection to the northwest, to Turkey, reviving the Silk Road connection to Europe. In 2001, the Mashhad-Bafq-Bandar Abbas line was completed, connecting landlocked Central Asia to the Persian Gulf. Iran also completed the Bafq-Kerman-Zahedan railway to Pakistan, connecting Iran to the Indian subcontinent.

There is also the strategic continental North-South Corridor, which goes from Russia to India. There is an agreement among Russia, Iran, and India to build a trade route through the Caucasus and Central Asia, through the Iranian railway network. This will tie in the port of Chabahar in southern Iran on the Gulf of Oman, on which development work is commencing. India is very interested in this, because shipping by sea takes about three weeks to the Black Sea, while the railway system through Russia takes one week.

Iran is connecting its south-north railway network to Russia through the Caucasus region through both Azerbaijan and Armenia in cooperation with Russia. In January 2013, the Armenian Ministry of Transport and Communication, Dubai-based investment company Rasia, and Russian Railways (RZD) subsidiary South Caucasus Railway (SCR) signed a trilateral agreement for the construction of the Southern Armenia Railway. The agreement covers the construction of a 316 km

FIGURE 1
Railways in Southwest Asia, 2012

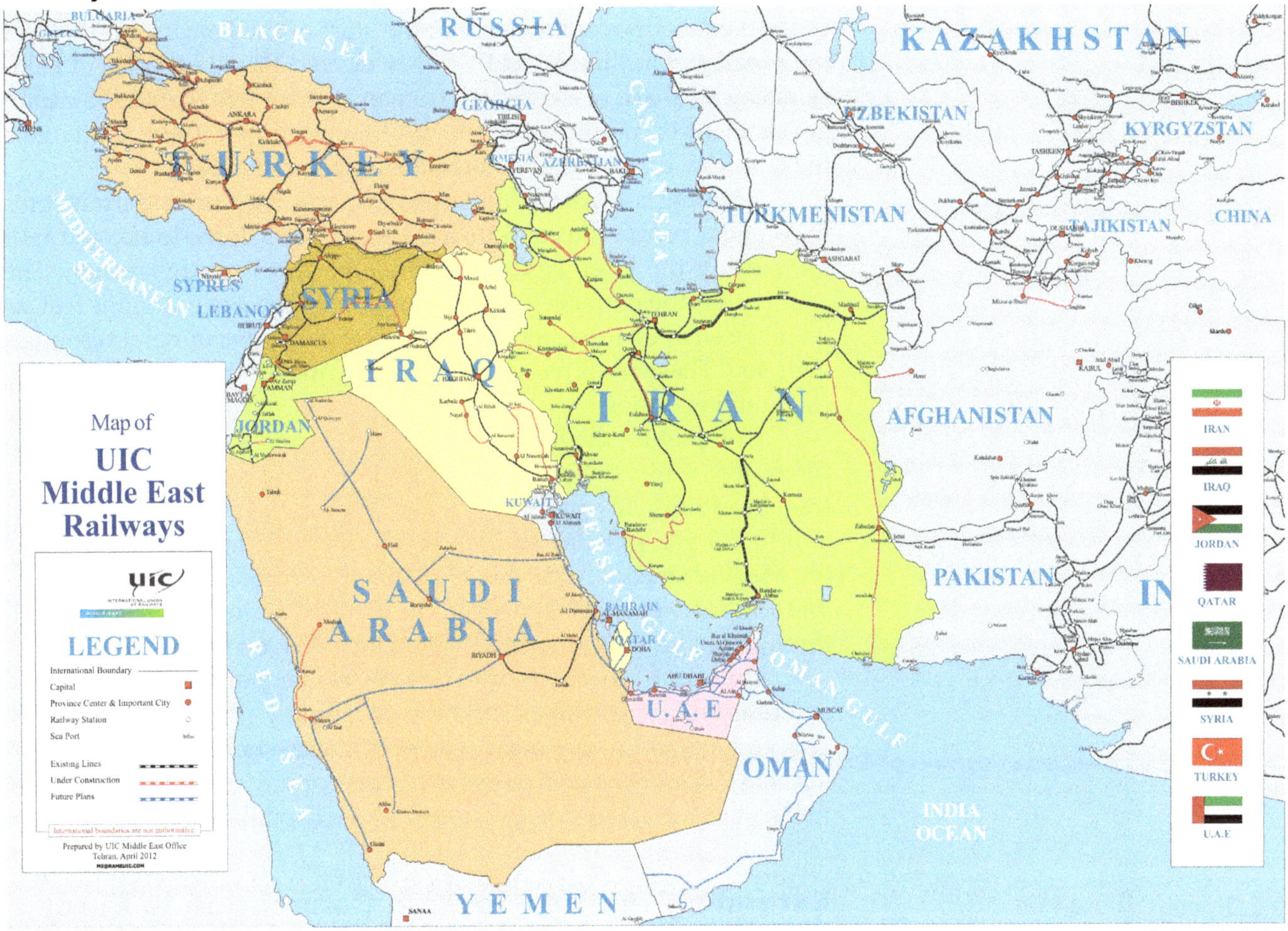

International Union of Railways, April 2012

railway linking Gavar, 50 km east of Yerevan near Lake Sevar, with the Iranian border near Meghri. The electrified, single-track railway line will be part of a new north-south corridor linking the Black Sea and the Persian Gulf, according to Armenian authorities. Interestingly, China is also involved in the project because the feasibility studies conducted by Rasia selected China Communications Construction Company to become the lead member of a consortium that will be responsible for the project.

The Iran-Pakistan gas pipeline, a vital economic endeavor for Pakistan's energy security and prosperity that was pursued by the Pakistani government despite American pressure, is crucial to bring Pakistan on board a regional solution for the situation in Afghanistan beyond the scheduled 2014 withdrawal of NATO forces from that country. On March 11, 2013, construction work on the Pakistani section of the pipeline was inaugurated by President of Pakistan Asif Ali Zardari and President of Iran Mahmoud Ahmadinejad. The pipeline in Pakistan is expected to be constructed in 22 months with the participation and financial backing of Iran. The Iranian part of the pipeline was completed earlier. However, internal economic and political instability in Pakistan itself, in addition to pressure from the Saudi-British-American axis to isolate Iran, is a major obstacle to this cooperation. Just two months after this ceremony, Iranian government officials expressed concern over the delay in the start of the Pakistani portion of the pipeline. On June 12, 2013, newly elected Prime Minister of Pakistan Nawaz Sharif reassured the Iranian government that the Pakistani government is committed to the fulfillment of the project and targets the first flow of gas from the pipeline in December 2014.

Building a modern Afghan nation will require modern institutions and a prosperous and thriving economy. China, India, and Iran are already the three largest economic partners of Afghanistan. The country has the potential of soon standing on its own economic feet, as it will be enabled to explore and exploit the massive mineral resources in its soil. These resources are estimated to make Afghanistan a world-class mining nation.

Completing the bridge to the west, Iran also has been active in promoting trade, transport, and economic exchange with Turkey and Iraq. In addition to a gas pipeline and railway to Turkey, Iran has been building a gas pipeline to Iraq, to be extended farther to Syria and the Mediterranean. A railway is planned to run adjacent to the pipeline and road projects. However, the destabilization of Syria and Iraq through sectarian violence has halted the work on the projects.

Thus, Iran will continue to be a key element of the New Silk Road or Eurasian Land-Bridge. Turkey is also a key player in connecting Asia to Eastern Europe across the Bosphorus—new bridges and tunnels have been built and others are under construction to connect the Asian and European sides of Turkey at the capital, Istanbul.

Extending the Iran-Turkish Connection to Europe

Rebuilding the Hijaz Railway from Turkey through Syria, Jordan, and Saudi Arabia to Eden in Yemen was under consideration before the events in Syria broke out in the spring of 2011. From Yemen, a tunnel or bridge to Djibouti on the Horn of Africa was also being considered at the time by the Yemeni government and corporations from the United Arab Emirates (UAE). From Yemen, through Oman and the UAE, a tunnel to the Iranian port of Bandar Abbas across the Hormuz Strait is a feasible transport corridor that would connect Asia to Africa directly. A railway and highway connection northward from the UAE through Saudi Arabia, Qatar, Bahrain, and Kuwait that will potentially connect to Iraq and Turkey, and through Syria to the Mediterranean, is under consideration by the GCC nations. From Jordan and northern Saudi Arabia, a bridge/causeway across the southern end of the Gulf of Aqaba to connect to Egypt was under study by the government of former President Hosni Mubarak in 2009.

Almost all these projects are shelved now due to the political and military destabilization of the region. Israel's connection to these networks, including gas and electricity networks, was openly discussed in the 1990s, but has been excluded since the collapse of the peace process between the right-wing Israeli government under Benjamin Netanyahu and the Palestinian Authority. All these projects, nonetheless, can be put back on the fast track, whenever a just international political order is established.

A Common Enemy: The Desert

What is striking about transcontinental regions where the Silk Road passes is that the landscape is a vast desert stretching almost continuously from the Atlantic coast of North Africa through the Arabian Peninsula, across the Zagros Mountains to Iran and Central Asia, and all the way to western China (**Figure 2**). The size of that transcontinental stretch of desert is about 13 million square kilometers. Areas that receive between 250 and 500 mm of annual rainfall are usually deemed to be semi-desert or semi-arid. In general, large parts of the area receive an average annual rainfall of 250 mm or less. Many parts of the great deserts get less than that, and sometimes no rainfall at all. The major deserts of the world are located within these regions. These deserts are currently expanding, due to not only the lack of adequate economic and sociopolitical measures, but also the destruction of existing green areas through the mismanagement of local resources. Long cycles of drought have also contributed to the expansion of the desert.

FIGURE 2

The vast desert that stretches almost continuously from the Atlantic coast of North Africa through the Arabian Peninsula, all the way through to western China, represents a major challenge for economic development for all the nations involved.

Sandstorms and dust storms are frequent events in Southwest Asia, especially in the Gulf region, but even extending to Iran, Afghanistan, Pakistan, and India. While sandstorms rise up to tens of meters, dust storms can rise to several kilometers into space. And they can cover whole countries. Their impact on urban areas can shut down airports, ports, hospitals, schools, and other vital facilities.

Attacking the Desert from the Fertile Crescent

The area stretching from Lebanon to Syria and down Mesopotamia to the Gulf is called the Fertile Crescent. Historians claim that agriculture in the world started in this region. However, it is not so fertile any more. By better managing the natural resources and creating new resources of land and water, green belts, trees, and vegetation selection and applying advanced agronomy and animal husbandry, and especially space-era science and capabilities, a new full array of productive activity can be established where once was desert.

In the 1970s and 1980s, "green belts" were being planned for eastern Syria, western Iraq, and parts of eastern Jordan. With successive rows of forests, the expansion of desert areas can be stopped and gradually reclaimed. In Iraq, for example, a plan was prepared decades ago to create a green belt in the west of the country (**Figure 3**). Due to the series of geopolitical wars and destruction of Iraq's infrastructure and agriculture, these plans were never implemented. Furthermore, the degradation of the soil and land has expanded the arid areas.

There now are active operations, although limited in scope, to return to this idea. Iraq and Iran signed an agreement in 2010 to invest approximately US$2.1 billion in projects to create green belts in the southwestern part of Iraq, especially in the region of the cities of Karbala and Najaf, which are frequently hit by sand and dust storms. One project, for example, established a 27-kilometer-long crescent lined with thousands of newly planted trees in a belt 100-200 meters wide. It is irrigated by 50 shallow wells (35 to 50 meters deep). The area is now the front line of Karbala's battle against increasingly frequent sandstorms and salinity and erosion of the soil. The project has involved planting more than 100,000 olive, palm, eucalyptus, and other trees, all of which were chosen for their resistance to heat and soil salinity.

FIGURE 3
Proposed 'Green Belt' To Stop Iraq's Desertification

Government of Iraq, 2012

Water Sources

The obvious question is, of course: Where will all the water come from to back this massive war on the desert?

The region is dominated by two major water systems—the Tigris and Euphrates rivers basin, and the Jordan River Basin. The latter is a relatively limited water system dominated by military and political conflicts for the control of the water between Israel and Lebanon, Syria, and Jordan. The Mesopotamian water system is larger and has greater potential. The general problem in this and other desert regions is that a great part of the rainfall disappears due to evaporation, transpiration, and run-off. To collect and use as much as possible of the precipitation, large-scale water infrastructure systems are required.

One of the most ambitious water infrastructure projects in the region, the South Eastern Anatolian Project (in Turkish abbreviation GAP), (**Figure 4**) has been under way during the past two decades. However, this project has created major problems for the countries downstream, Syria and Iraq, because it blocks the natural flow of the Tigris and Euphrates. What is needed is to establish a scheme of cooperation among the three countries, and even Iran, which shares some of the tributaries of the Tigris with Iraq,

FIGURE 4
The Southeastern Anatolian Project

EIRNS, 2012

The Southeastern Anatolian Project, called the GAP, would dam the Tigris and Euphrates rivers in Turkey, and create a huge reservoir behind the Ataturk Dam.

to make the entire Mesopotamian basin function most effectively as one unit. Legal and technical agreements have to be made to ensure the sound management of the system and a just share in the water. The GAP, begun more than 20 years ago and modeled on the Tennessee Valley Authority, envisions 22 dams to provide 7.4 GW of electricity, water management, irrigation, and flood control. Located in southeastern Turkey, the project covers 10% of the country's land area—75,000 square km and nine provinces in the Euphrates–Tigris basins and southeastern plains—and accounts for 20% of Turkey's arable land. The project includes the development of infrastructure of all types required for integrating the entire region, including transportation, power, tunnels, and canals. According to Turkish government estimates, when the projects are completed, 1.7 million hectares of land will be effectively irrigated. The region represents 28% of Turkey's total hydraulic potential, with the Ataturk Dam at its center.

There is a large number of dams in Syria and Iraq, but more can be done, especially in northern Iran, to build dams on the tributaries of the Tigris River that flow inside the Iraqi Zagros Mountains, such as the Greater Zab. Building new and maintaining the existing relatively modern system of dams, barrages, and canals in Syria and Iraq will rescue these two countries from flooding during the spring and drought during the summer.

Seawater Desalination

One thing that has become clear for the governments of the Gulf and other dry regions throughout the world is that the best solution to secure water for drinking, other urban usage, and industry is the desalination of sea water. Steps have been taken by the countries in the region to build conventional desalination plants on a large scale, investing heavily in the combined water desalination/power generation process with the use of fossil fuels such as natural gas and oil.

More than two-thirds of the world's production of fresh water by desalination occurs in the region. Saudi Arabia alone produces 25 million cubic meters of water per day, which is estimated to be one-half of the world's total. The UAE produces around 3 million cubic meters per day.

However, these countries will have to more than double that amount of desalinated water in the next decade and triple it in the decade beyond to meet projected demand. Water consumption will rise from 8 billion cubic meters in 2012 to about 11 billion cubic meters in 2016. Massive investments are already projected in this area.

A major problem with these projections is that the desalination of seawater is reliant on thermal power plants run by oil and gas. Reportedly, Saudi Arabia, for example, uses 1.5 million barrels of oil daily to produce the electricity and heat used for desalination. It is a net physical economic loss in the sense that valuable industrial raw material (oil and gas) that can yield many times their value if used as a base for petrochemical and other products rather than burned to achieve a relatively low energy flux-density compared with nuclear power.

Nuclear Desalination

One of the key solutions to this water shortage problem is the use of nuclear power for desalination and for increased industrial activities in the petrochemical field. As International Atomic Energy Agency studies show, medium-size nuclear reactors are especially suitable for desalination, often with cogeneration of electricity using low-pressure steam from the turbine and hot seawater feed from the final cooling system.

There are many new technologies being tested in this field, all of which point in the direction of higher temperature and pressure, something that can only be achieved efficiently through nuclear power. Fourth generation high-temperature nuclear power plants have long been proven as the most efficient, but almost no effort is being taken to invest in them.

At the moment, Iran is the only country in the region that has an operating large civilian nuclear power plant. The Bushehr plant, a product of cooperation between Iran and Russia, was inaugurated officially in September 2011, and reached its full power production capacity (1,000 megawatts) in August 2012. Iran is planning to build several new nuclear reactors, with the expressed aim of increasing the energy output of the country and desalinating seawater.

In December 2006, the GCC announced that the Council was commissioning a study on the peaceful use of nuclear energy. In 2007 the member states signed an agreement with the IAEA to cooperate on a feasibility study for a regional nuclear power and desalination program.

The UAE was the first of the countries in the study to launch its nuclear power program. The Emirates Nuclear Energy Corporation (ENEC) was established in 2009 in Abu Dhabi as an investment vehicle for the nuclear program. In December 2009, ENEC announced its acceptance of the bid offered by the South Korean Korea Electric Power Corporation (Kepco) to build four 1,400 MW nuclear plants by 2020 at

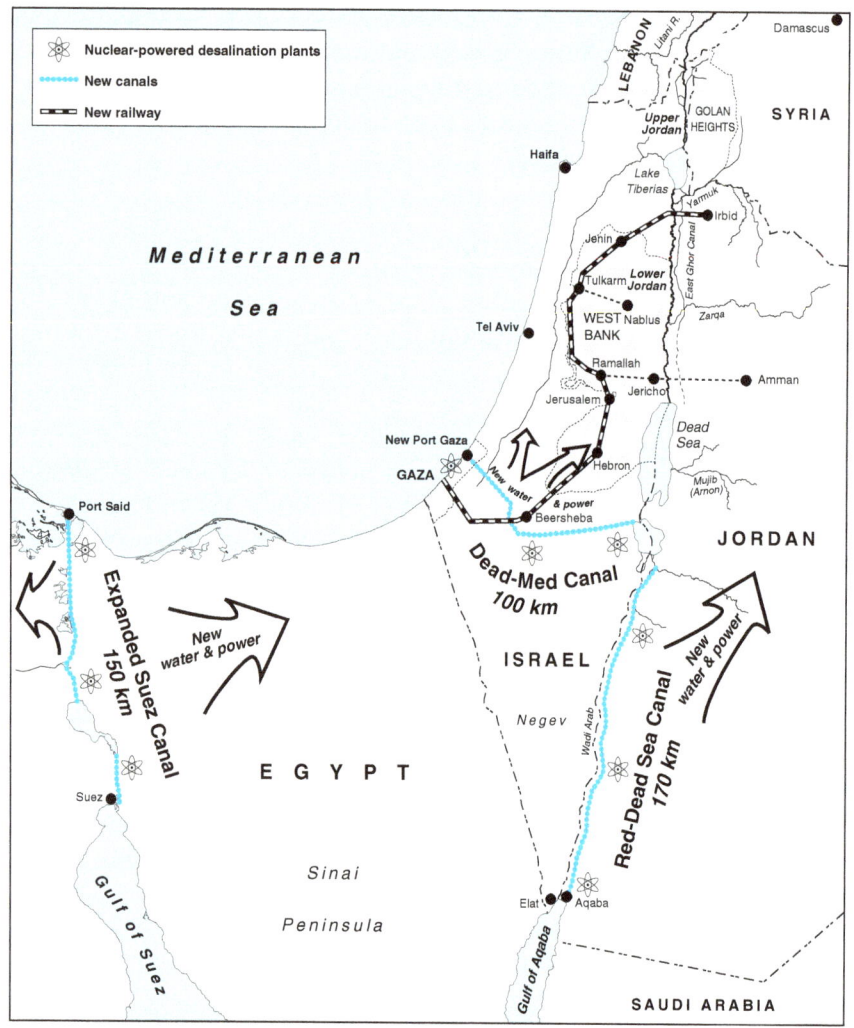

FIGURE 5
Features of the LaRouche "Oasis Plan", 1990

The roots of this plan for solving the water crisis in the Middle East go back to the mid-1970s, but it was codified in 1990, when LaRouche was campaigning to stop the first Gulf War.

the cost of US$20 billion. The construction of the first of the four plants was started in Baraka in July 2012, and the fourth and last will be completed in 2020.

Saudi Arabia announced through the royal decree of King Abdullah bin Abdul-Aziz Al-Saud in April 2010 the establishment of the King Abdullah City for Atomic and Renewable Energy. Shortly after that, the Saudi government announced plans to build 16 nuclear power reactors by 2030. Unlike the Iranian nuclear program, the GCC's programs are welcomed and approved by the United States and the West generally, for obvious geopolitical reasons.

In 2013 Jordan signed an agreement with Russia to build the first nuclear power plant in the country. The expressed purpose of the project was water desalination. Israel has similar plans for nuclear power development for desalination purposes. Lyndon LaRouche has argued since the 1970s that without turning to nuclear power to secure energy and water in this region, as prescribed in his Oasis Plan (**Figure 5**), no peace process can endure under the economic pressures including the shortage of water and power in the region. The Palestinian people especially, who have been deprived of even the little existing underground water by Israel, will have no chance of survival without large-scale seawater desalination. A United Nations report issued in 2012 determined that the Gaza Strip would be "unlivable by 2020" due to the complete depletion and contamination of the ground water aquifers.

Lebanon and the coastal cities of Syria are experiencing the same water crisis as the Palestinian areas. Only a combination of modern water management systems and nuclear desalination can reverse the catastrophic conditions in these two countries.

FOR FURTHER READING

"A Revolutionary Development Plan for the Near and Middle East," *EIR*, Dec. 12, 2012.

"The Persian Gulf: Peace and Reconstruction, or War and Destruction?", *EIR*, May 10, 2013. Also available in video at http://newparadigm.schillerinstitute.com

"The Keys to Peace: The LaRouche Doctrine," *EIR*, May 7, 2004. An interview with then presidential candidate Lyndon LaRouche by Hussein Askary.

PART 7
The Crucial Contributions of East and Southeast Asia

Japan Must Return to Leadership in Nuclear Energy

By Michael Billington

September 2014

Post-World War II Japan, like post-World War II Germany, became a world leader in high-end technological and industrial development, epitomized by its sterling machine-tool and engineering capabilities. Specifically, Japan developed a nuclear supply industry which, as other nations let theirs lapse, has become an indispensable component of the nuclear energy sector globally. If the vision of the New World Economic Order being put forward by the BRICS nations is to be carried out, Japan's nuclear industry capacity is a necessary component.

Yet today, all of Japan's nuclear power plants have been shut down, and thus its supporting industry is endangered as well.

Following the March 2011 Great East Japan Earthquake and tsunami, which killed more than 15,000 people, the Japanese government, then run by the green Japanese Democratic Party, chose to focus less on the overall tsunami disaster than on the damage done to several of the Fukushima nuclear plants. While the damage to the nuclear plants forced thousands of people to be temporarily relocated, it killed a grand total of *zero* people. All of Japan's 50 nuclear power plants were closed, and despite two attempts under the current LDP government of Prime Minister Shinzo Abe to begin reopening at least a few of the plants, they all remain closed as of September 2014.

Simultaneously, the green government of Germany under Angela Merkel is in the process of completely eliminating nuclear power in that nation, the preeminent industrial power in Europe. Significant political opposition to this suicide in both Germany and Japan has not yet reversed these policies.

The emerging conditions created by the BRICS, however, provide a crucial opening for Japan's role to be restored.

Two Japans

Lyndon LaRouche has long insisted that the world must recognize two Japans—one heavily influenced by the American System, the other by the British Empire. After the Meiji Restoration of 1868, Japan underwent a dramatic transformation and modernization, which was carried

Japan Must Return to Leadership in Nuclear Energy

An example of a forged steel component needed for nuclear fission reactors, made by Japanese Steel Works.

out in cooperation with several of the leading proponents of the American System of political economy associated with President Abraham Lincoln. The British were also active in Japan, arguing that Japan, like the British, is an island nation with few resources, and that it must thus follow the British Empire's militarist policies, aimed at conquest as the necessary means of obtaining the raw materials needed for an industrial economy.

Japan's Meiji leaders choose the American System of peaceful cooperation among nations, directed credit, and universal education, successfully transforming itself into a modern industrial state within a few decades. During the 1880s and 1890s, the British influence in Japan rapidly increased. Even though Britain virtually ruled China (after its two Opium Wars had subjugated China during the 19th Century), the British were at the same time actively preparing Japan for a war on China, including providing Japan with detailed intelligence on Chinese military capacities and weaknesses. In 1894, as part of Japan's war against China over the control of Korea, Japan crushed China's Northern Fleet in the Yellow Sea. The British proceeded to set up a military alliance with Japan, whose primary purpose was to use Japan to stop Russia's dramatic move to open up a land route to Asia via the Trans-Siberian Railway (a "Eurasian Land-Bridge" inspired by the same American System economists and industrialists who had built the Trans-Continental Railroad in the United States), which threatened to undermine British control of Asian trade via the sea routes.

Indeed, in 1905 Japan went to war with Russia, and in the process established a permanent military presence in China's northeastern provinces, including the last leg of the rail connections through Manchuria to the Trans-Siberian Railway. By the 1930s Japan had launched full-scale war on China, effectively launching World War II, with overt support from the British and their assets on Wall Street, led by J.P. Morgan's Thomas Lamont, who lobbied for U.S. support for Japan's war on China.

Only after that horrendous conflagration, and the useless and genocidal dropping of two atomic bombs on Japan by the British asset Harry Truman who took over the U.S. presidency after Franklin Roosevelt's untimely death, did Japan return to the American System approach, under American General Douglas MacArthur's direction, becoming one of the greatest industrial powers on Earth, and a global leader in nuclear power for peaceful purposes.

Now, that leadership in peaceful nuclear power is in shambles, while Prime Minister Abe has declared war on the pacifist Constitution which has guided Japan since World War II, pledging to "reinterpret" the Constitution to allow "collective defense" as a means of joining a potential U.S. war with China.

The World Needs a Peaceful, Productive Japan

Japan is suffering from its anti-nuclear policy, with the need to import massive amounts of fossil fuels to sustain its electricity supply. But also at threat are the many nations that depend on the output of Japan's nuclear-supply industries, and Japan's contribution to the needed leap worldwide to a thermonuclear-powered economy.

In fact, plans are in place to restart some of the nuclear power plants, but they keep being delayed. In 2013, Japan's major supplier of nuclear-power-generating equipment, France's state-owned Areva nuclear group, announced Tokyo's plans to restart six reactors by the end of 2013. "I think two-thirds of the reactors will restart" within several years, Areva's CEO Luc Oursel said. In the fall of 2014, we still hear the promise that a few reactors will be opened by the end of the year. Similarly, President Abe's commitment to restoring Japan's capacity to export nuclear power plants is active, but has not yet resulted in any contracts.

The greatest danger is that Japan's nuclear power production sector—the heavy engineering industry and research and development—becomes significantly degraded in terms of manpower and physical-technological capacity. The low-energy-flux-density "green" sources, such as solar, wind, and other so-called "renewables," can never meet the nation's (or the world's) growing need for energy. The following facts exemplify the dependence of the world's nuclear industry on Japan:

• Japan Steel Works (JSW) remains today the most prestigious and most prolific supplier of heavy forgings globally, and can manufacture 12 nuclear power plants sets a year. It produces large forgings for reactor pressure vessels, steam generators, and turbine shafts, and claims 80% of the world market for large forged components for nuclear plants. JSW has been manufacturing forgings for nuclear plant components to conform to U.S. Nuclear Regulatory Commission standards since 1974, and some 130 JSW reactor pressure vessels are in service around the world today. *The United States has no capacity to build such heavy forgings.* Russia builds its own pressure vessels, but they are of different design, and cannot be used except in Russian-built reactors.

The world's nuclear industry is inter-connected and inter-dependent on Japanese-manufactured engineering products. For example, India has a plan to set up 9,900 MW of nuclear-generated power in Jaitapur in Maharashtra, which would make it the single-largest nuclear-power-generating cluster, and France's Areva has secured a contract from India to supply two 1,600-MW reactors for installation in Jaitapur. However, Areva cannot deliver these large pressure vessels, which have to be forged by JSW. Moreover, India has plans to build 20 GWe of new nuclear reactors in the next 10-15 years. Some of those pressure vessels have to be cold-forged at JSW, no matter what firm in the West gets the contract to build the reactors. If the Abe Administration cannot revive nuclear power generation in Japan, and resolve the ensuing power shortage crisis, it is unlikely that JSW will be able to

Japan's status as a nuclear powerhouse is reflected in its nearly active 20 nuclear plants, as of 1999, when this sketch was made.

deliver any of these pressure vessels to India, China, or any other nation in the coming years.

• Along with the JSW, a leading Japanese industry that contributes significantly to the manufacturing of nuclear power reactors is the IHI Corporation. According to the company website, IHI Corp supplies reactor vessels, primary containment vessels, and piping systems. These components are for the traditional boiling water reactors and advanced boiling water reactors. IHI has also now started supplying components for pressurized water reactors, its website states. In 2011, it established a joint venture with Toshiba Corp. to manufacture steam turbine parts for nuclear power plants for domestic use as well as for export. The new name of the company, following the joint venture, is Toshiba IHI Power Systems Corporation. IHI is also a shareholder in Westinghouse.

• Another Japanese company that provides nuclear power reactor pressure and containment vessels, is the Yokohama-headquartered Babcock-Hitachi KK. The reactor vessels that they produce are for the boiling water reactors. Over its more than 40 years of existence, this company has reportedly supplied 15 such reactor vessels. In addition, it has also supplied equipment for fast breeder and high temperature gas-cooled reactors, as well as light water reactors.

The list goes on and on. Beyond Japan's heavy engineering capability, which allows many capital-rich companies in Japan, Europe, and the United States to receive orders and build nuclear power plants around the world, Japan has carried out an enormous amount of research and development in the field of nuclear technology. In fact, Japan has been a leader in the research and development of thermonuclear fusion, the necessary energy source of the future, since the 1980s. In magnetic fusion research, it is a partner in the International Thermonuclear Experimental Reactor (ITER) tokamak program, which also includes the United States, Russia, Europe, India, South Korea, and China. As a major contribution to solving one of the engineering challenges of fusion, Japan is taking the lead in developing the International Fusion Materials Irradiation Facility. The Japan Torus (JT)-60 tokamak, inaugurated in 1985, has undergone periodic upgrades, while the current JT-60SA (super advanced) upgrade, which is underway, will incorporate superconducting magnets for the tokamak's operation.

Japan has also excelled in inertial confinement fusion research. The GEKKO XII laser facility at Osaka University, completed in 1983, has also undergone upgrades to bring laser fusion closer to net energy production.

This is the Japan that the world needs, which must be restored—a leader in all areas of nuclear technology, for all mankind.

Ramtanu Maitra contributed to this article.

The Mekong Development Project—
A TVA for Southeast Asia

By Michael Billington

August 2104

In the immediate aftermath of World War II, the mighty Mekong River, rising in Tibet and flowing through China, Myanmar, Laos, Thailand, Cambodia, and Vietnam, nearly 5,000 km in all to the South China Sea, with a nearly 800,000 square km basin, was seen as a primary target for the kind of comprehensive development that Franklin D. Roosevelt had applied in his Tennessee Valley Authority (TVA) project, which transformed one of the poorest, underdeveloped regions of the United States into a center for advanced industry, agriculture, and scientific research. Several studies were carried out by the United Nations in the early 1950s, and in 1957 the Mekong Committee was created. Engineers from the TVA, U.S. Army Corps of Engineers, and U.S. Bureau of Reclamation conducted studies of the region, leading to plans for a cascading series of dams on the Mekong and its major tributaries. The resulting Indicative Basin Plan for 1970-2000 proposed a $12 billion long-term project to generate 17,000 MW of hydropower and provide for a ten-fold increase of irrigation during the dry season to about 6 million hectares of cultivatable land, doubling to tripling the productivity of that land.

These studies took place after the French colonial war against Vietnam, which ended with the French defeat at Dien Bien Phu in 1955. It was a time of great optimism, as the former colonial powers conducted the first international conference without the presence of the European colonial masters, at Bandung, Indonesia, in 1955, leading eventually to the creation of the Non-Aligned Movement. In the United States, the election of John Kennedy in 1960 promised to bring in a restoration of the international development policies intended by President Franklin Roosevelt in the post-war era. The Kennedy Administration tried (unsuccessfully) to recruit David Lilienthal, who headed the TVA, to a diplomatic position in Asia, to "create the atmosphere and steam behind the development of the Mekong River, a big Southeast Asian TVA," as Under Secretary of State Chester Bowles put it.

But that was not to be. The Mekong is still today one of the last untamed and undeveloped major river basins in the world (or, as the Greenies would say, an "unspoiled" river basin). This would not be so, if not for a combination of colonial wars throughout the 1960s-80s, and following that, a concerted, successful effort by the British Empire, through the synthetic environmental movement (the Greenies), to sabotage the development process at every stage.

The most recent report on the region, by the Mekong River Commission, admits that "management and development of the River remains limited today, in part due to unregulated river flows. The vast floodplains in Cambodia remain largely undeveloped and only a small proportion of the irrigation, hydropower, and navigation potential has been realized in the basin. The River remains mostly in its natural state." The Commission also points out that although poverty has been reduced through economic development, especially in Thailand and Vietnam, the region "remains among the world's poorest areas. Some 85% of its population is in rural areas. Safe water supply and all-weather roads reach only 50% of the households, and electricity consumption is only 5% of that of the industrial world. Floods and droughts claim lives and property and cause major economic losses."

Why?

The assassination of JFK in 1963 led quickly to the Indochina War, disastrous for Indochina and the United States. The development plans for the Mekong lay practically dormant for the next 30 years, as colonial wars left misery and poverty in their wake. Although President Lyndon Johnson in 1975 proposed a $1 billion project to build the Mekong, "which would dwarf even our TVA," he was thinking that Ho Chi Minh would give up in exchange for such a development proposal—a classic failure of U.S. understanding of Asia and of issues of national sovereignty. The great development plans for dams, roads, rail, and city building evaporated in the fire of napalm and carpet bombing across the region.

When the wars finally ended, the world had become infested with a Green disease. A diatribe against the development of the region published by the Stimson Center in 2010, with backing from two genocidal British-spawned fascist institutions, the Worldwide Fund for Nature (WWF) and the International Rivers Network (IRN), published a map (see **Map**) of the currently planned or constructed major dams on the Mekong today—8 in China and 11 on the lower Mekong in Laos, Thailand, and Cambodia. While five of the eight planned in China (where the Mekong is known as the Lanjiang) are now complete, with the other three under construction, none of the 11 planned dams in Southeast Asia have been built. The completion of the dams in China is considered a great disaster by the British imperialists behind the greenies, with China being denounced not only for building its own dams, but also for offering to finance and build several of the dams in the lower Mekong.

What is most interesting is that the Stimson report admits that these current plans, which they are demanding be totally stopped, are very close to the projects proposed back in 1970 by the U.S.-led Mekong Commission. They report that the U.S.-led team proposed "an immense project to exploit the hydropower potential of the basin to promote development and deter the spread of communism," to make the Mekong into a "much larger version of the Tennessee Valley Authority that supported the economic development of one of the country's most remote and impoverished regions."

What Happened to the Mekong Project

During the Indochina wars, much of the world was building major dams and water control projects. Although the massive North American Water and Power Alliance (NAWAPA) project was stopped with the death of JFK (for which the United States is now paying the consequences, with the drought in California, Texas, and the Southwest as a whole), the United States nonetheless has 8,000 major dams and 80,000 dams altogether, while virtually nothing was built in the Mekong region.

With the U.S. defeat in Vietnam in 1975, the United States ended all support for the Mekong project, and even though Vietnam removed the genocidal Khmer Rouge regime in Cambodia in 1979, the United States chose to support the Khmer Rouge claim to power and refused any support until 1992. By then it was too late—the greenies had taken over,

In 1995, the Mekong Commission finally was reconstituted under the new title of the Mekong River Commission, with Thailand, Laos, Cambodia, and Vietnam all participating. But the agreement was short on specifics, and a concrete work program was not agreed to until 2001, which, according to one analyst, was a "shift from a project-oriented focus to an emphasis on better management and preservation of existing resources." It totally eliminated all plans for mainstream dams and any other large-scale infrastructure development. The pressure for this came from the newly constituted environmentalist hit-squad called the Global Environment Facility, started by the World Bank but transformed into an independent body in 1992 at the Rio Earth Summit.

The next blow to the Mekong came in 1997, when the World Bank joined forces with the International Union for the Conservation of Nature (IUCN), perhaps the first overtly green-fascist institution, founded in 1948 by Sir Julian Huxley to re-create the pre-war racist eugenics societies under a supposedly more respectable name. Together the Bank and the IUCN created the World Commission on Dams. As the Stimson study proudly reports, as a result of this effort "the World Bank and the Asia Development Bank have largely abandoned the construction of large dams in the Mekong Basin and limited their role to supporting environmental mitigation and resettlement, funding scientific research [meaning global warming studies and related anti-scientific quackery], and in some cases providing risk insurance for

This map of the Mekong Delta highlights the fact that only in China have the necessary dams been built, or are under construction.

commercial loans that are contingent on environmental impact assessments and social impact assessments."

The fundamental lie used by these green fascists is that the dams and other large-scale infrastructure development would be harmful to the poorest of those families dependent on farming and fishing in the region. And yet they admit that the region is among the poorest regions in the entire world, while not admitting that this is entirely due to the lack of the development process they are dedicated to preventing. The Stimson study, in reference to the continuing efforts of the Mekong countries to develop their economies, says: "Mekong governments are making economic development decisions now that if not soon reconsidered based on more careful study, will imperil food security and livelihoods, threaten domestic political stability, and put great stress on still distrustful regional relationships."

Target—China

The Stimson study also reports: "The main threat comes from the construction by China of a total of 15 large to mega-sized hydropower dams on the river's mainstream in Tibet and its mountainous Yunnan Province, and the granting of concessions by downstream Southeast Asian countries to Chinese and other foreign development companies for the construction of 11 dams on the lower half of the same river."

The Mekong River Commission (MRC) itself at least admits that while "societal values shifted" to environmentalist, anti-growth policies in the West after they had already achieved their basic infrastructure development, "the situation in most poor countries is very different, and particularly in the Lower Mekong Basin due to its unique 20th Century history. Nevertheless, there are strong pressures today on the Lower Mekong Basin countries to adopt a different development path to growth to that taken by industrial countries."

The Asian Development Bank in 1992 formed a separate institution called the Greater Mekong Subregion (GMS), which included, in addition to the four members of the MRC, Myanmar and the southern provinces of China, Yunnan and Guangxi.

The GMS has been less focused on water development on the Mekong itself, and more on creating infrastructure for the region as a whole, and has made far more progress, especially in transportation infrastructure, than the MRC has on water development. The East-West Highway project has connected Vietnam's coast, through the Laotian panhandle, and through Thailand toward the Myanmar border. Plans are now underway to extend the route through Myanmar, Bangladesh, and India.

China, in the meantime, is building an "Oriental Express" rail line from Kunming to Singapore. As of September 2014, the Thai government has approved a plan for China to construct high-speed rail lines connecting Bangkok both to the North and to the Northeast of Thailand, while Laos is negotiating with China to build the portion of the North-South line that passes through Laos.

Thailand's Kra Canal, Keystone for South Asian Development

By Michael Billington

August 2014

In October 1983, *EIR* and the Fusion Energy Foundation, both founded by Lyndon LaRouche, held a conference in Bangkok, co-sponsored by Thailand's Ministry of Transportation and the Global Infrastructure Fund (GIF), part of Japan's Mitsubishi Research Institute, promoting the construction of a sea-level canal across the Isthmus of Kra in southern Thailand. A second conference on the same theme, also in Bangkok, was held a year later, in October 1984. Launching of this great project, which would represent a "keystone" for economic development throughout the Pacific region, was close to realization.

The process was subverted, by a combination of foreign intervention and opposition from certain powerful forces within Thailand. However, despite extreme instability in Thailand at the time of this writing, including a military seizure of power in May 2014, the potential for launching this project is definitively back on the table, with significant backing from China and Japan.

The Canal

Although the shipping distance saved by the construction of the Kra Canal is not comparable to that of the other two great canals, the Suez and the Panama—it will shorten the length of a trip from the Indian Ocean to the South China Sea by about 900 miles—it will nonetheless carry as much traffic as either of those canals, due both to the shorter route, and to the overcrowding of the shipping lanes in the Malacca Strait. That waterway carried more than 50,000 ships per year in 1983, but *EIR* projections at the time indicated, correctly, that economic growth in China and India would necessitate an additional route via a sea-level canal.

But the concept behind the Kra Canal goes far deeper than simply facilitating shipping. As Lyndon LaRouche told the 1983 Bangkok Conference: "The prospect of establishing a sea-level waterway through the Isthmus of Thailand, ought to be seen not only as an important development of basic economic infrastructure both for Thailand and the cooperating nations of the region; this proposed canal should also be seen as a keystone, around which might be constructed a healthy and balanced development of needed basic infrastructure in a more general way."

That conference, entitled "The Development of the Pacific and Indian Ocean Basins," presented the Kra Canal, together with construction of new deep-water ports at either end, and industrial zones in adjacent areas, as the central hub of an Asian-wide development approach based on projects including the development of the Mekong River basin, major water control projects in China, and water and power projects in the Ganges-Brahmaputra region of India. This, in turn, was part of a global "Great Projects" approach promoted by LaRouche, and by Mitsubishi's GIF.

The conferences also presented stark warnings that the failure to build the Kra Canal, and the industrial development parks associated with it, would lead inevitably to turmoil in the regions of southern Thailand, already suffering from underdevelopment and ethnic tensions between the Buddhist and Muslim populations in the region. Further, it was warned that the overcrowding of the Strait of Malacca would create a strategic crisis, because the Strait is a key bottleneck for oil and other trade for the Far East, especially a growing China, and thus vulnerable to sabotage and piracy.

Project Dimensions

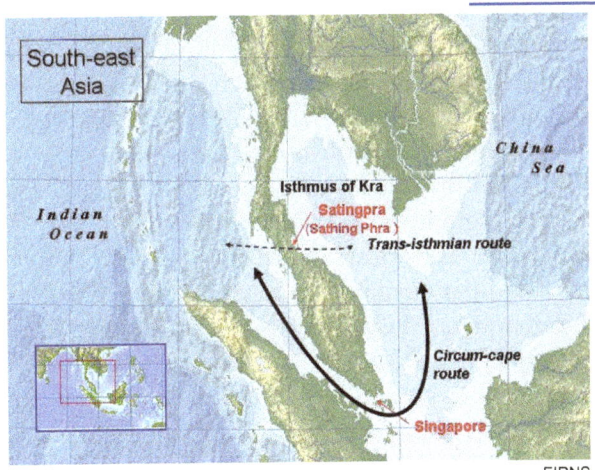

The proposed site of the Kra Canal

We can compare the dimensions of a proposed Kra Canal with other well-known canals. The width of the Kra Isthmus at its narrowest point is about 27 miles; compare this to the width of the Panama Canal—48 miles. The length of the various proposed locations for the Kra Canal range between 30 and 60 miles. The Suez Canal, for comparison, has a length of 119 miles.

The height of the interior mountain chain where the Kra Canal would be constructed is about 246 feet. Compare this to the height at the Galliard cut of the Panama Canal, which is slightly lower, at 210 feet.

The Strait of Malacca is not sufficiently deep for many large ships to pass through. The Strait is 620 miles long but very narrow—less than 1.6 miles at the narrowest and only 82 feet deep at the shallowest point. Currently, large ships are required to travel much further south, to the Lombok Strait, near Java, which has a depth of 820 feet. The Kra Canal would save about 1,200 miles in shipping transport in Asia; its depth, as projected in engineering studies, is expected to be 110 feet.

The Malacca Strait is by far the most heavily traveled of the world's strategic passageways, with more than twice the traffic of the Suez and Panama Canals combined. By a recent estimate, one-fifth of world trade goes through the Malacca Strait. Congestion or obstruction of the Strait, whether accidental or intentional, would dramatically increase the cost of trade and would cause severe danger to the economies of East Asian nations, which depend on oil from the Mideast.

Peace Through Development

A report on the 1983 Bangkok Conference, published in *Fusion* magazine (July/August 1984), addressed Thailand's security issue: "A major included strategic factor also deserves the attention of Thai policy makers. Contrary to some reported opinion and concern that a canal through the southern part of the Golden Peninsula would have negative security implications, severing the ethnically and religiously ill-integrated southernmost part of the nation from the rest of the country, the opposite consequence would be the projected outcome. The canal complex, as a major industrial growth-spot, would function as an integrating and unifying factor, joining together the southern, central, and northern provinces in a large common endeavor capable of inspiring the entire nation, uplifting the economic condition of the southern population, and thus reducing the potential for dissatisfaction and dissension."

General Saiyud Kerdphol, a former Supreme Commander of the Thai Armed Forces, in addressing the 1984 *EIR* Bangkok Conference, said: "Development and security must go hand in hand as a coordinated effort. We must recognize that economic, political, and social development all contribute to security—but that security, in itself, is not development."

Lyndon LaRouche addresses a conference promoting the Kra Canal in Bangkok, Thailand in October 1983.

Lyndon LaRouche addressed the importance of the Kra Canal for the entirety of the Asia-Pacific region in a May 2014 interview with the *Fortune Times*, a Chinese language journal in Singapore, saying that he had "specific, professional knowledge concerning the importance and feasibility of such an undertaking, and its profound implications for the entirety of major neighboring regions such as, most prominently, China and India, but also the entire region of the China-India development process throughout the Pacific region generally....

"Divide the maritime region of East and South Asia into three principal categories: China, a giant; India, a giant; and the maritime connection, throughout Southeast Asia's maritime regions, today. Add the impact of such triadic maritime and related connections, to the physical-economic relations to the Americas to the East, and the Middle East's underbelly and Africa. Then, the potency of a Kra Canal development appears not only as an eminently feasible feature, but a strategic political-economic force for the planet.

"The most common opposition to the Kra Canal, from within that region itself, is located precisely in Singapore. The chief source of resistance from Singapore, is entirely, global, British-imperial military-strategic interests. The completion of the Kra Canal, is not technologically difficult, if and when we take into account the massively beneficial impact of the creation of such a project; it would be principally the British imperial-strategic interests in the entire Indian-Ocean region, which has long remained the principal obstacle to the Kra during modern times. There are two truly great nations in Asia: India, and the more populous China. The sheer volume of maritime trade between the two great nations of Asia, and their connections through the South Asia maritime regions, represents the potentially beneficial, and also efficient project for the entire region of the Pacific and Indian Ocean regions, and the co-development of the major regions of Planet Earth as a whole. Water is still the most economically efficient mode of economic transport among the regions of the planet.

"On Singapore as such, Singapore itself, when freed from British strategic imperatives, will benefit far more from the success of the Kra Canal development, than that development would ever bring without the development of the Kra!"

The role of the British in preventing the construction of the Kra Canal goes back to 1897, when the British made a secret agreement with Siam (Thailand), which forbade the construction of a canal through the Isthmus of Kra without British consent. The agreement also brought exclusive commercial concessions in the area under British control. This Imperial policy of "no development" continued throughout the 20th Century. At the end of World War II, the Siamese Government, which had allowed the Japanese to occupy the country was forced to impose even stricter limits on its economic development, not to mention the heavy war reparations imposed on them.

Article Seven of the 1946 Anglo-Thai Treaty states that:

The Siamese Government undertake that no canal linking the Indian Ocean and Gulf of Siam shall be cut across Siamese territory without the prior concurrence of the Government of the United Kingdom.

Although this imperial treaty was eventually overturned, the British banking outpost in Asia, the City-State of Singapore, has remained the primary obstacle to the construction of the Kra Canal. However, Singapore agreed in August 2014 to become a Founding Member of the Asian Infrastructure Investment Bank (AIIB), initiated by China to provide a new source of funding for precisely this kind of regional infrastructure development, which may lead Singapore to lift its objections.

Centuries of Plans

The first call for a Kra Canal came from Thai King Rama I in 1793, who proposed a canal from Songkhla on the eastern shore on the Gulf of Thailand, to the Indian Ocean on the western shore, just above the Malacca Strait. The concept was taken up in the 1950s, and again in the 1970s, but a combination of instability internally and in the region, due to the colonial warfare in Indochina, prevented any significant regional cooperation.

However, a feasibility study, commissioned by K.Y. Chow of the Thai Oil Refining Company, was completed in 1973 by the American engineering firms Tippetts-Abbett-McCarthy-Stratton (TAMS) and Robert R. Nathan Associates, in collaboration with Lawrence Livermore National Laboratory. The study was updated by the Fusion Energy Foundation in preparation for the 1983 conference.

Peaceful Nuclear Explosives

A key aspect of the planning for the Kra Canal has been the potential advantages of the use of peaceful nuclear explosives (PNE) to carry out the excavations on the most difficult terrain. Today, the use of PNEs is completely left out of all discussions of the Kra Canal, due to the hysteria created by the enemies of development against anything nuclear. This particular form of anti-scientific brainwashing was not as extensive at the time of the 1983 conference in Bangkok, and the discussion demonstrated the enormous advantages for Thailand and the world in using this safe, controlled form of nuclear explosive.

With PNEs, the construction time and the cost of building the canal would be nearly cut in half. In addition, the assembly of the required advanced nuclear engineering and scientific manpower would facilitate the development of nuclear-related industries, as well as nuclear energy plants. A spokesman from Lawrence Livermore who attended the conference suggested that a major nuclear isotope separation plant could be constructed as part of the Kra Canal complex.

While some industrial progress was made across Asia in the 1980s and early 1990s, the speculative "globalization" bubble of the 1990s drew Asia in—with hot money and process industries substituting for basic infrastructure development—until the speculators pulled the plug in the 1997-98 crash, collapsing the Thai economy under hedge fund looting and International Monetary Fund (IMF) conditionalities.

One of the leading proponents of the Kra Canal has been former Prime Minister Gen. (ret.) Chavalit Youngchiyudh, who also founded the Thai-Chinese Culture and Economy Association. Thailand's close cooperation with China, now substantially stronger than in the 1980s, has created a new interest in the project, because China views infrastructure investment in foreign lands, especially in the Asian region, as mutually beneficial over the long term, rather than restricting investments to projects that promise immediate short-term profit to private investors, as is the dominant Group of Seven (G-7) policy today. This is the nature of China's proposed AIIB, which Thailand has agreed to join as a founding member.

In Japan, the Mitsubishi GIF is still committed to the project, while other leading economists are now deeply interested. Former Japanese Finance Ministry official and Representative to the IMF Daisuke Kotegawa has emphasized that Japan-China cooperation with Thailand in the construction of the Kra Canal, which would be of significant mutual benefit to the two Asian economic giants, represents precisely the kind of project that must be undertaken together, as a means of overcoming the mounting tensions between them.

With the election of Narendra Modi as Prime Minister of India, it is now likely that India will also be anxious to join in the Kra Canal project. Modi's campaign slogan was "development, development, development," and he intends to build on the extremely close economic relations he established with China and Japan as Chief Minister of Gujarat. The Kra Canal will thus become increasingly crucial to the expanding trade between East and South Asia.

Connecting Indonesia to the Eurasian Mainland

By Michael Billington

August 2014

The need to connect Indonesia, the largest economy in Southeast Asia, with the Malay Peninsula—as well as connecting its two major islands of Sumatra and Java—should be obvious. The lack of overland transport connections is a restraint on not only Indonesia, but all of Southeast Asia, China, and India. Projects for building these connections over both straits have been under discussion for decades, but now appear to be on the front burner.

Indonesia, the fourth largest nation in the world by population, is spread out over more than 17,000 islands, spanning more than 3,000 miles. Although separated by water from the Eurasian mainland, the two most populated islands, Sumatra and Java, could be relatively easily connected to mainland Malaysia by road and rail via bridges across or tunnels under the Malacca Strait and to each other via bridges across or tunnels under the Sunda Strait. These two straits are among the most strategically located maritime passageways in the world, connecting East Asia with South Asia, Africa, and the Persian Gulf by sea.

Further bridge and tunnel connections could eventually extend the overland connections from Java to Bali and other Indonesian islands all the way to Timor.

Malacca Strait Bridge

Denmark and China are actively involved in planning for construction of the 127 km Malacca Strait bridge, and a road from Malacca City in Malaysia to the town of Dumai in Riau Province in Sumatra. A 2006 study for the project found it to be feasible, at an estimated cost of $12.5 billion. Malaysia would construct the 48 km span over the Strait, while Indonesia would construct a 79 km highway connecting the bridge to Dumai.

The 48 km span over the Strait would be the longest sea-crossing bridge in the world.

In October 2013, China's Exim Bank agreed to finance 85% of the project, according to the Strait of Malacca Partners, appointed by the regional government to be the master planner and builder, but no fur-

Proposed bridges over the Malacca and Sunda straits would link Malaysia and Indonesia and the Indonesian islands of Sumatra and Java.

ther announcements have been forthcoming. In January 2014, Denmark indicated its interest in the project. Danish Ambassador Nicolai Ruge reported that Denmark had the expertise and engineering capabilities to undertake the project, based on the Øresund Sound Bridge and tunnel, which connects Denmark's Copenhagen with Malmö in Sweden.

The creation of the Asian Infrastructure Investment Bank (AIIB) under China's initiative could serve as a source of funding for this crucial project.

Sunda Strait Bridge

The outgoing Indonesian government of Susilo Bambang Yudhoyono has put a priority on building a bridge over the Sunda Strait, connecting Java and Sumatra, approving the project in 2007. It will entail a 27 km road and rail corridor and include pipelines for water, oil and gas, and fiber-optic cables, with roads and two suspension bridges connecting Java and Sumatra through two islands in the Strait, at an estimated cost of $10 billion. Here, too, the new AIIB could come into play.

In March 2014, the government announced that the project would proceed, despite the change of government in October.

Approximately 20 million people crossed the Strait in 2006, mostly by ferry, with that number expected to double by 2020, or even more with the completion of the bridge. Java, with 141 million people, is the most populous island in the world, while Sumatra, with 50 million people, is the fifth most populous.

With the Sunda and the Malacca Strait bridges, the huge cities of Java will have access by road and rail to all of Southeast Asia and China, vastly expanding the capacity for trade and communication.

PART 8
Australia—Driver for Pacific Development

A Vision to Bring Australia into the Land-Bridge Process

By Robert Barwick

August 2014

Australia is the empty continent, kept fallow by two centuries of British colonialism, lately in its environmentalist form. There is, however, a vigorous and growing political movement in Australia called the Citizens Electoral Council (CEC), allied with Lyndon and Helga LaRouche, which has a vision to develop and populate the vast expanses of the continent. Brought to life, this vision will contribute decisively to the future of the entire Pacific Rim growth area.

The CEC's vision includes reviving Australia's once-great agro-industrial capacity, developed in the emergency conditions of World War II. For a number of decades, post-war Australia boasted advanced engineering and manufacturing, centered on a world-class machine tool sector, which underpinned a highly skilled, high-wage work force, and established the nation as a resource for skills and capital goods for the surrounding region of developing economies.

From the 1980s onwards, the global drive for financial deregulation and free trade overran Australia, and systematically stripped away this capability. Free trade annihilated manufacturing, and reduced Australia's physical economy to a raw materials quarry; deregulation transformed the financial system into a casino of debt and derivatives. The CEC's development program will revive this lost capability, so that Australia can not only transform its own continent, but also contribute resources, high-end capital goods, and skills to a new era of development of the Asia-Pacific.

Currently Australia has only three people per square kilometer—23 million people on 7.6 million square kilometers. By comparison, the current population density of the United States, including Alaska, is 32 people per square kilometer; the state of California has 89 people per square kilometer.

From its colonial days, Australia has included people with a strong impulse to develop. With a few exceptions, the British largely suppressed this impulse. They parceled out huge tracts of land to pastoral houses controlled by London financial interests, which focused on grazing sheep for wool production, using cheap labor. The six separate British colonies that in 1901 federated into the nation of Australia all had different gauges of railway track, making a unified national railway system impossible. The British also kept close hold of the purse strings, as the colonies were dependent on London for financial capital. Of the very few development projects that London did finance, the standout was the Perth to Coolgardie pipeline, to deliver water to the booming Kalgoorlie gold mine in the middle of the Western Australian desert.

Machine Tool Manufacturing

Australia's contribution to the World Land-Bridge must be centred on reviving the world-class machine tool manufacturing capability that the nation had developed virtually from scratch during World War II. The story of Australia's machine tool industry is an inspiring proof that economic progress can be rapid and transformative.

Australia's pre-war economy was a colonial, agrarian backwater. It was ill-prepared to fight any war, let alone one for national survival. U.S. General Douglas MacArthur recorded in his memoirs that when he arrived in 1942 to organize the Allies' Pacific defense, Australia's poor military and economic capacity was the "greatest shock of the war." The reality of Australia's peril forced the government to launch an economic mobilization that rapidly transformed the nation into a productive powerhouse, in a way not dissimilar to what Franklin D. Roosevelt did in the United States.

Because of the pressing need for weapons and ammunition, one of the highest priority areas of the economic mobilization was machine tool manufacturing. At the beginning of the war there were only three machine tool manufacturers in the country, and only 15% of the machine tools in Australia were of munitions quality. An official history of the mobilization, *The Role of Science and Industry*, by D.P. Mellor, records the amazing results:

Under Prime Minister John Curtin, shown here (right) with General Douglas MacArthur, Australia built up a world-class machine tool industry, which is needed today.

"The years 1942 and 1943 witnessed an astonishing increase in the number and variety of locally-made machine tools. There was also a great deal of ingenious improvisation in the use of existing machines. Precision tools of a kind whose local manufacture would previously have been regarded as impossible became almost commonplace. At the peak of production in 1943 some 200 manufacturers employed 12,000 persons for an annual output of 14,000 machine tools. By the middle of 1944 what had been Australia's greatest single technological weakness had become a major source of strength."

Australia soon met its own machine tool needs, and exported to Egypt, South Africa, New Zealand, and India. In 1944, an observer noted with surprise, that 7 out of 10 of the 52,000 complex machine tools operating in Australia had been manufactured locally. Program head Colonel Thorpe boasted in 1943 that there was "no machine too large or too intricate for the Australian engineer to tackle."

At war's end, Australia's revolutionized economy allowed the government to pursue plans for post-war reconstruction and industrial development on a scale hitherto unthinkable, which in turn consolidated the nation's new engineering and manufacturing capacity. In 1948 Prime Minister Chifley established a machine-tool-intensive domestic car manufacturing industry, in partnership with U.S. automakers. The following year ground was broken on the enormous Snowy Mountains Scheme (see below). This project even included a brief flirtation with developing a domestic nuclear power industry, which did not come to fruition, for political reasons. Australian industries supported the development of rocket technology, which the Americans and British pursued at the Woomera rocket range in outback South Australia. By the 1960s, Australia was able to launch a satellite into space from its own soil, just the third nation to do so. In that decade agriculture and manufacturing production accounted for 40% of the nation's economic output, and 50% of employment.

Beginning in the late 1960s, this all started to break down under a concerted free trade and green assault. In 1967 Australian machine tool manufacturers lost their tariff protection, and cheaper machine tools from India increasingly displaced Australian production. The British Crown directed the rise of green fascism, which took over all aspects of Australian politics, society, and economy, and put a stop to any great infrastructure projects like the Snowy Mountains Scheme. In the early 1980s Australia adopted a radical program of financial deregulation, privatization, and free trade, which annihilated manufacturing while simultaneously expanding financial speculation, such that by the 2000s manufacturing and agriculture had shrunk to around 10% of GDP, but financial services had ballooned to 20%, from less than 5% in 1960. In the late 1990s the government scrapped the machine tool bounty, which was the final incentive for domestic production, bringing an end to any meaningful manufacturing in Australia. In 2013, Australia lost its greatest concentration of machine tools and skilled operators, when first Ford, then General Motors, and then Toyota announced an end to car manufacturing in Australia.

The Australian economy now consists of a raw materials quarry, shipping mountains of iron ore and coal to Asia; a derivatives-addicted financial sector; and a speculative bubble in housing. However, Australia's own experience in World War II proves how rapidly this industrial decay can be reversed, when national survival demands it.

The Requirements for Development

The first prerequisite for Australia to play its rightful role in the global development process is for it to regain its sovereignty. This means leaving the British Commonwealth, becoming a republic, and nationalizing its huge raw materials deposits, some of the richest in the world, from British Crown cartels such as Rio Tinto and BHP Billiton.

In addition, Australia must return to its tradition of national banking, which had been introduced to the nation by an American immigrant named King O'Malley in 1911. O'Malley modeled his banking proposal on Alexander Hamilton's First Bank of the United States, declaring in a 1909 speech to Parliament, "I am the Alexander Hamilton

FIGURE 1
Snowy Mountain Scheme

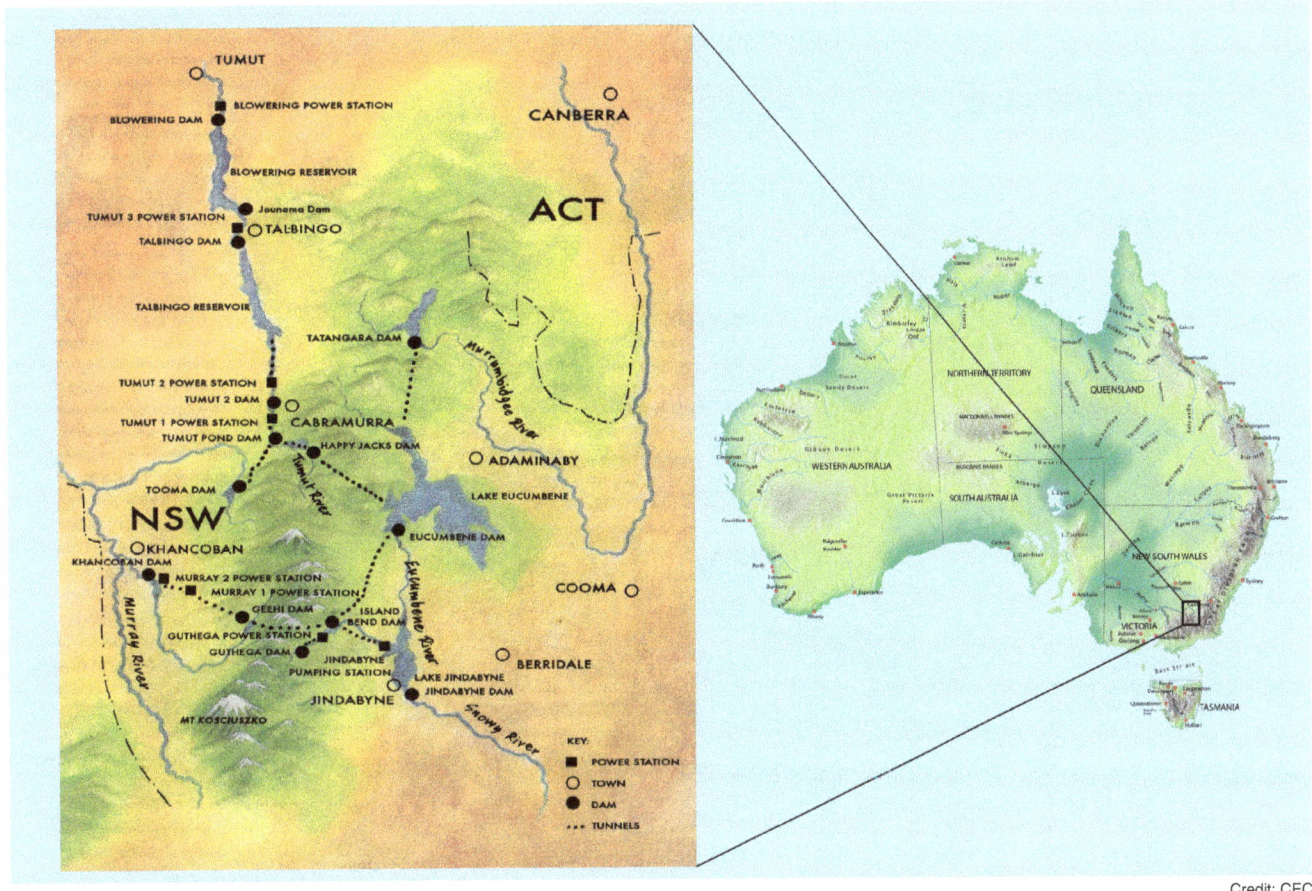

Credit: CEC

The Snowy Mountain Scheme, a major source of water for irrigation and electricity production, was built on the model of the TVA.

of Australia. He was the greatest financial genius to ever walk the earth and his ideas have never been improved upon." Thanks to O'Malley, national banking became a cornerstone of the Australian Labor Party's platform for many decades, and survived to provide the credit for the nation's greatest infrastructure project, the Snowy Mountains Hydroelectric Scheme (**Figure 1**).

Started in 1947 and completed in 1975, the Snowy transformed Australia, creating its agricultural breadbasket, and it remains a model for what Australia must now do on an even grander scale. In 1967, the American Society of Engineers hailed the Snowy Scheme as "one of the seven engineering wonders" of the modern world. It covers an area of 7,780 square kilometers, with 16 dams, seven power stations, 145 kilometers of tunnels, and 80 kilometers of viaducts. It diverts the headwaters of the Snowy River and two other rivers, westward across the Australian Alps into the irrigation areas of the Murray and Murrumbidgee rivers, thus creating one of the most productive irrigated areas in the world—the Murray-Darling Basin.

The Snowy Scheme, which was developed by engineers trained in the United States, and in collaboration with the Americans, is a model

FIGURE 2
Australia and the Water Projects It Needs

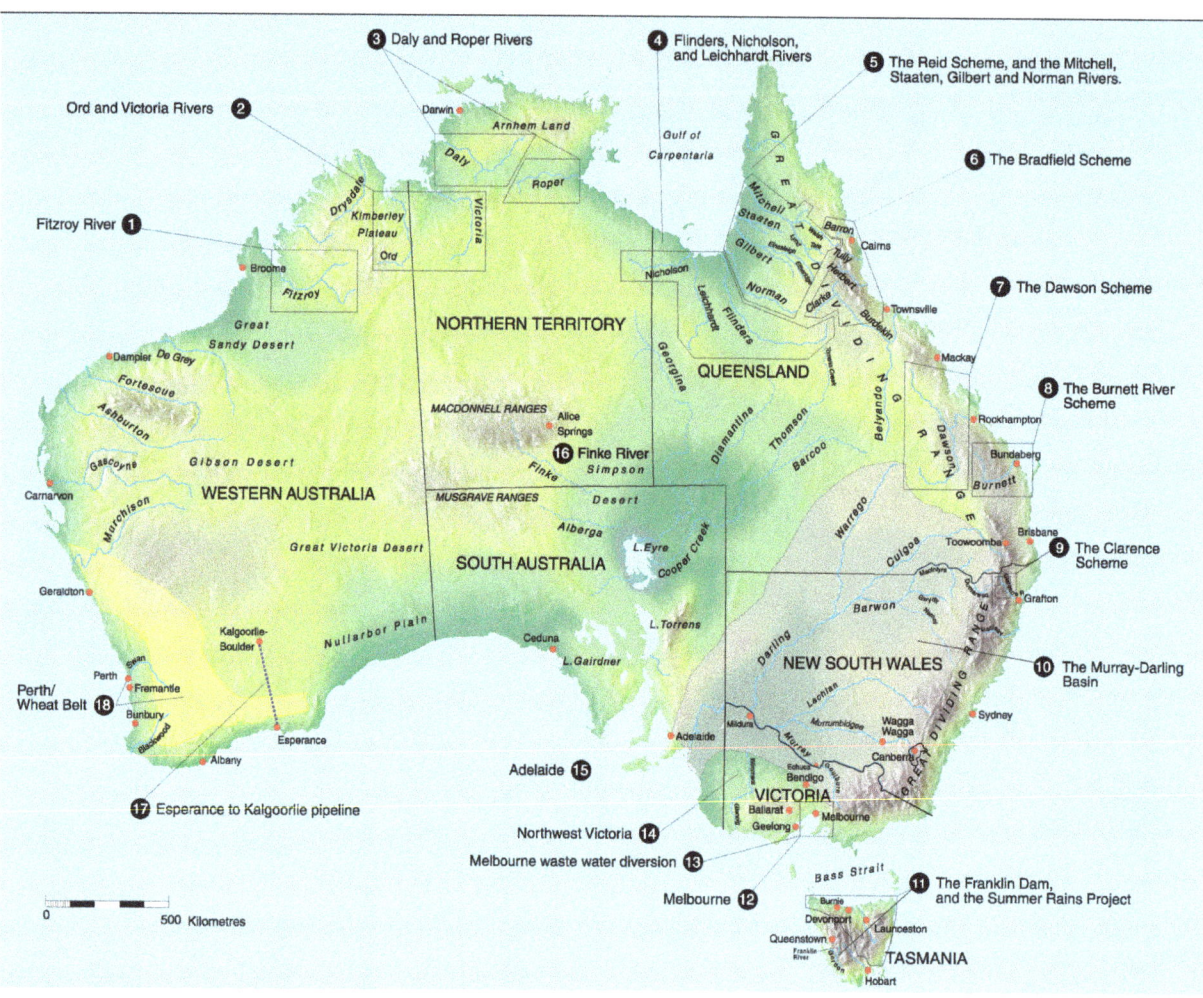

Credit: CEC

This map is the centerpiece of the CEC's "Infrastructure Road to Recovery" plan, devised in conjunction with engineer Lance Endersbee.

for a series of other projects, 18 in all, which are urgently needed today, to green the desert and provide adequate water for the continent. When constructed, these projects would transfer water from rivers in coastal mountain ranges, such as the Clarence River in New South Wales on the East coast and the Fitzroy River in the West, inland to the arid regions.

The Bradfield Scheme

One of the most exciting of these projects is the Bradfield Scheme for Watering the Inland (**Figure 2**). It was drafted in 1938 by Dr. J.J.C. Bradfield, the engineering genius who built the famous Sydney Harbor Bridge. Updated in the 1980s, this brilliant project for biospheric engineering has been gathering dust on the shelf for decades. We will take it as a case study for what must be done.

The highest rainfall area in Australia is in tropical North

Queensland, where warm, moist air from the Pacific Ocean is pushed up sharply by the Great Dividing Range, which runs the length of Eastern Australia. The Tully and Herbert Rivers, only 123 and 340 kilometers long, respectively, carry this rainfall straight back out to sea. Bradfield proposed a series of dams and tunnels to divert the headwaters of these two rivers through the Great Dividing Range, down into the fertile but dry plains of central Queensland, and eventually into Australia's inland salt lake, Lake Eyre. The project will reach from the far northeast of Australia, down across the continent to Lake Eyre in south-central Australia.

Bradfield envisaged not only the diversion of water for irrigation, but the transformation of the now-arid inland climate, through establishing vegetation and an increased permanent cycle of rainfall.

High-Speed Rail: The Asian Express

Australia is also in urgent need of an upgrade of its transportation system. Before his death in 2009, leading Australian engineer Professor Lance Endersbee designed a program he called the Asian Express.

A high-speed rail grid uniting Australia's major cities, in combination with a high-speed national shipping line, the Asian Express would place any part of Australia just one to four days from any part of Asia, including the world's two largest ports, in Singapore and Hong Kong, and many other major ports (**Figure 3**). The concept started as a Melbourne to Darwin fast-freight railway, which could transport high-value horticulture products from the southern states to Australia's northernmost port in just 24 hours. Endersbee explained that the Asian Express would conquer the "tyranny of distance" that has always limited agricultural production in the areas along the route to crops with a long shelf-life, such as wool, grains, beef, and canned and dried fruit. The fast freight service would open up the land along the route to higher-value production including dairying and intensive horticulture, to supply the populations of Australia.

Professor Endersbee later expanded his concept into the Australian Ring Rail, to go around the top end of the continent and connect in Perth to Australia's existing east-west trans-continental railway, the Indian-Pacific.

To achieve the full potential of this project, Australia must establish high-speed shipping services into Southeast Asia. The distance from Darwin to Singapore is equal to the length of the Mediterranean, and except for monsoon season, the sea state is mostly calm, which will enable fast ferry services from Darwin to the huge ports of Singapore, Java, Hong Kong, and Koahsiung in Taiwan. Australia is a world leader in constructing high-speed catamarans: Tasmania's Incat and Western Australia's Austal Ships hold many speed records, including the fastest trans-Atlantic crossing and the longest distance travelled by a commercial passenger ship in one day. Both companies have developed their largely passenger/vehicle catamarans into fast freight carriers using container and roll-on, roll-off technology.

FIGURE 3
Container Ports in Asia

Credit: CEC

With high-speed rail connecting to high-speed shipping, all of the great Asian container ports, among the biggest in the world, will be within 1-4 days transportation from factories and farms in Australia.

Nuclear power

Australia contains one-quarter of the world's known reserves of uranium—more than any other nation—and one-quarter of the world's reserves of thorium, similar to India. Thus the nation represents a key resource for the development of a nuclear power grid that can supply cheap, abundant power and desalinated water to the nation. Currently Australia has one small research reactor, and a hysterical green movement determined to suppress nuclear power. However, there is a growing movement in Australia's universities to train engineers and scientists in fission and fusion technologies. With Australia's abundant domestic reserves of nuclear fuels, it has the potential to develop economies of scale in nuclear power generation that will contribute to breakthroughs in fission and fusion worldwide.

Space

Australia has participated very closely with the United States on the Apollo manned Moon missions, and is well-positioned to pick up this role again. In the 1960s and 1970s, more than 2,000 American, British, and European rockets were launched from the Woomera site. It ceased operating in 1976; since then, the nation's potential as a base for space exploration has not been matched by its political leadership.

There are a number of sites that are serious contenders to host a space base. Woomera is the tested site, but it is not as close to the equator as other options; proximity to the equator takes advantage of the centrifugal force created by the Earth spin, which enables the launch of heavier payloads.

Cape York Peninsula at the northern tip of Queensland is an excellent option, just 12 degrees south of the equator. Various politicians over the years have proposed its development as a space base, but the government has not aggressively pursued this opportunity.

The most interesting recent proposal has come from the Russian government's Aviation and Space Agency, which has expressed interest in developing a commercial launch facility on the Australian territory of Christmas Island in the Indian Ocean. Russia is attracted by the proximity of Christmas Island to the equator. An international consortium called the Asia Pacific Space Centre, which includes participation from Australian and South Korean investors, garnered a commitment from the Australian government in June 2000 to contribute $52 million to the project to upgrade the transport and infrastructure on the island.

BIBLIOGRAPHY

For further reading on projects needed to build up Australia's capabilities, see *The New Citizen*, "Australia's Blueprint for Economic Development," April 2006, available at www.cecaust.com.au.

PART 9
Europe — Western Pole of the New Silk Road

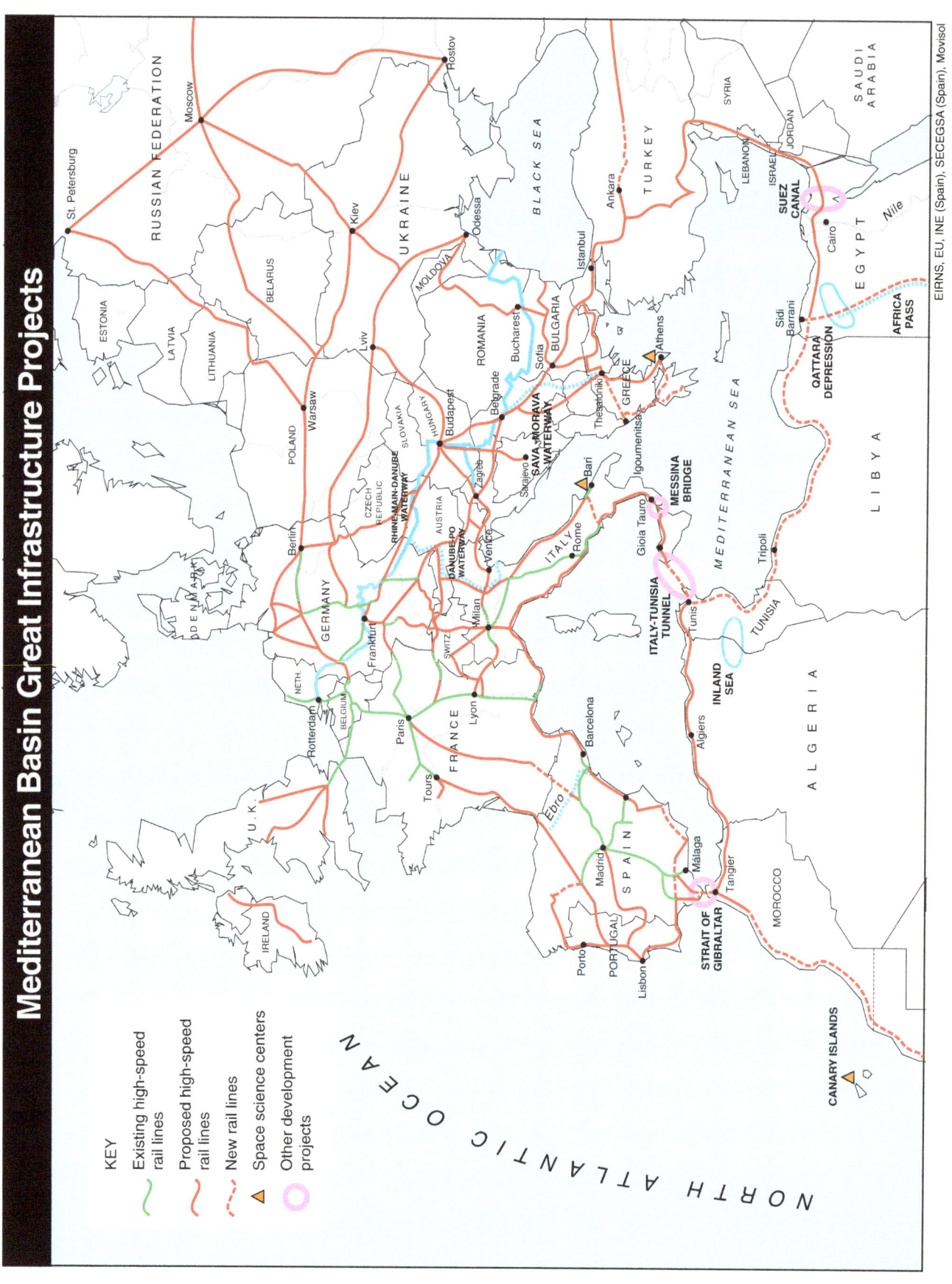

Germany: The Key to European Integration into the New Silk Road

By Bruce Director

August 2014

The key to European integration into the New Silk Road development corridor is the Federal Republic of Germany, which since Roman times has been at the center of trade and cultural development between Asia and Europe. The ancient east-west trade route in Europe, the Via Regia, which extended from Spain to Kiev, passed through Germany, and connected to the Silk Road through Russia and Central Asia to China. The north-south road from the Baltic Sea to Italy also crossed Germany. Thus German cities such as Leipzig and Frankfurt became international trading centers. German centers of mining, industry, culture, and education have historically been mixing bowls for collaboration between Europe, Russia, and Asia. With its industrialization under Bismarck, Germany became the strategic center for the economic development of the Eurasian continent, thus drawing the ire of the British Empire, which orchestrated the dumping of Bismarck and more than 70 years of warfare, from 1895 until the fall of the Berlin Wall 25 years ago.

Since 1989, Germany, whose post-war economic miracle was based on the production and export of high-technology machine tools and industrial goods, has increasingly turned toward the decaying zero-growth environmentalist and monetary policies of the British Empire. German financial institutions have been integrated into the City of London and Wall Street, and have joined the ranks of the world's "too big to fail" zombie banks. Germany has shut down nuclear energy and spoiled its countryside with silly, non-performing windmills. It stopped developing frontier technologies such as magnetically levitated (maglev) trains, only to see the happy development of these technologies in China. Once the pinnacle of classical culture, the nation of Bach, Beethoven, Brahms, Schumann, and Schiller, Germany now sees its stages taken over by Bertold Brecht's Regietheater, the avant-garde, and worse.

To fulfill its rightful role as a key component of the World Land-Bridge, Germany must reverse these disastrous policy trends and implement the following reforms:

1. Introduction of the Glass-Steagall separation of the banking system

Just as Franklin Roosevelt, with the original Glass-Steagall Act in 1933, put Wall Street effectively on a leash and thus defended the common good, today the business operations of normal commercial banks must be hermetically separated from those of investment banks. Then speculative losses will hit the authors of such schemes, while depositors and taxpayers will no longer be liable for more losses, as is currently the case under the "bail-in" clause in federal German and European laws, to which the German finance ministry has committed itself.

2. No to "bail-in"—no expropriation of savers and the middle class

The expropriation of all accounts of more than 100,000 euros (as happened in 2013 in Cyprus) would not only affect the savers, but also municipalities and somewhat larger concerns—that is, all those who have major liquidity requirements in order to meet their ongoing obligations (e.g., salaries, suppliers, and other costs).

3. State credit creation by establishing a National Bank and regional development banks

If Germany is to achieve full and productive employment, it needs a credit system that provides the means, at low interest, for developing infrastructure and for research and development (see below). In contrast to the current practice by the central banks of printing money in an unsuccessful attempt to rescue ailing banks, this form of credit expansion would not be inflationary, because the funds would be coupled to productive capacity, and thus increase tax revenue. A newly structured Kreditanstalt für Wiederaufbau (KfW), the German version of the Reconstruction Finance Corporation, could once again, as in the time of the 1950s Economic Miracle in West Germany, provide long-term, low-interest loans through state development banks for specific projects. Some states, such as Saxony, already have such banks; back in 1991, after its founding, the Development Bank of Saxony had such a perspective.

4. Concentration on infrastructure and high-technology areas

Infrastructure and high-technology industry increase society's overall productivity by physical standards (output/km²), and from the standpoint of the common good. These include, above all, infrastructure projects to link up with the New Silk Road and Eurasian Land-Bridge, as well as science and high-technology projects that overcome the existing limits of human knowledge and allow an increase in our knowledge of the universe (e.g., nuclear fusion, plasma physics, and space flight).

5. Dialogue of cultures and cultural renaissance

The unifying cooperation in economy, science, and culture is the best guarantee of a long-term perspective for peace. Germany can make a significant contribution because of its own rich cultural history, a tradition which must be revived. In the state of Saxony alone lived such

Overview of Main Rail Routes

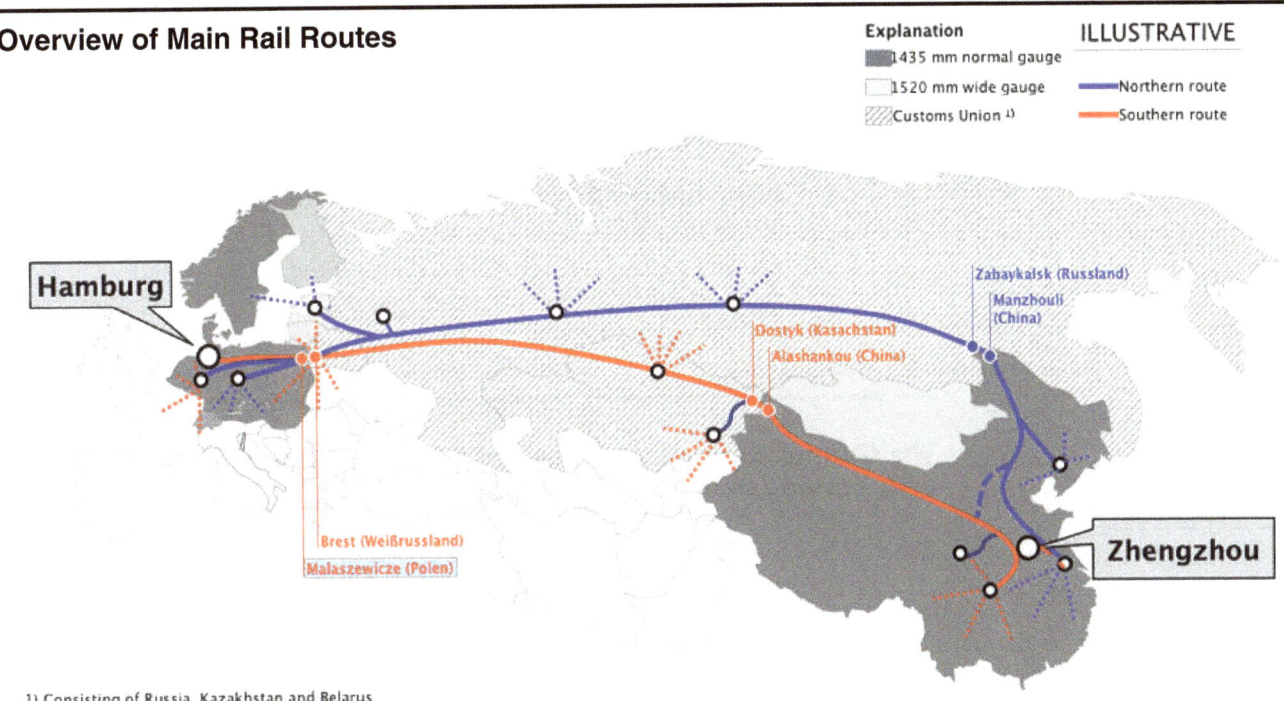

1) Consisting of Russia, Kazakhstan and Belarus

Source: *Railway Gazette*

This map shows the two regular rail routes from Germany to China. Besides Hamburg, the other two German rail hubs for the traffic are Duisburg, in the Ruhr region, and Leipzig, as described in the article.

spirits as the great polymath Gottfried Wilhelm Leibniz, one of the most important German influences on Russia, China, and the United States. Friedrich Schiller, the poet of freedom as he was called by the former slave and friend of U.S. President Lincoln, Frederick Douglass, worked in Saxony and left his mark. But also, the tradition of the classical music of Johann Sebastian Bach, Felix Mendelssohn Bartholdy, Robert and Clara Schumann, and Johannes Brahms is still a unifying force for humanity and should be promoted much more. Germany thus has a special responsibility to realize an image of man that places human creativity at the center, and contributes to the universal "pursuit of happiness."

Transportation Infrastructure Requirements

But Germany can only make a productive contribution if the physical lifelines of necessary infrastructure are in good condition, or are created. Railways, roads, and waterways urgently need to be renovated nationwide. Almost every year the ADAC, the German affiliate of the American Automobile Association (AAA), and other institutions warn that Germany's entire road and bridge network is outdated and urgently requires repairs. The problem of closure of railway stations and routes is well known. There is also the lesson to be learned from the major floods that have devastated parts of Germany since 2002—now that rivers are no longer sensibly maintained, their flow-capacity decreases, and they cannot accommodate water in the same quantities as before. Therefore there is an urgent need to address this in the context of transnational projects for water management.

Two areas are of paramount importance—the rail network, which will play a particularly important role as part of the Eurasian Land-Bridge, especially with the involvement of maglev technology, and the necessary expansion of waterways.

1. Construction and Modernization of Rail Transport

Since 2008, the Trans Eurasia Logistics (TEL) has existed as a joint project of the German Bahn AG with the Russian Railways RZD. A logistics network came into being along with the railway, reaching from Duisburg to Beijing, so that today, goods can be transported

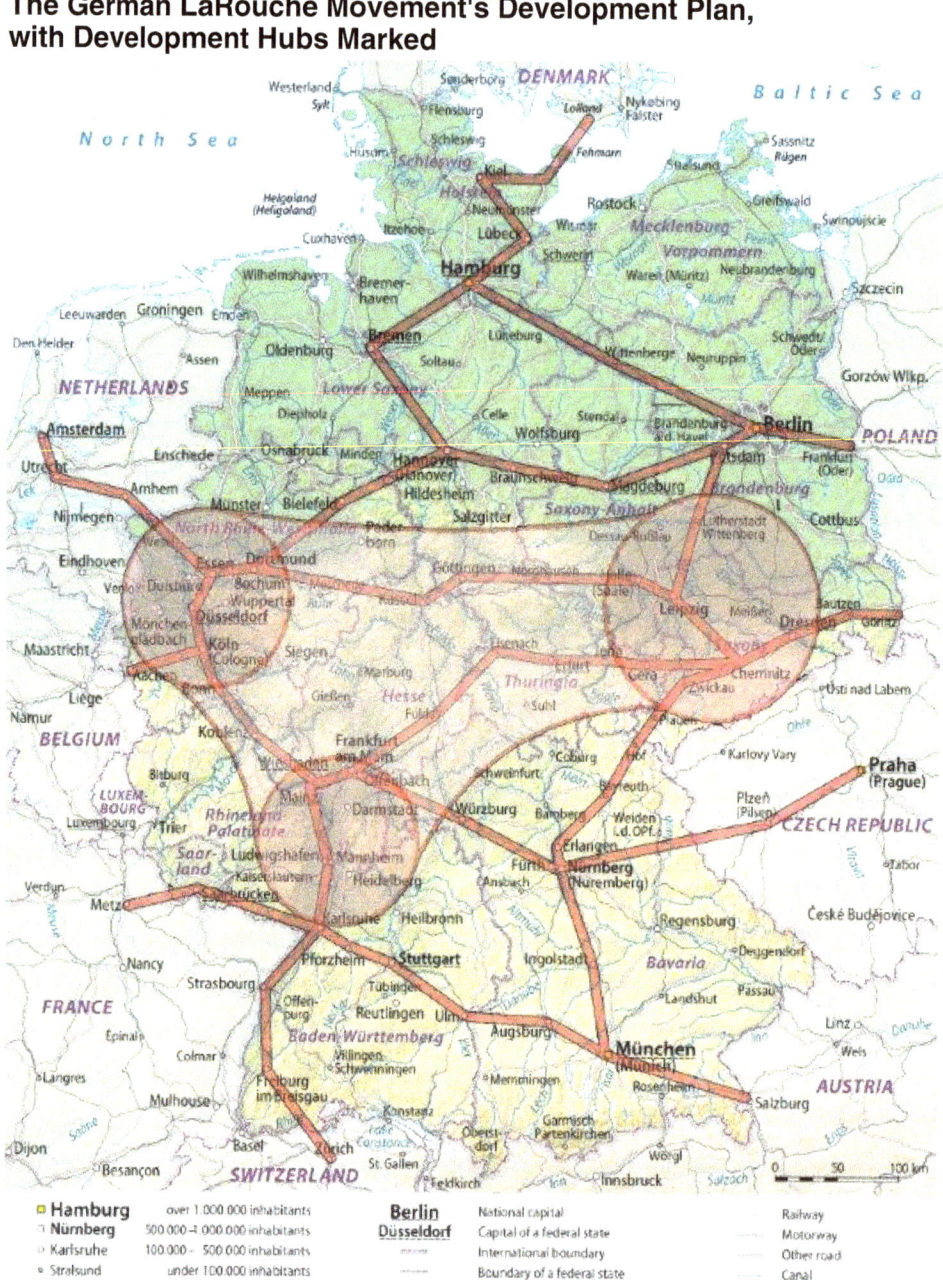

The German LaRouche Movement's Development Plan, with Development Hubs Marked

and distributed on seven rail routes across all of Eurasia:

• Three routes between Europe and Russia: the Moscovite, Tubeteika, and Matroschka lines

• Two routes between Europe and China: New Silk Road and Tiger Train

• Two routes between Russia and China: China Russia Multinet and Central Asia Express

All seven sections travel over two main routes—one in the north on the Trans-Siberian Railway, and one in the south through Kazakhstan. These are main infrastructure corridors of the Eurasian Land-Bridge: Corridor A in the north is already fully existent, and the Asian part of Corridor B is at our disposal.

Here, the approximately 12,000 km of the northern route can be traveled in 19 days and the 10,000 km of the southern route in 18 days. At a price of 2,200 euro per 40-foot container, rail transport is one-third more expensive than by ship, but more than twice as fast.

Since 2011 there has been a train with 36 containers traveling every day from Leipzig-Wahren to Shenyang in China. At first, it only carried quantities of the 8,000 components from BMW to its new factory in China. But these trains can also handle 50 containers. BMW has required a 63,000 m² logistics center in this context, and thus created 600 jobs. Dr. Karl-Friedrich Rausch, chairman of the board of DB Mobility Logistics AG & Transport, described this connection as a "significant impulse for the Eurasian Land-Bridge."

Vladimir Yakunin, head of Russian Railways, has explained that currently 80% of trading volume moves from east to west. He welcomed the Leipzig logistics center, because it represents the first step toward improvement in changing this ratio. The increase in goods exported from west to east would mean a significant boost to the German economy.

In the context of these developments, the Leipzig region will become a major European logistics center and thus one of the most important gates to the East. As a centuries-old trade fair city, such a role corresponds to the character of Leipzig much more than does the role of a post-industrial shopping city.

In addition, one might create a rail corridor along the route from Dresden-Prague-Bratislava-Uzhhorod-Lviv-Kiev. This could link not only to Odessa, but also to Crimea. From there an important strand of the New Silk Road could come into being, over the bridge that Russia plans to build across the Azov Sea, and a continuation to the southern Russian growth center Krasnodar, then north along the Caspian Sea along and through Kazakhstan to Western China.

Russia, whose railway chief Yakunin announced plans on May 22 in St. Petersburg to build a new trans-Siberian connection from Moscow to Vladivostok using maglev technology, also would gain a great advantage from this second route via Krasnodar. All the regions along this route can benefit from the fact that a maglev container freight version will allow goods to be transported by rail at 200 km/hour instead of by truck at 80 km/hour over longer distances.

Discussions are already under way in China and Russia on emphasizing maglev technology instead of trains with wheels in the future. Germany should finally promote this technology and incorporate it in the proposed projects, wherever it makes sense. Görlitz, as Saxony's carriage-building site, can be a central location for the production of Transrapid trains, which will function best when combined with a maglev line connection to Dresden, so that the trains from the Görlitz plant can travel directly to the new long-haul network and be deployed on the future routes from Dresden to Prague, Dresden to Kiev, Cracow to Dresden, and in the triangle Berlin-Leipzig-Dresden.

2. Development of Inland Waterways

In the Czech Republic and Poland, the construction of the Danube-Elbe-Oder canal is being discussed. Germany should participate. At the same time, the Elbe River should be expanded; to implement this, the water level at Dresden must be regulated through a series of weirs and locks, so that even in the summer months, it will be high enough to permit the passage of ships. For all three countries, Germany, Poland, and the Czech Republic, new conditions will arise for a flourishing of industry, thus raising the productivity and the living standards of the population.

A second major water project for Germany will be the completion of the Elster-Saale canal for freight and passenger transport. The construction of this canal, first proposed in 1856, was stopped due to the war in 1942, and should be completed now. For this purpose, the harbor and a canal about 8 km long must be completed to the Saale River, where a lift lock should be installed. This will require the construction of three locks, at Halle, Planena, and Werder (Merseburg), and a straightening of the watercourses at Calbe.

The Rhine-Main Canal

Source: EIR, 1992.

The Rhine-Main canal, traced here, first conceived by Charlemagne in the 9th Century, was finally completed in 1992.

Thus with 1.2 billion euros, divided between the two states Saxony and Saxony-Anhalt, Leipzig could be connected via the Saale to the Elbe. Leipzig is already a factor in Eurasian commerce, with its new rail network links to China. This position could be enhanced through this canal, and lead to a flowering of industrial prosperity in the Halle-Leipzig region.

Greece and a Marshall Plan for the Mediterranean

by Dean Andromidas and Marcia Merry Baker,
adapted from EIR's "An Economic Miracle for Southern Europe and the Mediterranean!"

September 2014

The future of the economies of Greece and of all the Balkan countries lies in taking advantage of their geostrategic location in the eastern Mediterranean, to make them the economic development gateway to Eurasia to the northeast, Southwestern and Southern Asia to the east, and Africa to the south. For Greece, this is an historic role. For the Balkans, this is a critical mission, to serve as a corridor of peace and development, and to uplift a war-torn area, stretching from the north Adriatic, eastward through Ukraine, and southeastward across to Southwest Asia.

On June 19, 2014, Chinese Premier Li Keqiang met in Athens with Prime Minister Antonis Samaras of Greece, after which Li announced specific joint projects, but stressed overall that the two ancient civilizations of the East and West—China and Greece, respectively—enjoy a glorious history and culture. They made special contributions to civilization, and now they will collaborate on the future. Specifically, they will build the port of Piraeus into a regional transit center and gateway for trade to all Europe.

The principal, intercontinental vectors of this region's connections are evident in the 1997 east-west "three corridors" on the Eurasian Land-Bridge (see Part 1). The Balkan peninsula is at the Mediterranean Basin juncture of these routes, and with full intermodal development for trade and transit—rail, road, waterways, air, ports, and sea—the critical geo-position of Greece and the Balkans can be maximized for the benefit of all.

First, consider these intercontinental corridors more closely; next, look in brief, at a few of the priority regional corridors across the peninsula itself.

Greece and the Balkans connect to the north, into the full Eurasian east-west land-bridge development corridor. To the west, via the Rhine-Main Canal, there is the connection to the international ports of Antwerp, Rotterdam, and Hamburg. To the east, via the Danube corridor, there are links into the Black Sea Basin. This continues eastward to the Dnieper River, the Don-Volga Canal, and deep into central Asia and western Siberia, via the Caspian Sea. Greece and the Balkans will thus be integrated into the trans-Eurasian rail corridors spanning the landmass.

Greece and the Balkans also connect to the east/southeast by railway corridors leading into Turkey, across the Anatolian peninsula, then branching eastward into South Asia, through Iraq and Iran, to the Indian subcontinent.

This Anatolian peninsula overland route also proceeds southward through the trans-Jordan, across the Sinai, into north and east Africa.

The connections by sea in the Mediterranean and outward worldwide are self-evident, but an entirely new horizon of transportation and trade is embodied in the current expansion of the Suez Canal.

Peninsular Corridor Priorities

A quick overview of the priority transportation/development routes and regions across the Balkans and Greece can be obtained by starting with the picture more than 20 years ago, of what were identified as ``priority corridors'' for modernized rail-lines (and implied, related road, water, and other infrastructure), stipulated by transportation ministers at the March 1994 Second Pan-European Transportation Conference on the island of Crete. There were 10 European corridors designated, of which four traverse Greece and/or the Balkans.

Map 1 shows a May 1994 European Community Transport Infrastructure map, from the Crete meetings, presenting an ``Outline Plan for a European High-Speed Train Network—2010.'' Besides a high-speed rail line shown for Greece itself, vector-arrows elsewhere in the Balkans show the direction of other routes to be worked out.

Needless to say, very little indicated on the envisioned ``2010'' map has materialized, with one of the few exceptions being the historic completion in 1992 of the Rhine-Main-Danube Canal, creating a waterway corridor all across Europe, from the Black Sea to the North Sea, as first envisioned over a millennium ago by Charlemagne (see Germany section).

Today, this transportation development perspective must be reactivated on an emergency basis. The specific Balkans priority transportation links, as first proposed at the 1994 Pan-European conference, out of the 10 designated corridors are:

Corridor 4. On the major west-east link across Europe, going from Berlin to

MAP 1
The Pan-European Transport Corridors

Wikimedia commons

Nine of the 10 corridors, first identified in 1994, are shown on this map, although very few have been completed.

Istanbul (Berlin/Nuremberg-Prague-Bratislava-Gyor-Budapest-Arad-Craiova-Sofia-Istanbul), there must be branch links between Sofia and Thessaloniki.

Corridor 5. On the major west-east link between northern Italy and Ukraine, there are important branch links into the Balkans. The main corridor is Venice-Trieste/Koper-Ljubljana-Budapest-Uzhgorod-Liviv, extended through Rijecka-Zagreb-Budapest and Ploce-Sarajevo-Osijek-Budapest.

Corridor 8. The Adriatic Sea to the Black Sea, from Albania to the ports of Varna and Burgas on the Black Sea (Durres-Tirana-Skopje-Sofia-Plovdiv-Burgas-Varna).

Corridor 9. Going from Greece to Moscow, beginning at the easternmost Greek port of Alexandroupolis to Dimitrovgrad-Bucharest-Chisinau-Lyubaskeva-Kiev-Moscow.

Corridor 10. From Salzburg to Thessaloniki (Salzburg-Ljublijana-Zagreb-Belgrade-Nis-Skopje-Veles-Thessaloniki). The ancient Roman Via Egnatia, from the Adriatic to the Bosporus, is a priority redevelopment route.

Aegean, Adriatic North-South Axes

From Greece northward, two development axes are evident—the Aegean on the east, and Adriatic on the west.

The Aegean north-south axis, beginning in the south with the port of Piraeus, and proceeding northward, via Thessaloniki (second largest city in Greece) to the Danube valley, encompasses the routes designated above in Corridors 4 and 10, and is a powerhouse for development.

The port of Piraeus, at Athens, was until the current crash, the tenth-largest container port in Europe and its largest passenger port. Up until now, it has been Greece's only major port, with little transshipment. But its potential to serve as an international entrepôt is clear. The China Ocean Shipping Co. (Cosco) has leased one of the port's two container terminals for 35 years. Piraeus serves as a hub for China's exports into Central and Eastern Europe.

The BRICS impetus throws into focus the need to upgrade the entire rail and road grid of Greece and the Balkans, to allow fully intermodal freight traffic. This flies in the face of the killer intent of the 2012 Troika austerity memorandum order to Greece, to shut down its rail service going outside of the country!

There is a need to double-track the entire length between Athens and Thessaloniki, which requires construction of several tunnels through the mountains. High-speed rail between these two cities will cut travel time from six hours to less than three.

The development corridor continues northward from Thessaloniki along the Axios River, which becomes the Vardar in the Former Yugoslav Republic of Macedonia (FYROM). Via the northward-flowing Morava River, the way proceeds to Nic and Belgrade. To the east, a corridor reaching to Thessaloniki provides Sofia, Bulgaria, with even closer access to the sea than the Black Sea ports of Burgas and Varna.

Turning to the west is the axis known as the Ionian/Adriatic

MAP 2
Railway Axis of the Ionian/Adriatic Intermodal Corridor

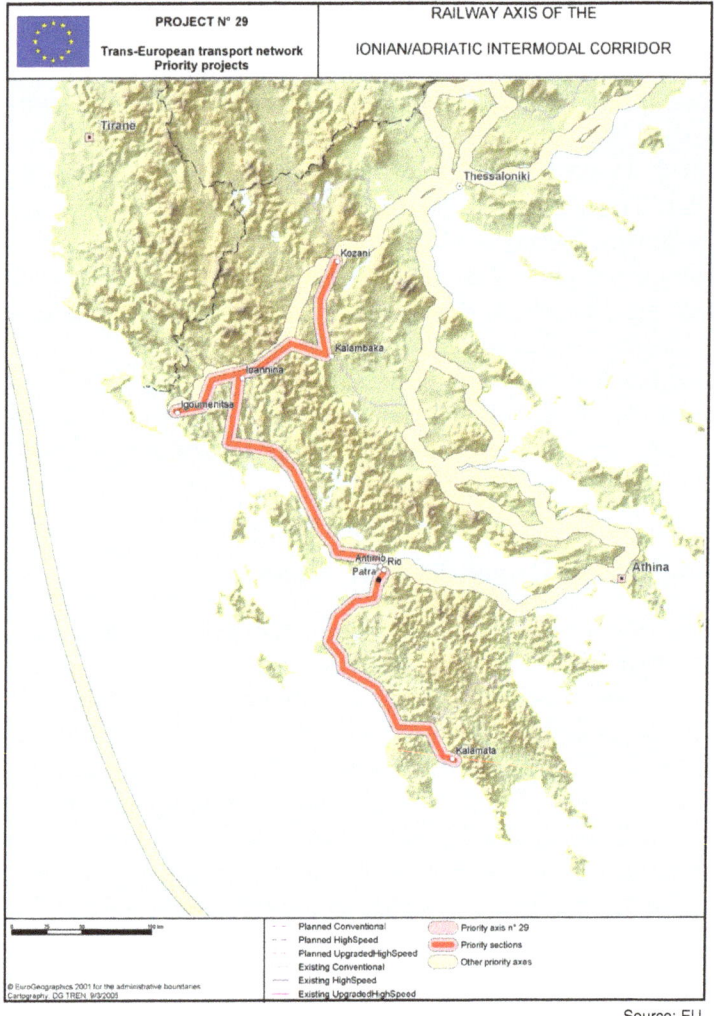

Source: EU.

Intermodal Corridor (shown in **Map 2**). The Pan European Plan (Corridor 7 described above), traces the priority route for modernized rail, to connect this region into Eurasia.

On the Adriatic coast of Greece is the port of Igoumenitsa, one of the most important ports in the region. There is a project underway to further develop the link between the port of Taranto, Italy's second-largest, and Igoumenitsa, and then through the Egnatia Odos Motorway, across northern Greece, linking it with the ports of Thessaloniki, Kavala, and Alexandroupolis, and then with Istanbul. Thus it would provide access to all the Balkans, including Albania, the FYROM, and Bulgaria.

To the south, there is the port of Patras on the northwestern tip of the Peloponnese, with its recently completed South Port, strategically located to serve the new traffic flows from the enlarged Suez Canal.

Power, Water, Agro-Industry

Along with the outlined transportation infrastructure, the necessity of upgrading power, water, health systems, and other basic infrastructure is obvious. The top of the agenda is nuclear power. There are no nuclear reactors in Greece, nor the republics of former Yugoslavia. Nuclear-powered desalination is particularly required in the many Greek islands. This economic platform will then serve the vast improvements required in agro-industry. Other projects pending, include the South Stream gas pipeline project being promoted by Russia, which will traverse the Black Sea and supply gas to all the countries of the Balkans, as well as Italy and Western Europe.

As early as possible, the food import-dependence of Greece and the Balkans can and must be reversed. This condition results from globalization of agriculture imposed under the EU and World Trade Organization regime, and the Troika diktats. As much as 40% of Greece's food is imported. One immediate policy matter, is to shift the cotton-for-export production, into food crops.

In total, only 20% of the land area of Greece is suitable for agriculture, but there are pasturelands, hill farms, delta lands, and coastal plains in the inventory, all of which can be made highly productive. In this respect, there must be full application of space-based infrastructure—GPS/GLONASS precision farming—with all needed inputs and mechanization. Greek agronomists have already done all the groundwork. What is required is full-scale deployment of these potentials.

Greece, Seafaring to Spacefaring

Greece has the world's largest merchant fleet. In addition to what this signifies for general economic activity, it also involves a precious resource of skilled labor in the maritime, industrial, and machine-tool sectors, and a shipbuilding capacity capable of tooling up for high-tech tasks. This resource is vital for the Mediterranean development drive overall.

One notorious, negative feature of the Greek shipping sector should be identified, however. The industry is historically an integral part of the financial complex of the City of London, in service for decades to the British Empire. But now, with the crash of the monetarist system, this British-centered nexus of insurance, shipping, and commodity control is in chaos.

Under a new Mediterranean Basin Marshall Plan approach—in the worldwide BRICS context, the valuable Greek shipping capacity can be redirected into heavy-duty service for development.

The final note on Greece, is that its seafaring future extends to the skies. The ancient Hellenes were the original ``Peoples of the Sea,'' as immortalized by Homer's *Iliad*, but today, that legacy can be expressed by transforming a nation of seafareres into a nation of spacefarers.

For example: The Elefis Shipyard not only builds state-of-the-art vessels, rail cars, and other craft, but also it built the Delta-Berenike, a self-propelled special purpose vessel that is being used as a stable platform to build the Cubic Kilometer Neutrino Telescope, one of only four that exist in the world today. It has been erected at a depth of 5,200 meters, the deepest point in Europe. The site of the telescope is 17 km off the coast of the Peloponnese. The headquarters of the project is in the small city of Pylos on the Bay of Navarino. Ancient Pylos, which is a few kilometers from the modern city, is the location of the palace of Nestor, of *Iliad* fame, giving the name Nestor to the project.

Greece and the Balkans can move to the front lines of the world extraterrestrial imperative. The institutions are in place, although under assault from the EU and Troika, to the point of some 12,000 Greek scientists having left their homeland for work abroad. Restoring the mission of development, means that Greece is well placed to further research by scientists from throughout the region. There are centers at the National Observatory in Athens, the National Center of Scientific Research at Demokritos, institutes at the big universities of Athens and Thessaloniki, and several others. One critical specialty is remote-sensing for earthquakes.

The National Center of Scientific Reseach was founded in the 1950s as the Nuclear Research Center of Demokritos, with a boost from the U.S. Atoms for Peace program, from which it received an experimental reactor. The founding of this institute initiated a wave of repatriation of scientists at the time, who had been conducting research abroad because of the lack of opportunities in Greece. Today, with the BRICS initiatives, this science mission can be resumed to proceed with achievements never before imagined.

Italy: Build the Mezzogiorno, and a New Renaissance

by Claudio Celani

This is adapted from the June 2012 EIR Special Report on "An Economic Miracle for Southern Europe and the Mediterranean."

Speaking about a ``Marshall Plan'' for Southern Europe, the mind goes naturally to the original Marshall Plan which, although only a shadow of Franklin Roosevelt's real intention for postwar world reconstruction, supplied urgently needed credit for the reconstruction of Europe. Italy owes its reconstruction to that credit, but also to the competence of its ruling class of that time, which was able to use it for policies and institutions modeled after the best experience of FDR's New Deal.

The Fund for the Development of Southern Italy (Cassa per il Mezzogiorno), established in 1950, was perhaps the experience closest to the New Deal on the European level. The Cassa is still today a model for the development of Southern Italy and other underdeveloped regions of the Mediterranean area.

Italy's Mezzogiorno, with a population of 20 million, includes the regions of Molise, Campania, Basilicata, Puglia, Calabria, and the islands of Sicily and Sardinia.

This region saw strong sustained development from 1950 to 1965, and less so until 1975, when it was interrupted. Italy today could have the highest productivity in Europe, with a northern part that is as productive as Germany, and a southern part that is exactly one-fourth less productive than the North. Accordingly, whereas unemployment in the North is around 10%, in the Mezzogiorno it is more than 25%. The rebirth of the Mezzogiorno means the rebirth of Italy.

Thanks to the Cassa, the development of Italy's Mezzogiorno took off, going through a decade (1950-60) in which, for the first time, the income of southern families grew at the same rate as the income of northern families.

Private land ownership, the figure of the ``independent farmer,'' appeared in Southern Italy only in 1950, with the De Gasperi[1] land reform that distributed 30% of the latifundia to farmers. The Cassa was fundamental in ensuring that the new farmers would get credit

---

1. Alcide De Gasperi (1881-1954) served as prime minister during 1945-53; and as foreign affairs minister and interior minister before that.

and means for productive improvements, including irrigation, seeds, machines, and livestock.

In the 1950-60 decade, the Cassa was flanked by the role of the state conglomerate IRI in building infrastructure and industries throughout Italy, and by the state oil company ENI (Ente Nazionale Idrocarburi) in providing cheap energy through the discovery of large gas reservoirs in the northern Po plain. The steady 7% yearly growth was called the ``economic miracle''; inflation was defeated, and for a short time, even became negative. The national currency, the lira, was recognized for its stability. In 1959, full employment was reached.

When, in 1975, the role of the Cassa was abruptly downsized through the devolutionary introduction of regional governments that took over jurisdiction of long-term investments, the Cassa per il Mezzogiorno had created 2 million hectares of irrigated land; built 62 dams, 52 aqueducts, and numerous sewage systems; modernized 20,000 km of roads, and built 6,000 km of new ones; electrified railway lines; and started numerous industrial centers. However, the job was only half done.

After the model of the Tennessee Valley Authority and the New Deal projects for the Appalachian region, the Cassa was given unprecedented technical competence and power, including funding to finance a 10-year program that was drafted and executed by the Cassa itself, under approval of a special government committee composed of the Minister for the Mezzogiorno, and the ministers of the Treasury, Finance, Public Works, and Labor.

In addition to the long-term projects, which the Cassa leaders drafted with an integrated approach, new projects could be adopted yearly, according to the changed situation. The Cassa's structure allowed it to move funds it had earmarked for a project, to another project, if priorities changed. Local authorities were forced to collaborate with the Cassa and put their competencies at its disposal. As the Cassa's long-time president, Gabriele Pescatore, often said, the aim of the Cassa was to create ``a process of self-subsistent capital accumulation.''

The regional devolution meant a shift from a unitary integrated approach for infrastructural development of the entire Mezzogiorno, to local approaches and views, breaking up the unitary vision and ending the development process, which degenerated into localism and clientelism.

Today, the original approach of the Cassa per il Mezzogiorno must be revived, if we want a rebirth of Southern Italy, and a locomotive for the entire Italian economy, and that of all the Mediterranean.

Springboard for Development of North Africa

If we want to plan a rebirth of the Mezzogiorno, we have to consider its geographical role in the center of the Mediterranean, and its potential land connection from central Europe through the Italian peninsula, to Africa.

From its northeast to its southernmost point, the Island of Lampedusa stretches for 1,291 km as a natural ``bridge'' between Northern

Africa and Central Europe. It is 140 km from the coast of Tunisia, and 70 km from the Albanian coast. There are projects to connect at least one of these two distances through an undersea tunnel.[2]

Italy is the only country belonging to ``Southern Europe'' that has a self-subsisting industrial capacity, able to provide capital goods for itself as well as other countries. Italy has the second-largest manufacturing sector in Europe, after Germany. The problem is that this industrial base is concentrated in Northern Italy, and partly in Central Italy, whereas Southern Italy is underdeveloped.

Italy's industrial potential is now blocked by its loss of sovereignty. The euro system vetoes the creation of credit for development, and forces industrial companies to outsource production. These two main problems must be eliminated through re-establishing monetary and credit sovereignty, and protective measures of commerce.

If this is done, Italy can go back to the FDR-style methods used in the postwar reconstruction, and use its huge scientific and industrial potential to develop its southern part, while helping develop neighboring countries, such as Greece, Spain, Portugal, and Northern Africa.

By extending its capacities in the Mezzogiorno, northern Italian industry will enjoy the unique advantage of being closer than any competitor to its export markets. Italy's Mezzogiorno must become the production site for capital goods for itself and for the entire Mediterranean region.

Defeating the Environmentalist Mercenaries

Any development program in Italy must reckon with an occupation force called the environmentalist movement. After 1987, environmentalists have, for two decades, successfully prevented any major infrastructure from being built in Italy, inducing anti-science and technology psychosis in the population. An attempt by the central government to bypass this in 2001, with a bill called ``Legge Obiettivo'' (Objective Law), which made local approval not binding for a list of strategic infrastructure projects, has been only partially successful. A program for an Italian economic recovery must therefore involve a war against the foreign occupation force, the environmentalist movement, steered by London. This must be conducted both at a cultural level, by organizing the population with cultural optimism, bringing forward the real values of the Italian culture rooted in the 15th-Century Renaissance, as well as on the political-intelligence level, exposing and destroying the foreign intelligence networks controlling the environmentalist operatives.

Here are the main projects to be implemented.

Energy

Energy is the main deficit item in the Italian trade balance. Italy imports 78% of the energy it consumes, both as electricity, and as fuel

---

2. See Dr. Nino Galloni, ``The Sicily-Tunisia Tunnel: Link to Africa,'' *EIR*, Feb. 25, 2011.

for industrial and domestic consumption. Of its electricity, 12% (43 TWh) is imported from France, Switzerland, and Slovenia. Of the electricity produced domestically, 66% (230 TWh) comes from imported natural gas. Coal is 18% and oil is 16%.

This causes energy prices for production to be on average 30% higher than Italy's industrial competitors. To stay in the market in today's insane system of free trade and globalization, Italian producers are thus pressured to reduce labor costs. Due to this and to the higher taxation (more than 50% of the gross wage), Italian wages are among the lowest in Europe.

This is the result of the demolition of Italy's nuclear capability which, in 1966, was the third-largest in the world after the United States and Great Britain; in 1987, when that capability was shut down, Italy was the technology leader in Europe. A solution to Italy's energy problems will come through a massive comeback of nuclear energy.

Italy's nuclear tradition goes back to Enrico Fermi, the father of the first nuclear reactor, built in Chicago in 1942. Enrico Mattei built the first Italian commercial reactor in 1958. After the first oil crisis in 1973, Italy had four active nuclear plants and the government pushed a plan to build six new reactors. A massive British-led economic and political assault against Italy, using the newly-born environmentalist mob, brought the Italian nuclear program first to a stop, then to a shutdown, with a national referendum conducted in 1986 under the emotional shock of the Chernobyl accident.

When the government resumed a nuclear program, planning to build eight new plants in order to achieve 25% of its electricity from nuclear, the same forces organized another referendum in 2011. Destiny had it that the referendum coincided with the Fukushima accident following the Japanese tsunami in February 2011. The massive Goebbels-like media propaganda resulted in another plebiscite against nuclear energy, and the nuclear program was cancelled.

The new nuclear reactors can be built in Southern Italy, starting with one per region—Campania, Basilicata, Puglia, Calabria, Sicily, and Sardinia. With a mixed system of European Pressurized Reactor (EPR) and High-Temperature Reactor (HTR) complexes, production of about 10 GW (10,000 MW) can be reached with the first shot. Antiseismic and other considerations will lead, in some cases, to building the plants on floating platforms off the coast. At the same time, four plants can be built in central and Northern Italy, in Trino Vercellese, Latina, Caorso, and Montalto di Castro, on the same site as the old plants, for a total capacity of about 16 GW. In a second phase, this capacity can be doubled.

Although due to the nuclear moratorium, Italian industry has not built any nuclear plants since 1987, companies such as ENEL, ENI, and Ansaldo (Finmeccanica Group) have continued to participate in international consortia, so that the know-how has been maintained. This means that Italy could start exporting nuclear technology after the first phase of its own nuclear program is completed.

Transportation Networks

A revolution in freight transport is indispensable in Italy, and will produce a great boost in productivity. Currently, only 10% of commercial goods are moved on rail, 0.1% on barges, and 0.6% on coastal waters, despite Italy's 7,750 km of coastline. The huge remainder goes by truck on the roads, with a great expense for gasoline and rubber, and creation of massive traffic congestion. Producers do not use rail because it is slow and inefficient. It takes a container less time to go from Milan to Berlin than from Palermo to Rome. An effort to change this involves upgrading the rail network, making it faster and more efficient.

Currently, Italy is completing its sections of three Trans-European corridors of high-speed rail that connect most of the country's major cities—Corridor 6 (Lyon-Kiev), Corridor 1 (Berlin-Palermo), and Corridor 24 (Genoa-Rotterdam). The Milan-Salerno part of Corridor 1 (**Map 1**), which involved major engineering work in its Bologna-Florence Appenine part because of 73 km of tunnels, is already functioning. The Turin-Venice section of Corridor 6 is being completed. The Milan-Genoa section of Corridor 24 is being developed.

The Italian sections of Corridors 6 and 24 are opposed by environmentalist groups, which are often violent, and backed by the media. The environmentalist mobilization against the Turin-Lyon section, which includes a new 57-km-long tunnel under the Alps, has developed into violent clashes with the police and against the construction site. Recently, prosecutors in Turin arrested 24 leaders of the insurgents, among whom were two former members of the Red Brigades terrorist group.

The same groups oppose the new Genoa-Milan high-speed project.

Once implemented, however, these three lines will not be sufficient. Italy has 55.4 km of rail per 1,000 km², about half the density in Germany (94.5 km). Italy has 238 km of rail per 1 million inhabitants, as compared to 481 in France, and 412 in Germany. The high-speed section is currently 13 km per million inhabitants, as

MAP 1
Italy and the Trans-European Project 29

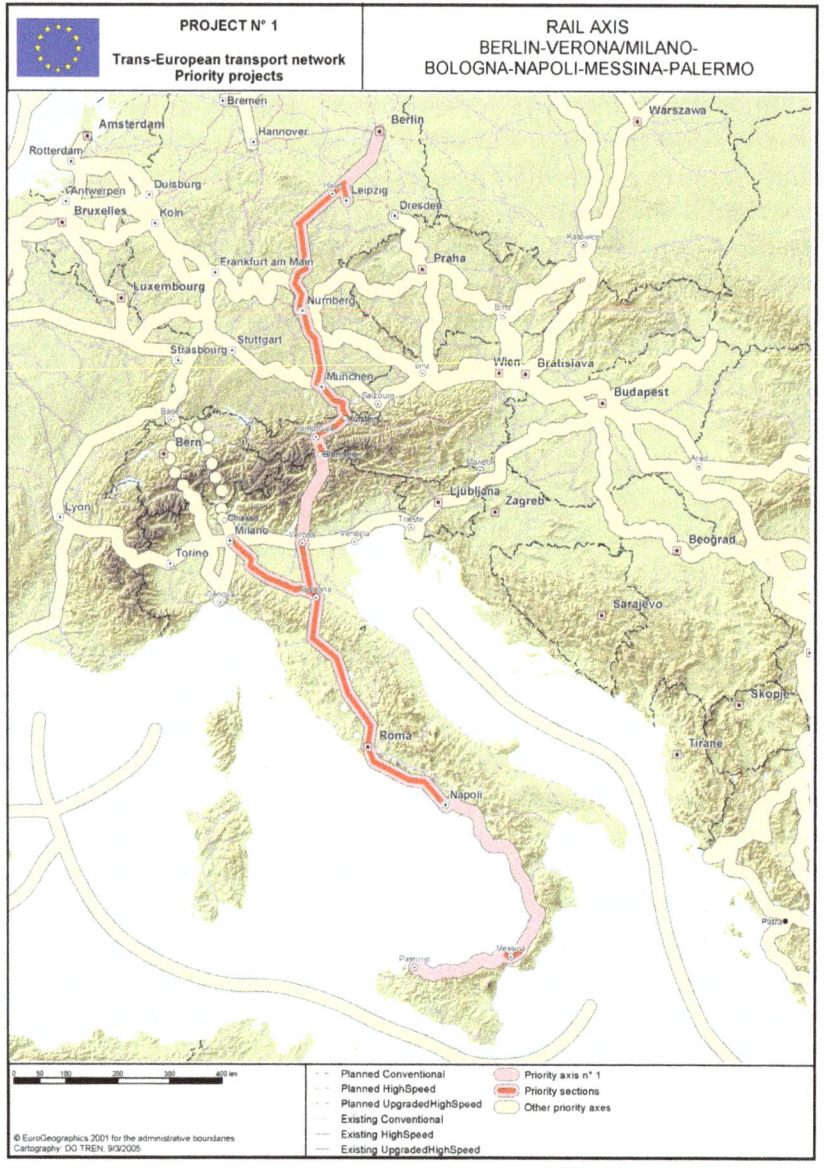

Source: EU

opposed to 16 in Germany, 30 in France, and 35 in Spain. Furthermore, if we take the conventional lines, only half of the total 22,935 km are electrified, and 9,213 km are single rail. The latter case dominates, for instance, in Sicily.

These figures, however, supplied by the national railway company, do not reveal that a large portion of the secondary lines is in a state of decay. This involves connections among minor centers and lines used by commuters.

Thus, an effort to modernize the Italian railway system means double-tracking the single-track lines, electrifying half of the current network, and doubling it on a national scale.

In the Mezzogiorno, railway lines must be quadrupled, and high-speed rail lines must be extended beyond the current southern terminal, Salerno, to the tip of the ``Boot'' and, over the future Messina Bridge, to Palermo.

From Palermo, the line will be continued to the small town of Pizzolato, in the province of Trapani, where an undersea tunnel will connect with Capo Bon, in Tunisia.

The bridge over the Messina Strait will be a major engineering enterprise. At 3.3 km, it will be the longest single-span suspension bridge in the world (**Map 2**).

The bridge will connect the cities of Messina and Reggio Calabria, creating a single, large urban conglomerate, with more than 2 million people. This urban center will be connected by the high-speed line to Central and Northern Italy, to Central Europe, and, via the same line and the Sicily-Tunisia tunnel, to North Africa.

Part of this center, on the Calabrian side, is the deep-sea port of Gioia Tauro, which could become the main port receiving cargo ships coming from the Suez Canal. Currently, 30 million containers per year (20-foot equivalent units, or TEU) move through the Mediterranean, and Italy handles fewer than 4 million, 3 million of which are in Gioia Tauro. At least 20 TEU head to Gibraltar, circumnavigate the Iberian Peninsula, and dock in Rotterdam in order to reach Central Europe. It

MAP 2
Italy: Messina Bridge and Tunnel to Tunisia

Source: Movisol

would be much easier to unload the freight in Gioia Tauro, put it on a train, and ship it to the North, but this is not convenient now because of the inefficiency of the rail connection.

After Gioia Tauro is efficiently connected via rail, starting by making the current conventional rail efficient, while building a high-speed connection to Salerno, freight would take 30 hours or less to reach Berlin, as opposed to the current time of one week.

The high-speed rail must be extended over the Messina Bridge to Palermo and beyond, so that Corridor 1 can be projected all the way into Africa.

This will be achieved with the undersea tunnel to Tunisia, a project of the Italian national research agency ENEA. The distance between the coastlines is about 155 km, and would be reached by five tunnels constructed between four intermediate artificial islands that will be built with the excavated material. There would be two tunnels in each direction, plus one service tunnel.

The tunnel will provide a fast commercial railway route to export capital goods for the development of North Africa, not only from Italy, but also from Central Europe (**Map 2**).

The Maglev

The Italian railway industry has been weakened by the lack of demand resulting from the slow modernization over the past two decades. Thus, Fiat has sold its Fiat Ferroviaria division to the French Alstom, leaving Ansaldo-Breda (Finmeccanica Group) as the only company able to produce modern locomotives. However, the current government is planning to privatize Ansaldo-Breda too, because its balance sheet is in the red. Ansaldo-Breda produces the newest version of the Italian high-speed train, ETR-500, which was designed in the 1980s.

The new private French-Italian company NTV has now been allowed to run on the Italian high-speed line with the most modern version of the TGV (built by Alstom) which has a better performance than the ETR-500. Thus, things are looking grim for the Italian rail industry.

However, the Italians could outflank such problems by going for magnetically levitated trains (maglev), after the Chinese model. The Chinese have obtained a license to build the Transrapid, a Siemens technology, on the condition that they do not sell it abroad.

Waterways

Italy has a very poor internal waterway system. Basically, only the Po River is partially navigable, along with a network of channels in the Veneto-Emilia Romagna region that go back to the time of the Republic of Venice.

And yet, the Lombardy region is studying plans for making the Po entirely navigable from the Adriatic coast to Milan.

At the same time, a major waterway could be opened in North-East direction, connecting the Adige River with the Inn River, creating a waterway that goes from Venice to Passau, connecting the northern Italian network to the entire central European waterway system.

The project, developed by the company Tyrol-Adria AG, foresees a canal-tunnel between the Inn River in Austria and the Adige in Italy, which come within 70 km of each other on the plain. The tunnel-canal would be 78 km long, and would be large enough to allow the passage of barges of the EU Class V. The water would be pumped into the tunnel, creating an artificial current that would push the ships, thus avoiding the use of engines and pollution of the tunnel. The energy for pumping the water is produced by hydroelectric plants built along the Inn.

Space

Italy has a long tradition in aerospace, starting with Leonardo da Vinci's studies of the flight of birds, and in modern times, with the aerodynamic school in the 1930s, to the early phase of the space programs in the 1960s. In 1964, Italy was the third country in the world to put its own satellite in orbit, after the Soviet Union and the United States. Using a platform built in international waters in the Indian Ocean, at the Equator, Italy launched the ``San Marco'' project of five satellites, which were sent into orbit thanks to vectors provided by NASA.

The author of the project was Luigi Broglio, the founder of the Department of Aerospace Engineering at the University of Rome, in 1956.

Since then, Italy has developed its own aerospace industry, which is today part of the state-owned Finmeccanica corporation, and in 1988 established its own space agency, Agenzia Spaziale Italiana. ASI contributed, along with others, to building major parts of the International Space Station. Italian astronauts participate in European Space Agency (ESA) programs and have been on board several missions of the U.S. Space Shuttle.

Recently, ESA completed the project of the European launch vehicle Vega, with an Italian design, and 63% Italian participation, as a coronation of Luigi Broglio's original vision. Vega, able to carry a payload of 1,500 kg into lower orbit, is currently one of the three available launch vehicles worldwide, together with the French Ariane 5 and the Russian Soyuz.

Italy is therefore best fit to play a leading role in a European-Russian-American-Chinese effort for the colonization of the Moon and outer space.

Spain: The World Land-Bridge's Bridge to African Development

by Dennis Small

This report is adapted from the June 2012 EIR Special Report "An Economic Miracle for Southern Europe and the Mediterranean."

FIGURE 1
Economically Active Population (EAP) and Employment
(millions)

	2008	2009	2010	2011
TOTAL EAP	22.8	23.0	23.1	23.1
Employed	20.3	18.9	18.5	18.1
Productively Employed	10.7	9.7	9.4	9.0
—Productively Employed as % EAP	47%	42%	41%	39%
YOUTH EAP (16-24)	2.4	2.2	2.0	1.9
Employed	1.8	1.4	1.2	1.0
Productively Employed	0.8	0.6	0.5	0.4
—Productively Employed as % EAP	35%	27%	24%	20%

Sources: INE, EIR

Spain, today notorious as the epicenter of the current disintegration of the trans-Atlantic banking system, and for having the highest rates of unemployment in Europe—an official 24.4% overall, with shocking youth unemployment of more than 50% (**Figure 1**)—tomorrow will be one of the key geographic and economic bridges from Europe to Africa, in a recovering world economy. It will play a central role in providing crucial science-driver programs, infrastructure, engineering, and capital goods to North Africa in particular; and in the process it will productively employ and re-employ its own massively un-, under-, and mis-employed labor force, most especially its youth, in high-productivity jobs.

In order to create 10+ million new productive jobs in Spain, and to help create further millions of jobs throughout the Mediterranean Basin, Spain—along with its sister nation on the Iberian Peninsula, Portugal—will develop major projects in the following areas:

• **Rail:** Spain will build high-technology industrial corridors on either side of some 15,000 kilometers (9321 miles) of new, high-speed rail lines (including magnetically levitated systems) that will crisscross Spain and Portugal, and link up with the World Land-Bridge in southern France.

• **Strait of Gibraltar Tunnel:** A 40 km (25 miles) tunnel built under the Strait of Gibraltar, from Spain to Morocco, will allow European rail corridors to be connected to future North African rail systems. This will be a project on the scale, and of the significance, of the Bering Strait tunnel and the Darién Gap project, because like them, it will link an entire *continent* into the World Land-Bridge.

- **Water:** Spain will dust off existing, viable water-transfer projects, such as the Ebro River project, to transfer about 1 cubic kilometer of water per year to the semi-arid Mediterranean coast; and it will also produce some 1.5 km³ (.36 mi³) of fresh water yearly with nuclear-powered desalination plants.
- **Nuclear energy:** In addition to the nuclear plants needed for desalination, Spain will build modern nuclear power plants to produce about three times the 7,500 MWe per year that the country currently gets from its eight aging nuclear plants. This will allow Spain to rid itself of the economically destructive (and scientifically incompetent) emphasis on wind and solar power, which has been imposed on it by the British Empire's fascist Greenie movement, led by the World Wildlife Fund (WWF). Where is Don Quixote when we need him?
- **Space science:** The Canary Islands is an ideal location for a new Euro-African space center, including a major satellite-launching facility and related science city. This will be coordinated with critical work being done in Greece, Italy, and other nations around earthquake precursor detection and other endeavors involved in the Strategic Defense of the Earth program, in furtherance of the common aims of mankind.

This will not be the first time in its history that Spain will play a catalytic role at the crossroads of cooperating civilizations. Under the personal guidance of Alfonso X, "The Wise," King of Castile and Leon from 1252-82, the Castilian capital of Toledo was built into Europe's most important scientific center of the time, and the nexus for the transmission of the Greek Classics and the highest achievements of the Arab Renaissance into continental Europe. Alfonso was especially known for his work in astronomy, and for his Toledo school of translation, which brought together the outstanding scholars of the world's three major monotheistic religions—Islam, Christianity and Judaism—to render the most advanced religious and scientific texts of each culture, into the languages of the others.

It is past time for a new "Alfonsi Era."

Great Water Projects

Spain's precipitation produces about 112 km³ (27 mi³) of water per year, which comes to about 2,700 m³ (3531 yd³) per capita per year. That compares to an average of 10,600 m³ (13864 yd³) for Europe as a whole. Of that total available, the amount actually used (withdrawals) is 875 m³ (1144 yd³) per capita per year, which is pretty much on a par with the rest of Europe. But again, the average conceals the fact that the central mesetas and Mediterranean coast of Spain are desperately short of water. As a result, there is serious over-exploitation of aquifers in these drier regions.

Average precipitation in Spain as a whole is 650 mm (26 in.), but most of the central mesetas and Mediterranean coast get under 500 mm (20 in.), and much of that under 300 mm (12 in.) (**Map 1**). The province of Almería in Andalusia is probably the most arid region in all Europe; its Cabo de Gata area receives barely 125-150 mm (5-6 in.) of rain a year. (Arid or desert areas are conventionally classified as

MAP 1
Spain: Annual Rainfall and the National Hydrological Plan (PHN)

Sources: INE (Spain); EIR

receiving 0-250 mm (0-9.84 in.) of precipitation per year; semi-arid is 250-500 mm (9.84-19.7 in.))

Spain has constructed a significant number of dams (the total went from 60 at the beginning of the 20th Century, to about 1,000 today), and has reservoirs capable of storing some 54 km³ (13 mi³) of water—almost half the annual runoff, which is the highest proportion in all Europe. About 80% of all Spain's water withdrawal is used in agriculture, especially in the more productive southeast. About 20% of the agricultural land area is irrigated, and it is estimated that that land produces about half of the country's total food output.

In June 2001, the Spanish government proposed to implement a very modest National Hydrological Plan (PHN), which would have transferred about 1 km³ of water per year from the Ebro River in the northeast of the country, down the Mediterranean coast, complemented by about a half-dozen desalination plants. But it was stopped dead by the British Monarchy's World Wildlife Fund (WWF), and their Greenie allies inside Spain.

Of all Spain's rivers, the Ebro has the highest discharge rate. The average discharge registered in the Tortosa gauging station, located 48 km from the river mouth, was 13.8 km³ (3.3 mi³) per year from 1960 to 1993 (equivalent to an average flow of 425 m³/s) (102 mi³), which is, however, highly irregular over the course of the year. That amount has also been reduced over the years, as more water has been withdrawn upstream, with the 2000-08 average at Tortosa reportedly being 8.8 km³ (2.1 mi³) per year, down from 13.8, two or three decades earlier.

Since the 1930s, 138 reservoirs have been constructed in the Ebro River basin, with a total storage capacity of 6.8 km³ (1.6 mi³)—more than half the average annual discharge from 1960-1990.

The idea of the PHN (see **Map 1**) was to transfer 1.05 km³ (.25 mi³) per year, or about 12% of the Ebro's current annual discharge of 8.253 km³ (1.98 mi³). Of this total amount, 0.19 km³ (.046 mi³) was to be transferred northwards to Barcelona; 0.315 km³ (.076 mi³) south to Valencia; 0.45 km³ (.108 mi³) south to Murcia; and 0.095 km³ (.023 mi³) south to Almería. About 120 new dams were to be built, along with canals and 10 pumping stations. Other than the northward portion for the urban area of Barcelona, the remainder of the transfers were intended for primary use in agriculture.

But by 2004, the Spanish government of José Luis Rodríguez Zapatero had shelved the PHN, and put in its place a program for providing a lesser amount of water to the Mediterranean coast (0.715 km^3 (.17 mi^3)) by desalination plants—a project which predictably never materialized. The prime mover in the sabotage of the Ebro project was Prince Philip's WWF, which is explicitly opposed to any water transfers from one basin to another, anywhere in the world.

The WWF published a report in 2004 which classified Spain as among the three worst countries in Europe in terms of water management, and in a press release headlined "Seven reasons to stop the Spanish National Hydrological Plan," denounced the PHN as "illegal under EU legislation," "not economically justified," and—of course—"environmentally damaging." This led to a European Parliament inquiry (i.e., inquisition), which likened the planning involved "to the old Soviet-style of water management," and demanded the Spanish government answer the WWF's accusations. The upshot was that the project was shelved.

Under our Marshall Plan for the Mediterranean Basin, Spain will immediately restart the stalled National Hydrological Plan's Ebro water transfer project, which will require expelling the WWF, and its influence, from the country. This will produce numerous side benefits, such as ending Greenie mental pollution of the youth, as well as possibly putting an end to the Spanish monarchy—after all, King Juan Carlos is also the honorary president of WWF Spain.

However, the Ebro project alone is insufficient to put a serious dent in the water shortfall in most of the country. An ambitious nuclear desalination project should also be initiated, with which fresh water will literally be manufactured.

The most efficient power source to drive desalination plants is nuclear power. One leading type of reactor is a modular High-Temperature Gas-Cooled Reactor (HTGR), capable of producing 350 megawatts. One "island" of four modular HTGR reactors can produce a total of 1,400 megawatts of power. This level of power, when transmitted to a multi-stage flash distillation desalination plant, will generate about 145 million cubic meters of fresh water per year. It will also generate, beyond that, 446 MW of net electrical output.

If Spain were to build, initially, 10 such nuclear islands, principally along the Mediterranean coast, each hooked up to water desalination plants, it will generate about 1.5 km^3 (.36 mi^3) of new fresh water per year—50% more than the amount to be transferred from the Ebro. That will allow for high-technology agriculture to really take root in the country, along with the numerous downstream industries that this implies.

In this way Spain will become a net food exporter not only to Europe, but to Africa as well.

Full Tilt for Nuclear

Spain will never develop unless it rids itself of the British Empire's green ideology which has taken over the country, especially the youth, and has transformed Spain into a world leader in the clinically insane policy of fostering solar panels and windmills.

Spain got a good start in nuclear energy, beginning construction on its first nuclear plant in 1964, which went into operation in 1968. Over the course of the 1970s and early 1980s, eight nuclear reactors were put into operation. But then in 1983, a moratorium on further nuclear plant construction was adopted under the London-run government of Prime Minister Felipe González (1982-96), which reaffirmed the moratorium in 1994 and abandoned five units that were then under construction.

Today, the country has eight aging nuclear power plants, which in 2010 provided 21% of the country's electricity generation. Natural gas produced 32%; coal 9%; and a stunning 15% came from windmills, and 5% from solar and other so-called renewables (see **Figure 2**). In other words, wind and solar—with their destructively low energy-flux densities—today produce as much electricity in Spain as nuclear energy!

FIGURE 2
Total Electricity Generated, 2010
(thousand GwH)

	Amount	% of Total
Natural Gas	96	32%
Nuclear	62	21%
Wind	44	15%
Solar, Other Renewables	17	5%
Hydroelectric	39	13%
Coal	26	9%
Fuel Oil-Gas Oil	16	5%
TOTAL	300	100%

Source: INE

Over the last few years, vast financial subsidies to wind and solar led to huge increases in the installed capacity in these sectors. But in 2010, the government reneged on its rate subsidies for solar, when budget austerity became the order of the day.

Total electricity consumption in Spain had been rising steadily until 2008, but since then has declined to the current level of about 5,600 kwh/year per capita. Total energy consumption also peaked in 2007, and since then has fallen by 15% per capita. In terms of energy self-sufficiency, Spain is extremely dependent on oil imports: Oil is 47% of total energy consumption, and natural gas another 23%, and in both categories it is all imported. Nuclear is 12% of the total energy consumed, and it is 100% produced in Spain. All in all, Spain only produces about one-quarter of all the energy it consumes.

Under our plan, nuclear energy will replace the insane current emphasis on windmills and solar power, which produce neither the energy output nor the energy-flux density levels required by modern society. Even the addled Don Quixote knew that it made sense to get rid of windmills.

Currently, nuclear produces about 7,500 MWe per year, one-fifth of the total electricity produced in the country. The proposed 10 nuclear islands required for desalination are a good start on improving that situation, generating about 14,000 MWe per year, which will nearly triple the current level. Of this, 9,500 MWe will be "earmarked" for desalination, and 4,500 MWe will be available as net electrical output. A dozen or more fourth-generation nuclear plants will also be built in the interior of the country, to produce some 20,000 MWe per year. This will allow Spain to immediately phase out the economically destructive wind and solar emphasis, and to gradually reduce Spain's enormous dependence on imported oil and natural gas.

In Portugal, at least three such nuclear islands will also be built along the southern coast, to similarly desalinate water and produce net electrical energy.

Building the Bridge to Africa…

One of the bright spots of Spain's physical economy is its rail sector, both in terms of existing infrastructure as well as world-class engineering and production capabilities. High-speed trains now run on 2,600 km of track in Spain, with significant additional lines under construction. The existing government plan—which can never be executed within the euro straitjacket—projects having 10,000 km of high-speed track by 2020.

Historically, Spain has had a different gauge (1,668 mm (64.7 in.)) from most of Europe (1,435 mm (56.5 in.), also called the UIC gauge), which has created major bottlenecks requiring—until relatively recently—transfer of passengers and freight at the French border. Portugal's slightly larger gauge of 1,774 mm (69.8 in.) is inter-operable with Spain's, so the two are often referred to as the "Iberian gauge." This is also a major problem as you move east into Ukraine, Belarus, and Russia, which have a third gauge (1,520 mm (59.8 in.)).

The very raison d'être of the World Land-Bridge, especially as you move into maglev and other high-speed rail lines, demands a solution to this problem. New lines can and should be standardized, but interim solutions to link existing rail networks of different gauges are also required. Rather than transferring passengers and cargo between trains (and switching out locomotives), which is highly inefficient, there is now technology, pioneered by Spanish companies, to automatically change the gauge of the existing axles while the cars are in motion (at about 15 kph (9.3 mph)). This requires axles specially constructed for this purpose.

Spain's Talgo company pioneered work internationally in this area, developing the first commercial application of a track change-over system in 1969. A second Spanish company, CAF, developed its own system in 2003. Other countries now producing similar systems include Poland (SUW 2000, in 2000), Japan (in 2007), and Germany (Rafia, no commercial application yet).

In 1988, Spain decided to construct all of its new high-speed rail corridors at the European (UIC) gauge. There are currently four principal high-speed corridors: Madrid-Barcelona; Madrid-Valencia; Madrid-Valladolid; and Madrid-Sevilla/Málaga (**Map 2**).

There are a number of Spanish companies involved in high-speed rail today, including Talgo, Renfe, CAF, AVE, etc. CAF recently signed contracts for building five high-speed rail lines in Turkey. And Talgo has built and runs rail lines in Kazakhstan, Argentina, the United States, and the Portugal-Spain-France-Switzerland-Italy corridor in Europe. They also just sold 17 cars and one locomotive to Russian Railways, which will now be able to run continuously between Moscow (standard gauge) and Berlin (UIC gauge). Existing high-speed rail lines also link Berlin to Paris and Perpignan, and from there they will go under the Pyrenees Mountains through a new tunnel, to Figueras on the Spanish side, and down to Barcelona and Madrid.

The success of the entire Marshall Plan for the Mediterranean Basin will rely on Spain building on strength, and assuming a leading role in engineering, building, and exporting high-speed rail systems. It will simultaneously develop related downstream industries, including construction, steel, metalworking, electrical and electronic components, telecommunications, etc., while leapfrogging ahead into magnetic levitation (maglev) technologies. The new, productive, high-

MAP 2
Spain and Portugal: High-Speed Rail Lines (EU Project 19)

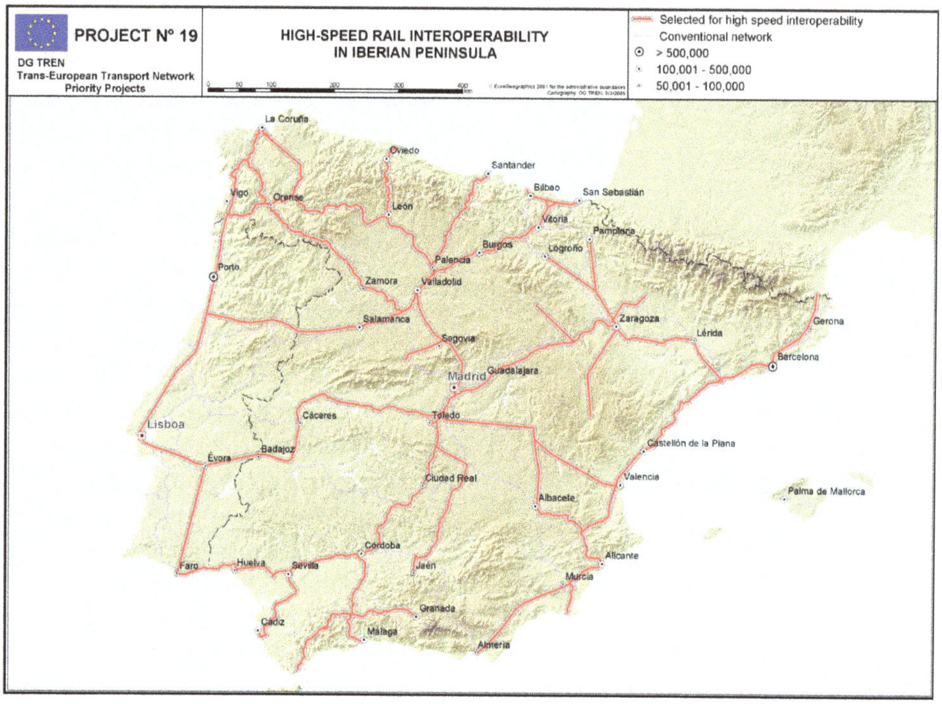

Source: EU

technology jobs so created will make a serious dent in today's unemployment problem.

There are some existing rail links connecting Spain and Portugal with the rest of Europe, and these will be improved and broadened (see **Map 2**). In addition to the Barcelona-Madrid corridor (which is operational), this will include:

- An Atlantic branch: Madrid-Valladolid (operational)-Burgos-Vitoria-Bilbao/San Sebastián-Dax-Bordeaux-Tours (Paris).
- An Iberian branch: Madrid-Lisbon-Porto.

Similarly, the EC's Priority Project 16, for a freight railway axis Sines/Algeciras-Madrid-Paris, links the key ports of Sines (southwestern Portugal) and Algeciras (southern Spain), with the center of Europe (**Map 3**). This requires the construction of a high-speed freight corridor, including a new high-capacity rail link for freight across the Pyrenees, which would involve a long-distance tunnel through the Pyrenees.

Although technically viable, these EU projects are financially and politically frozen, and will never be implemented under the current Maastricht diktat.

As for Portugal, the agreement with Spain to build a high-speed rail line from Madrid to Lisbon was suspended by the current government of Passos Coelho in 2011, on Troika orders. Not only should that line be built, but existing Spanish plans to link the two countries with four high-speed rail lines (Vigo-Porto; Salamanca-Porto; Madrid-Badajoz-Lisbon; and Seville-Huelva-Faro) should go ahead, and internal Portuguese high-speed lines connecting Lisbon with Oporto, and

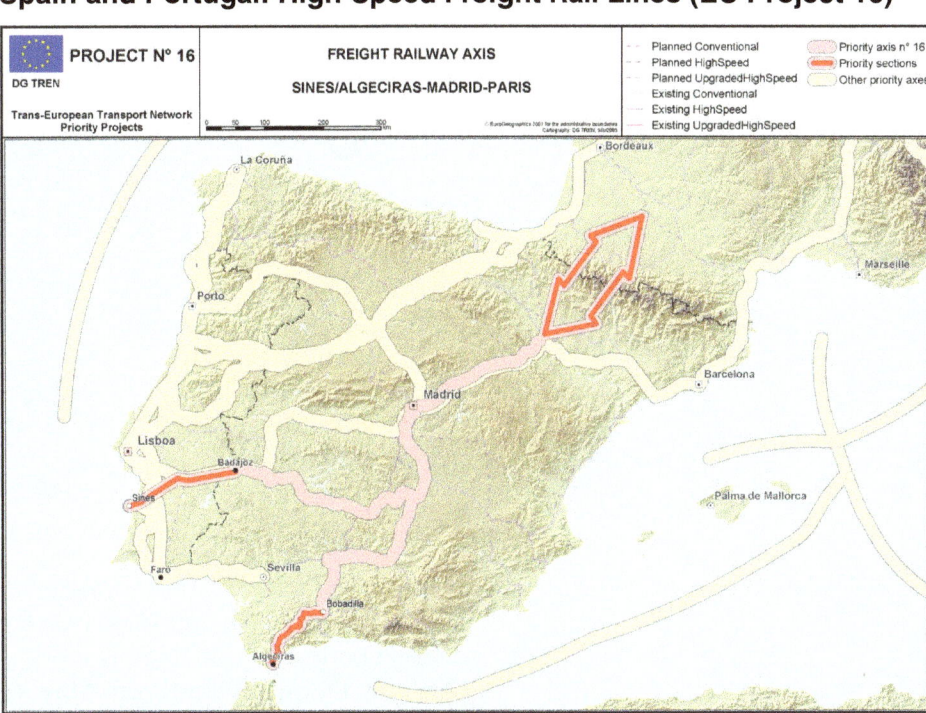

MAP 3
Spain and Portugal: High-Speed Freight Rail Lines (EU Project 16)

Source: EU

Lisbon with Faro—all at international UIC gauge—must also be built (see **Map 2**).

The southernmost point of this network in Spain is Algeciras. From here, a new high-speed rail line will be constructed to Tarifa and Cádiz, since Tarifa will be the Spanish terminus of a tunnel with high-speed rail going under the Strait of Gibraltar to Tangiers, Morocco, and from there will link to the whole Africa leg of the World Land-Bridge.

The idea of a tunnel was first proposed in Spain in 1930, and since that time, various options have been considered, including a fixed bridge (ruled out because of the impossibility of building supporting pillars in 300 meters, or more, of water), a floating bridge (discarded because of the strong cross-currents in the strait), and a tunnel bolted to the seabed (not viable, both because of the strong currents and the seabed's instability in that region).

In 2003, Spain and Morocco agreed to explore the construction of a fixed tunnel, and in 2006, their SECEGSA (Spain) and SNED (Morocco) state companies hired the renowned Swiss tunnel engineering company Lombardi to draft a design for the project. In 2009, the Lombardi proposal was presented to the EU—after which absolutely nothing has been done, because the entire Eurozone and world financial system is collapsing.

The Lombardi plan considered the option of a bridge at the narrowest point between the two continents (14 km (8.7 mi)), but since the seabed there is a very deep 900 meters (2953 feet), it was discarded as impracticable. The selected route instead runs at a more western

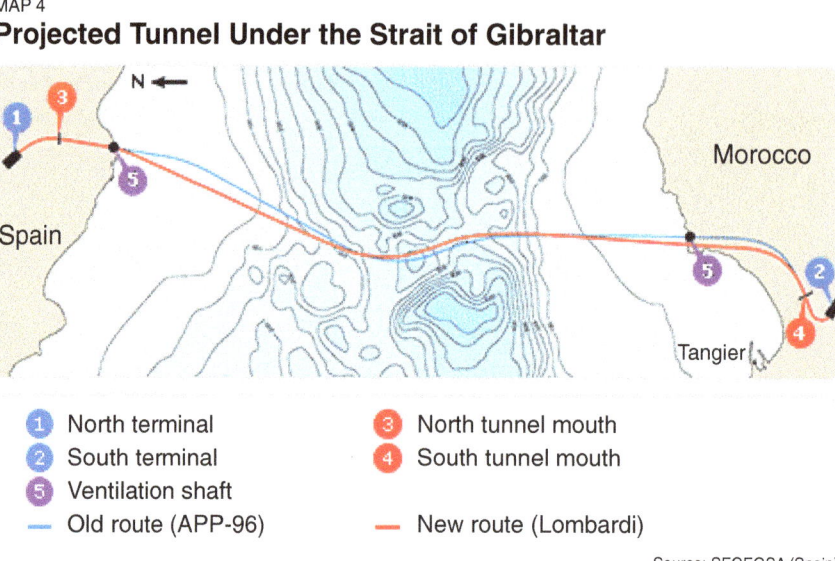

MAP 4
Projected Tunnel Under the Strait of Gibraltar

- ① North terminal
- ② South terminal
- ⑤ Ventilation shaft
- — Old route (APP-96)
- ③ North tunnel mouth
- ④ South tunnel mouth
- — New route (Lombardi)

Source: SECEGSA (Spain)

point, from Tarifa, Spain, to Tangiers, Morocco, a route where the sea floor is "only" 300 meters deep (984 feet)—which would make this the deepest undersea tunnel in the world. The length of the tunnel would be about 40 km (24.9 mi) (see **Map 4**). It would consist of two tubes for train lines for both passengers and freight, with an emergency or service tunnel running between them.

Lombardi estimated that it would take about 15 years to build, given the engineering problems to be solved—including the fact that it would run through a highly active seismic area (the Azores-Gibraltar Transform Fault), and difficulties in the stratification of the seabed there, described as a virtual "cocktail of sand, stone, and mud that make for a digger's nightmare." In fact, engineers have had to invent new boring methods just to drill exploratory holes, given the rock formations and the fierce underwater currents.

For purposes of comparison, the Channel Tunnel is only 50 meters (164 feet) below sea level, and is 49 kilometers (30.4 miles) long. The Bering Strait Tunnel would be at about the same depth (54 meters (177 feet)), and run 85 km (52.8 mi) in total, but it would make use of the Big Diomede and Little Diomede islands as "stepping stones," making the longest stretch only about 35 km (21.7 mi) long.

Once completed, and linked to high-speed rail lines, SECEGSA/SNED calculate that it would take 1.5 hours to get from Casablanca to the tunnel terminus in Tangiers; 30 minutes to cross to Tarifa, Spain; under 3.5 hours to then get to Madrid; and then 2.5 hours more to get to Barcelona. In other words, it would take less than 8 hours to get from Casablanca to Barcelona!

The joint SECEGSA/SNED website summarizes their concept of the project as follows: "The Fixed Link Through the Strait of Gibraltar can be considered the decisive connection between two continents and two great seas, which will articulate a heretofore unknown system of transportation between Europe and Africa and the Mediterranean surroundings."

As part of this project, it would be appropriate to return Gibraltar to Spain, from which the British stole it in the 1700s.

On the Morocco side, the Strait of Gibraltar tunnel will link up with high-speed rail lines in North Africa. The French are already helping to build high-speed rail lines in Morocco, and the entire North Africa rail project is a perfect area for French-Spanish cooperation.

...and on to Other Planets

Achieving these ambitious projects on planet Earth, however, depends on inspiring coming youth generations with mankind's true mission, his extraterrestrial imperative. The scientific breakthroughs, and the related cultural optimism, that is so sorely lacking today, will only come with such a focus and mission.

With that in mind, our Marshall Plan for the Mediterranean Basin will also construct a world-class Euro-African spaceport and associated science city on the Canary Islands. This location—100 km (62.1 mi) off the western coast of Morocco, at the same latitude as the U.S.'s Cape Kennedy—is ideal for such a project.

There is, in fact, already advanced scientific work underway in the Canaries. The Canary Islands are the site of a number of observatories, the latest and biggest of which, the solar telescope GREGOR, was inaugurated on May 21, 2012 on Tenerife. There, on the plateau at the foot of the 3,718-meter-high (4,066 yards)Teide volcano, the telescope, Europe's biggest, is being run by a consortium of researchers from the Kiepenheuer Institute for Solar Physics, the Astrophysical Institute Potsdam, the Institute for Astrophysics Göttingen, the Max Planck Institute for Solar System Research, and other international partners, who began constructing the GREGOR solar telescope there in 2000.

Scientists at GREGOR will not look directly at the Sun; this will be done using electronic detectors, such as spectrographs, polarimeters, interferometers, and cameras. GREGOR's rotating-fold mirror deflects the bundled beam generated by the adaptive optics system to the various instruments. Their purpose is to measure various physical solar parameters with an unprecedented level of precision, in particular, the Sun's magnetic field, and in doing so, reveal small structures down to a scale of 70 kilometers (43.5 mi)—an astounding resolution capacity, given that the Sun is located approximately 150 million kilometers (93 million miles) from Earth.

Tenerife is already the site of numerous astronomical observatories, and will become the site of a larger scientific complex, a space city, which will be connected to the existing airport by a maglev train—especially since the area is mountainous and not suited for traditional train systems. A feasibility study for a maglev track connecting the south and north of the island has already been done by the German Railway Research Institute in Berlin.

The island of Lanzarote, a lava-dominated landscape that strikingly resembles the surface of the Moon and of Mars, could serve as a testing site for coming Euro-African space missions—mankind's true destiny.

DOPE, INC.

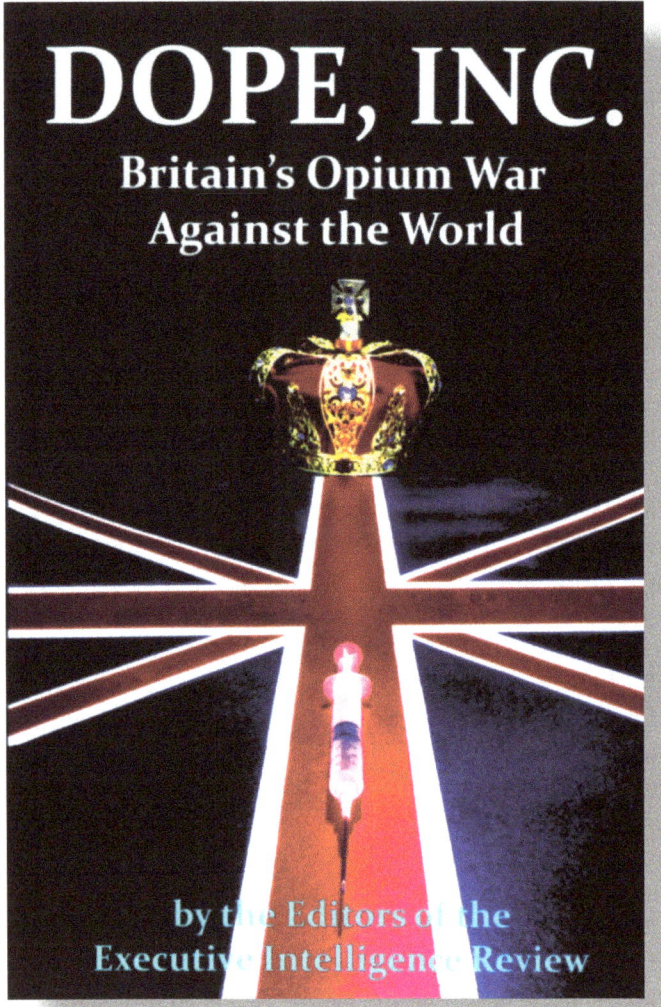

Is Back In Print!

Dope, Inc., by the Editors of the Executive Intelligence Review, is back in print, by popular demand, for the first time since 1992. Commissioned by Lyndon LaRouche in 1978, the book immediately "went viral" before the term was even invented!

The new, 320-page paperback, subtitled "Britain's Opium War Against the World," includes reprints from the third edition, and in-depth studies from EIR, analyzing the scope and size of the international illegal drug-trafficking empire known as Dope, Inc., including its latest incarnation in the drug wars being waged out of Afghanistan, and against Russia and Europe today.

FROM THE BACK COVER:

"Dope money supports the bankrupt world financial system. A trillion dollars goes through the Cayman Islands, the Isle of Man, Dubai. Speculation makes it trillions more. It sucks the blood of the real economy; and the dope destroys mankind's powers of reason."

This edition, published by Progressive Independent Media, is currently available in limited numbers, so there is no time to waste in ordering yours today.

PRICE: $25, plus $5 shipping and handling

available through www.larouchepub.com and EIR at 1-800-278-3135.

PART 10
Africa — Test for Global Progress

A Nuclear-Based Infrastructure Platform Is Necessary for Africa's Future

By Lawrence K. Freeman

August 2014

As Helga Zepp-LaRouche has consistently insisted, the continent of Africa is the conscience of the world. If the nations of the world permit the rampant genocide currently ravaging most of Africa to continue, the rest of the globe will ultimately follow.

That genocidal process is writ large today in the Ebola outbreak in West Africa, a lawful outcome of the scourge of poverty, underdevelopment, and warfare that has devastated the continent persistently, despite its nominal "decolonization," and represents a global threat.

But today, thanks to the BRICS process, there is also an alternative future visible for African nations. It is represented first and foremost by the shift undertaken by the Egyptian government of Abdel Fattah el-Sisi, toward crash programs of major infrastructure development, which will not only lift up the condition of the Egyptian people, but also link up Africa with the Eurasian continent. BRICS member South Africa, with its recent major commitment to nuclear power development, provides another focus for escape from the genocide paradigm.

In the following pages, we review the basic infrastructure needs of the African continent, combined with certain keystone projects, many of which have been on the drawing boards for decades. We assert that the only way to overcome the hideous deficit of investment in Africa over the past centuries, is a leap to a nuclear-based infrastructure platform, immediately.

Egyptian President Abdel Fattah el-Sisi

Overview

The 54 nations that constitute Africa have a population of 1.2 billion, which is expected to double in less than two generations by the year 2050, due to the high fertility rate of Sub-Saharan Africa. According to projections, by 2050 Africa will have 25% of the world population, will be home to almost 1 billion people under 18 years old, and will be the youngest continent with a median age of 25.

Average life expectancy in Africa is the lowest in the world, at 58 years. It has the lowest percentage of manufacturing exports at 1.5% and the lowest contribution to global trade at 2%. Africa has 255 million people who do not have enough to eat—25% of its population, the highest prevalence of hungry people in the world. The highest infant mortality rates and maternal mortality rates at birth worldwide are in Africa, and shockingly, it remains the only continent in the world where cholera is still endemic.

Africa is suffering from a catastrophic lack of all types of infrastructure, both hard and soft—energy, water management, rail transportation, educational and healthcare facilities, and social services.

Belying all the monetarist-statistical nonsensical propaganda from the International Monetary Fund (IMF) and sundry financial institutions, mindlessly repeated in Western capitals, that Africa is prospering with some of the fastest growing economies in the world, is the horrifying reality of this new lethal outbreak of the Ebola virus in West Africa that has claimed more than 2,000 victims officially as of this writing, with tens of thousands more affected. This outbreak is a deadly reminder of the complete failure to develop stable healthy African economies after they achieved their independence from the European colonial powers. Impoverished countries, with no healthcare infrastructure, a lack of hospitals, and a shortage of doctors, and where the majority of its citizens live in abject poverty, are proof of the complete lack of development of African nations over the past half-century.

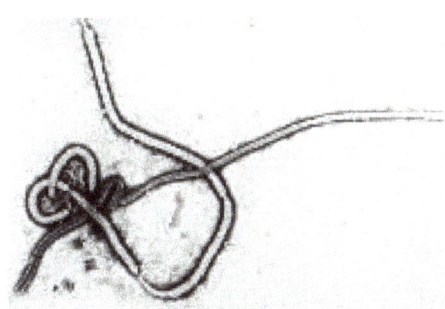

An electron microscope picture of the Ebola virus now ravaging Africa.

There are no objective limitations to Africa's potential. There is not now, nor has there ever been any legitimate objective reason for hundreds of millions to perish from disease, hunger, and war, with 4 million children dying each year. Africa not only has more than sufficient unused fertile soil and water systems to feed its entire population, but the resources to become a breadbasket for the world.

Given the population explosion underway, especially the number of youth and new entrants to the labor force, Africa will face an existential crisis unless it immediately embarks on a crash nuclear-powered science-driver infrastructure program. Only regional and transcontinental transformative infrastructure projects will provide the level of new high-skilled jobs to absorb Africa's expanded labor force and provide the energy, transportation, and controlled flows of water to grow the food necessary to finally eliminate famine and hunger.

Sub-Saharan Africa has the largest deficit of infrastructure per capita and per land area in the world. Infrastructure is the platform of any functioning—i.e., growing—physical economy. Despite the obsession by Western leaders and financial institutions to blame the lack of development of African economies on the lack of good governance, transparency, accountability, and not being business friendly, even the World Bank in its 2013 report was forced to admit that the "negative impact of deficient infrastructure is at least as large as that associated with corruption, crime, financial markets and red tape restraints."

Nowhere is this more dramatic than the lack of energy, access to electricity, per capita use, and industrial consumption throughout the Africa continent.

Energy

Half of the continent's population, 600 million people, has no access to electricity; of those who have access, brownouts and blackouts are common. The 48 countries of Sub-Sahara Africa are estimated to consume 70,000 MW, with South Africa using almost half of that. This means a per capita consumption level of less than 100 Watts, with Nigeria, for example, at 20-25 Watts per capita, compared to average per capita U.S. consumption of 1,400 Watts. If each African were to enjoy an energy living standard equivalent to the U.S. levels, Africa would require 1,680 GW (1,680,000 MW).

Given Africa's numerous river systems, much is being done to generate power from hydroelectric dams. Sudan's Merowe Dam, completed in 2009, generates 1,250 MW, and Ethiopia's Gibe III, expected to come online in 2015, will add 1,870 MW of generated electricity. Mega-projects such as the construction of the Grand Renaissance Dam in Ethiopia, which is expected to bring online an additional 6 GW (6,000 MW) by 2017, and existing plans to build the Grand Inga Dam near the mouth of the Congo River in the Democratic Republic of the Congo, which will have a capacity to generate 40 GW (40,000 MW) of electrical power, are necessary and exciting. However, these positive efforts by themselves will be insufficient to bring Africa into the 21st Century, and provide the present and future generations with an advanced standard of living commensurate with a progressive agricultural-industrial society. This can only be accomplished with a nuclear energy (fission)–driven economy, followed by the next level of energy technology, a thermonuclear fusion–based society.

The Republic of South Africa (RSA) is the only one of 54 African countries to use nuclear power as part of its energy grid, and it is only the RSA that has made nuclear energy part of its future, as President Jacob Zuma made clear speaking at the August 4-6, 2014, U.S.-Africa Summit in Washington, DC. Currently the RSA, which already has the highest per capita consumption of electricity on the Sub-Saharan continent, is in discussions with Russia's Atomic Energy Agency (Rosatom) for the construction and financing of six nuclear power plants, which would have the capacity to generate 9,600 MW of electricity. This is a very important initiative, because many African leaders have been conditioned to accept the view that nuclear power is too advanced or unnecessary.

Issoufou Mahamadou, the President of Niger, a country in the Sahel that is 75% desert, publicly advocates the need for nuclear energy to be part of his country's future energy grid. Nuclear energy, unlike hydroelectric power, does not require rivers and dams. Each and every African nation must make nuclear energy part of its future. With 1,000 nuclear plants producing 1 GW (1,000 MW) of power each, Africa can make great progress in meeting the energy needs for its rapidly growing population.

Senegalese scholar Cheikh Anta Diop in the 1960s and 1970s advocated for African economies to be powered by nuclear energy, and thermonuclear fusion energy, and wanted to establish training centers for Africans to master these technologies.

Diop wrote in 1978: "However, if that source of energy control [fusion] were to become available, with effective control of thermonuclear reactions, the energy needs of the planet would be answered for a period of a billion—repeat, one billion—years. The future instruments that produce this energy, whether called thermonuclear reactors or tokomaks ... will be fed in their final and truly operational stages by heavy hydrogen, obtained basically through electrolysis of sea water."

Thinking into the future, Diop demanded that thermonuclear energy be studied in Africa, and called for the creation of "a pilot fusion center in an appropriate African country, open to all qualified African researchers willing to follow this line of pursuit."

Transportation

The lack of rail transportation, much less high-speed trains, is appalling in the second decade of the 21st Century, and severely affects the economies of all 54 nations. Transcontinental high-speed rail lines connecting all countries on this vast continent are imperative. This requirement goes far beyond constructing intra-country transportation systems and paved roads. Chinese Premier Li Keqiang, speaking at the African Union headquarters in Addis Ababa, Ethiopia, on May 5, 2014, emphasized that one of China's goals is to help connect all African capitals by high-speed rail. This project would of course advance many other features of development. To bring prosperity to all African nations, it will be necessary to build this interconnected transportation network across the continent. Its construction will require putting

Existing Railroads in Africa

Proposed Railroads for Africa

EIRNS/Fusion Energy Foundation, 1980

to work a "mega army" of potentially unemployed youth to lay hundreds of thousands of kilometers of track.

Other transformative transportation projects have been studied or are being built.

President Zuma has been leading the effort to construct a South-North transportation corridor from Durban on the east coast of South Africa, to Dar es Salaam in Tanzania, and then north to Cairo, Egypt, which would affect the entire Eastern half of Africa. The Cape to Cairo rail and road corridor consists of 204 projects, of which various stages of progress are being made on 81 road projects, 48 rail projects, and 6 bridge projects.

A year before the formal launching of the BRICS New Development Bank, President Zuma made the funding of this trans-continental project a major topic of discussion at the first meeting of the BRICS Business Council in Johannesburg on August 20, 2013: "I warmly invite you to collaborate with us in realizing the delivery of infrastructure on the continent…. We specifically champion the North-South corridor with its particular emphasis on road and rail infrastructure, initially from Durban to Dar es Salaam and ultimately Cape to Cairo. There is a lot of scope indeed for mutually beneficial partnerships with the BRICS community, which will create much-needed infrastructure in Africa." This theme was repeated by President Zuma when he spoke at the national Press Club luncheon in Washington, DC on August 4, 2014.

Another transformative transportation infrastructure project is the Dakar to Port Sudan Railway Line, approved by the Organization of Islamic Conference-(OIC).

This rail line, conceived more than a century ago, would go from Dakar, Senegal on the Atlantic Ocean across the continent to Port Sudan on the Red Sea. This main line will cross through Sudan, Chad, Niger, Mali, and Senegal and include branches to Djibouti, Libya, Uganda, Cameroon, Nigeria, and Burkina Faso and spurs into Ethiopia and Kenya, connecting the Indian and Pacific Oceans to the Mediterranean and Red Seas. The total rail network will extend for 14,000 km, and will cause a revolutionary transformation of all the countries involved across the girth of Africa and begin the taming of the Sahel desert. The feasibility of this project has been determined, and it only lacks the political will to finance it.

A third transportation project called the African Pass, developed by Egyptian Engineer Aiman Rasheed, calls for a high-speed rail and modern highway be built from Sidi Barrani in northwest Egypt, south to the Great Lakes Region, crossing through Sudan, South Sudan, Rwanda, Burundi, Uganda, the Democratic Republic of the Congo, and Central Africa Republic. (See Appendix–Part 10.)

To connect Africa's new expanded rail system to Eurasia, the construction of a bridge/tunnel across the Strait of Gibraltar that has been on the drawing board for decades, and thoroughly studied for feasibility, should finally begin. Again, all that is lacking is political will to finance this project.

Water

The major river systems of Africa represent a major resource. The Niger River in West Africa flows from Guinea on the Atlantic Ocean, northeast to Mali, then south into Niger and continuing through Nigeria to form the Niger Delta on the Bight of Benin. Although a long river, it is not large in terms of water flow. The Nile River is the longest in the world, but not the most voluminous; added to that, it has many countries to serve. The Congo River is the second largest river in the world, by volume of water discharged. Ethiopia has several river systems that flow into Sudan and South Sudan. While much more can be done in managing the river waters and transporting water from wet basins to dry basins, nuclear desalination will be necessary to provide for Africa's growing population and agricultural production, particularly in arid regions such as Egypt and Sudan.

The Transaqua Project, developed by the Italian engineering firm Bonifica, is a grand water infrastructure project that has been studied since the 1980s, and if it had been built decades ago, would have already transformed the Great Lakes Region. The proposal is to transfer 100 billion cubic meters per year, 5% of the Congo River's 1.9 trillion cubic meters per year that flow into the Atlantic Ocean, through a newly created canal across the Central African Republic into River Chari, which empties into the currently disappearing Lake Chad. In addition to refurbishing Lake Chad, on which the lives of more than 30 million farmers and fisherman from Niger, Cameroon, Chad, and Nigeria depend, Transaqua would reverse the encroachment of the Sahel desert and create an economic Renaissance for all the countries of the region. (See Appendix–Part 10.)

A second feature of Engineer Rasheed's African Pass that parallels Transaqua, is to construct a 3,800 km long canal from the Congo River that will travel north through the Central Africa Republic, South Sudan, Sudan, and into Egypt to fill the Qattara Depression.

Both projects apply the same principle—human intervention to improve nature for mankind's benefit by transporting water from moist regions to dry ones, and simultaneously creating new economic wealth.

Food

As stated above, Africa is the potential breadbasket for the world. A few examples indicate the potential.

The Democratic Republic of the Congo (DRC), more than six decades after its independence from Belgian colonists, has recently begun an aggressive agricultural program cultivating large areas of its unused land, thus taking advantage of its plentiful rainfall. According to a report from Harvard Business School in December 2013, the DRC has 80 million hectares of arable land, of which only 1% is being cultivated. However, the DRC, among the most deficient countries in basic infrastructure per capita and per land, will require massive investment in energy, transportation, and healthcare, if it is to realize its full potential level of food production.

It has been known for decades that the concentration of fertile arable land in Sudan and South Sudan, with their many rivers, if developed, could feed 750 million to 1 billion people. Is it not a crime against humanity that the Sudanese people live in such hardship and poverty, when they could be employed in growing food to feed the whole continent? Since Sudan freed itself from British colonial rule in 1956, there has been no strategy to develop what was, prior to July 9, 2011, the largest country in Africa.

Mali's "inland delta," emanating from the Niger River just north of Segou, which extends further north to Timbuktu, is the largest body of water in the Sahel. Yet, this "jewel in the desert" is completely under-utilized. When the French controlled the Office of Niger before Mali's independence, they projected that the naturally irrigated 1.9 million hectares could potentially yield 2.5 million tons of rice; yet today only 100,000 hectares—5% of the total available—is being farmed.

Raw Materials

Following President Nixon's August 1971 decision to put the final nail in President Franklin Roosevelt's Bretton Wood System, his then National Security Advisor Henry Kissinger and de facto British agent promulgated that U.S. policy was not to economically develop "Third World" countries, but rather to have unrestricted access to their natural resources, and find ways to reduce the rate of growth of their populations. Kissinger's December 1974 report, "National Security Study Memorandum 200-Implications of Worldwide Population Growth for US Security and Overseas Interests" (NSSM-200), infamously asserts what U.S. policy has been for the past four decades:

"Whatever may be done to guard against interruptions of supply and to develop domestic alternatives, the U.S. economy will require large and increasing amounts of minerals from abroad, especially from less developed countries. That fact gives the U.S. enhanced interest in the political, economic, and social stability of the supplying countries. Wherever a lessening of population pressures through reduced birth rates can increase the prospects for such stability, population policy becomes relevant to resource supplies and to economic interests of the United States."

Included in NSSM 200 was a list of 13 countries that were particular targets of this policy, due to their rapid rate of population growth. Among them were three African nations—Egypt, Ethiopia, and Nigeria.

The Kissinger/British anti-population policy has, in fact, never been repudiated.

In line with this policy, it should be no surprise that Africa has remained primarily a raw materials producer. One indicative statistic: Exports from Sub-Saharan Africa in 2012 were $400 billion, of which $300 billion were from extraction of natural resources—oil, natural gas, precious metals, and diamonds.

The Egyptian Model

For decades, many regional infrastructure projects have been discussed, reviewed, studied, and put into the planning stage by the African Union and regional associations. One of the early ones was the Organization of African Unity's Lagos Plan of Action, adopted in April 1980, in response to which economist Lyndon LaRouche and *EIR* wrote a book-length critique. The Lagos Plan had numerous fatal flaws, among which were the promotion of "soft technology" and "alternative energy sources," and a reliance on the "positive role" of the World Bank and the IMF. Subsequent plans have varied in quality, but the best ones have usually lacked the necessary political backing to force them through against opposition from the international financial institutions, and internal sabotage operations.

The exception today is what is going on in Egypt. Egypt's President Abdel Fattah el-Sisi, in the image of General Abdul Nasser, the founder of modern Egypt, announced on August 5 an era of great projects to rebuild the nation. He announced the construction of a New Suez Canal, the government's intention to complete the Toshka Agricultural Project, and the determination to construct Egypt's first nuclear plant, the 2,500 MW Dabaa power plant, within five years.

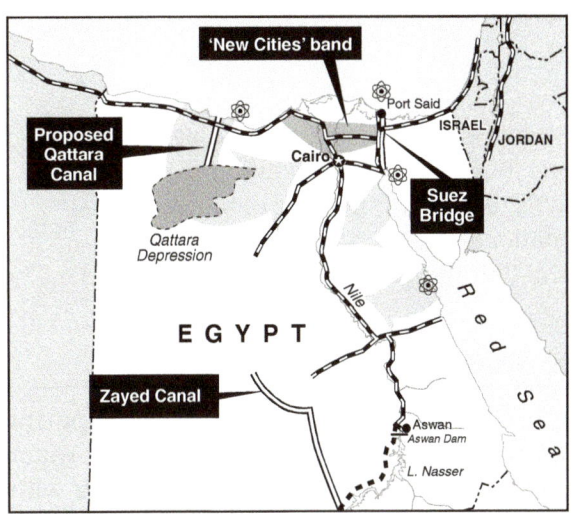

This schematic shows a number of the major projects proposed for transforming Egypt. The Sheikh Zayed Canal, a part of the Toshka project, now has more than 50 km completed.

President el-Sisi did not wait for permission from anyone to proceed with these projects, which have been on the drawing boards in some cases for decades. On August 6, the Egyptian Army Corps of Engineers began work moving earth to build the New Canal. They have called on all able-bodied youth and men under 45 years of age to apply for job vacancies. The president also insisted that the project be funded by Egypt itself, through the sale of government-issued debt certificates, sold only to Egyptian citizens. Thus, the project will be free of the possibility of external sabotage, and be a true *national* project in which the citizenry is invested.

This triad of mega-projects will form the core of a network of projects, including roads, ports, airports, industrial-manufacturing centers, and canals, which are designed to transform Egypt, and allow its citizens to prosper. *EIR* has featured a number of detailed studies on the current projects, which will be updated as they proceed. Egypt can indeed be a model for nations in Africa, and throughout the world, of how to turn around an economy for the general welfare of its citizens, present and future.

FOR FURTHER READING

"The Rebirth of Africa," *EIR*, January 1, 1993, which includes Lyndon LaRouche's ``Critique of the Lagos Plan: Stop Club of Rome genocide in Africa.''

"Peace through Development in Africa's Great Lakes Region," Proceedings of a Seminar in Walluf, Germany, April 26-27, 1997, *EIR* Special Report.

APPENDIX A

The Transaqua Project

> "The measure of investment costs does not lie in the millions of dollars alone, but also in the liberation from wars, millions of individuals rescued from dying of hunger, social peace, and an international conscience."
> —Dr. Marcello Vichi, 1992 (Originator of Transaqua Project)

For Example: The Transaqua Project

The project, dating back to the 1970s, was part a development perspective for the Sahel region of Africa involving massive water management projects interconnecting with trans-African transport corridors to provide the infrastructure for modernizing the economies of all central African countries and ending the genocidal conditions of life imposed by centuries of colonial oppression. The Transaqua project itself involves diverting 100,000 million cubic meters per year of freshwater from the Congo River to fill up Lake Chad. That's approximately 5% of the freshwater of the entire Congo basin. The project would involve construction of 2,800 km of navigable canals, "river highways", each 25 meters deep and 100 meters wide, an inland port, a series of hydroelectric power plants that will generate approximately 4 gigawatts of electric power, a crossing with the Lagos-Mombasa road axis connecting two ports on the Atlantic and Indian Oceans, and connection to the Mediterranean Sea via the Lagos-Algiers Trans-Saharan road (see maps).

Once the water has reached Lake Chad, it can be moved further north providing irrigation for an estimated 5-7 million hectares of land in the Sahel, helping to halt the southward expansion of the Sahara Desert, and providing food for the nearly 20 million Africans in the region who are continually threatened with starvation if the Sahara's

Transaqua will bring approximately 100,000 million cubic meters (81 million acre feet) of freshwater from the Congo River basin through 2,800 km of navigable canals to Lake Chad.

The New Silk Road Becomes the World Land-Bridge

growth cannot be stopped. Compare African agricultural output estimated maximally at 500 kg per hectare to European production of approximately 10,000 kg per hectare. This discrepancy is entirely due to lack of infrastructure – water, roads, equipment, fertilizer, etc. Transaqua, Africa Pass (see accompanying article), and other large regional infrastructure projects will lay the foundations for a prosperous future for all of Africa, and give humanity the benefits of the discoveries sure to come from a continent of happy, productive souls.

Elliott Roosevelt reported in his 1945 book *As He Saw It* that his father, Franklin Delano Roosevelt, in his discussion with Winston Churchill, said, "Divert this water flow for irrigation purposes? It'd make the Imperial Valley in California look like a cabbage patch! The Sahara would bloom for hundreds of miles.... Wealth! The Imperialists don't realize what they can do, what they can create! They've robbed this continent of billions, and all because they were too shortsighted to understand that their billions were pennies, compared to the possibilities! Possibilities that *must* include a better life for the people who inhabit this land..."

APPENDIX B

Africa Pass

The Africa Pass plan has the potential to revolutionize the economies of Sub-Saharan and North Africa, as well as relations within the continent and across the Mediterranean into Europe. The author, Aiman Rsheed, made the first public presentations of his Africa Pass in the aftermath of the Egyptian revolution of January 2011 and presented a draft to the office of Prime Minister Kamal Al-Ganzouri in February 2012.

Summary of the Project

The Africa Pass will include two major components:

Transport: In the first phase, it includes the building of a major modern seaport in Sidi Barrani in northwestern Egypt near the border with Libya, which will be connected to the Great Lakes nations (Rwanda, Burundi, Uganda, Democratic Republic of Congo, the Central African Republic, and South and North Sudan) by high-speed rail and modern auto highways (**Map 1**). In the second phase Somalia and Ethiopia will be connected. In the third, Egypt will be connected to Asia through a tunnel underneath the Suez Canal, and a bridge from south Sinai to Saudi Arabia across the Tiran Island in the south of the Gulf of Aqaba. In the fourth a high-speed rail network across North Africa

MAP 1
Africa Pass: Four Phases of Transport Corridors

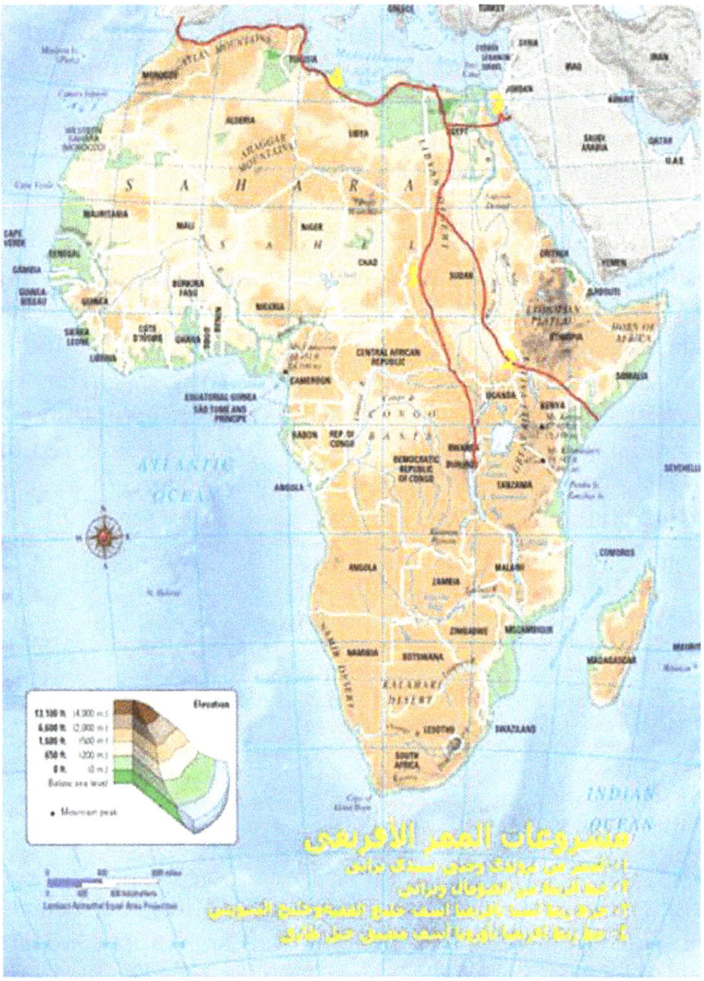

westward will connect to Europe through the planned Gibraltar tunnel.

Inside Egypt alone and along the Africa Pass corridor, five large cities are envisioned to be constructed like a string of beads with 250 km between each city, in an area that is practically only desert now.

The building of the Sidi Barrani port, a modern container-handling and industrial center on the Mediterranean with a large international airport, is the first and easiest part of the project to accomplish, according to the study. The large industrial zone and tourist zone in the area will attract industries, skilled Egyptian labor, and investors, and immediately provide work for large numbers of Egyptians, who are currently unemployed.

Water: The more impressive water project presented by Rsheed is similar to the Transaqua Canal Project (presented thoroughly by *EIR* and the Schiller Institute based on the work of Italian engineer Marcello Vichi). An irrigation canal, 40 meters wide and 15 meters deep and about 3,800 kilometers long, will extend from the highlands in eastern Congo, where the mighty Congo River originates, and flow northward through the Central Africa Republic, South and North Sudan, into Egypt to fill the Qattara Depression west of Cairo with fresh water (**Map 2**). Seven hydropower stations will harvest the power of the flow of water from a height of 1,500 meters above sea level in the south and down into the Qattara Depression, which lies 80 meters below sea level.

The canal will be constructed parallel to the rail lines and roadways. Electrical and electronic communication lines will accompany the Africa Pass to allow for building agricultural and urban centers along it. Oil pipelines can be added to the corridor to allow the landlocked countries to export their oil.

Around the Qattara Depression alone, millions of acres of agricultural land can be created, turning Egypt into a breadbasket, rather than being, as is the case now, dependent on imports of food. The freshwater Qattara lake and the green areas around it will have enormous hydrological effects, moderating the weather in the desert and increasing the hydrological cycle in the region with ever greater rainfall, diminishing the size of the desert.

MAP 2

Africa Pass: Canal from Eastern Congo to the Qattara Depression

Objectives of the Project

1. Development of nine African nations through real economic development projects.

2. Turning Egypt and the other African nations through which the project passes, into industrial, labor-attracting centers instead of labor-flight disaster areas.

3. Opening an export outlet for the agricultural products of the nations of the Great Lakes region, which are now wasted for lack of storage and low-cost, rapid means of transport. It is estimated that agricultural and other products will be ready for shipping from their place of origin to the Mediterranean within two days, with the help of the high-speed rail and Sidi Barrani port. This project will open up new agricultural sectors in the region which have lain dormant and isolated, such as the enormous potential of livestock and dairy production in both parts of Sudan. It will also lead to the elimination of hunger and starvation in many parts of Africa, especially the Horn.

4. Redistributing the population, especially of Egypt, into new cities, towns, and service centers in a fertile environment that is aspiring to grow.

5. Re-establishing Egypt's leading role and connection to Africa with renewed economic and diplomatic cooperation, which was launched by Egypt's new government and Foreign Ministry after the revolution.

6. Developing the water resources of all the nations included in the project and ensuring the production of large amounts of clean hydroelectric power. Inside Egypt, the Africa Pass will be complementary to the New Nile Valley project, which will start at the Toshki Canal in the south near Aswan and run parallel to the Nile northwards, opening new agro-industrial centers in the desert.

7. The cultivation of millions of acres of land around the Qattara Depression, and generation of power.

KNOW YOUR HISTORY!
America's Battle with Britain Continues Today

The Civil War and the American System: America's Battle with Britain, 1860-1876
W. Allen Salisbury, ed.
$15.00
PDF download

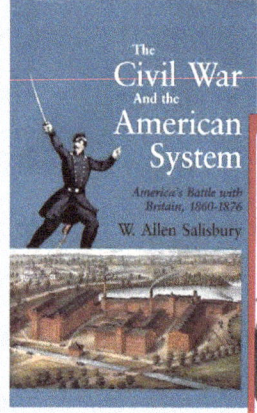

How the Nation Was Won
America's Untold Story 1630-1754
by H. Graham Lowry
$9.99
PDF download

The Political Economy of the American Revolution
Nancy Spannaus and Christopher White, eds.
$15.00   PDF download

Order online from
store.larouchepub.com

EIR News Service, Inc. P.O. Box 17390 Washington, D.C. 20041-0390
1-800-278-3135

GENOCIDE
RUSSIA AND THE NEW WORLD ORDER

Russia in the 1990s: "The rate of annual population loss has been more than double the rate of loss during the period of Stalinist repression and mass famine in the first half of the 1930s . . . There has been nothing like this in the thousand-year history of Russia."
—Sergei Glazyev

Paperback, with a **preface by Lyndon H. LaRouche, Jr.**

Economist Dr. Sergei Glazyev was Minister of Foreign Economic Relations in Boris Yeltsin's first cabinet, and was the only member of the government to resign in protest of the abolition of Parliament in 1993.

Now available in PDF format from the LaRouche Publications Store.   **$20**

Order by phone: **800-278-3135**
Online: **www.larouchepub.com**

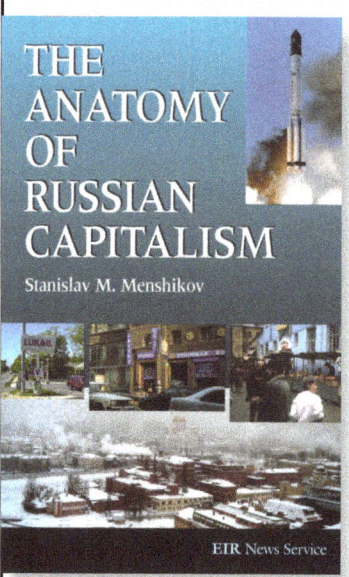

This English translation of the work of Russia's authoritative economist, Stanislav Menshikov, presents a critical analysis of the complex economic processes in Russia following the collapse of the Soviet Union.

Available through
Executive Intelligence Review

Order by calling 1-800-278-3135, or at the EIR online store, at www.larouchepub.com.

$30 plus $4.00 for shipping and handling
VA residents: Add 5% Va. sales tax

PART 11
Bringing the Western Hemisphere On Board

Rediscovering the Americas

by Dennis Small

September 2014

It was Cardinal Nicholas of Cusa, the revolutionary 15th-Century philosopher and founder of modern science, who played an instrumental role in the project that led to the 1492 "discovery of the Americas" by Christopher Columbus, working off a map provided to Paolo dal Pozzo Toscanelli by circles closely associated with Cusa. It was also Cusa who famously wrote in his 1450 *The Layman: About Mind*: "Mind is a living substance. . . . Its function in this body is to give it life and because of this it is called soul. Mind is a substantial form of power."

If the Americas, like the rest of the world, are to be saved from the economic disintegration and New Dark Age now besetting it, it will have to be on the basis of re-discovering—and rebuilding—the Americas, based on applying that "substantial form of power" which Cusa invoked. That process is now underway with the momentous July 15-16, 2014 summits of the BRICS nations and UNASUR, sparked by Argentina's courageous fight against the criminal vulture funds.

The entire region of the Americas, from the tip of Alaska in the north to Tierra del Fuego in the south, contains enormous challenges to biospheric engineering which will require major scientific and technological breakthroughs to conquer them. In the very north, there is the Bering Strait tunnel (see **Figure 1**), probably the single most decisive project for the World Land-Bridge, as it will connect not only Russia and the United States, but in fact all of Eurasia to the entirety of the Americas.

Moving southward, there is the Great American Desert (see **Figure 2**), a swath of arid and semi-arid land that covers a large part of the national territories of the United States, Canada, and Mexico, which can only be revived by massive bioengineering, beginning with mastery of the huge flows of water, not on earth, but in our atmosphere (see article, Part 2). Then there is the

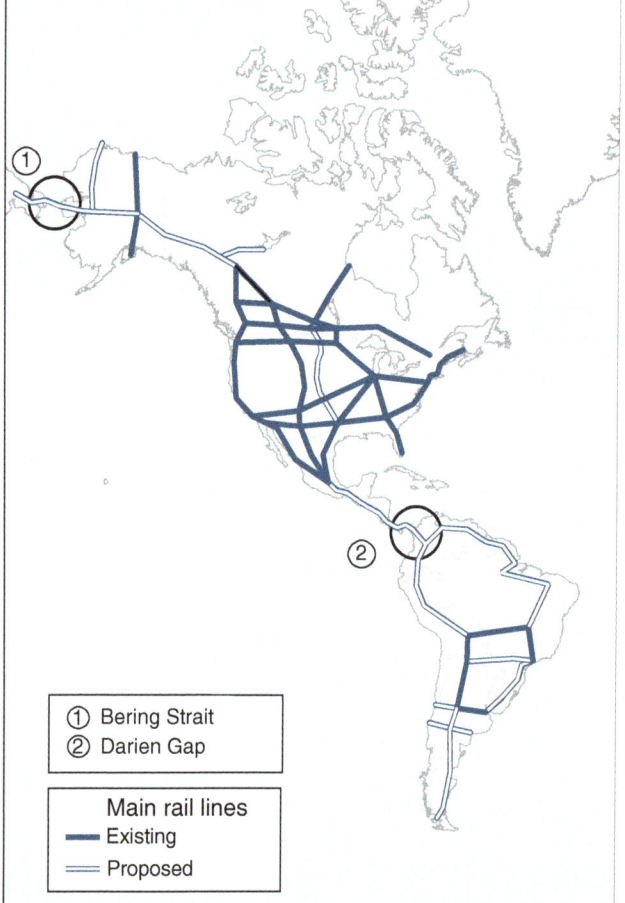

FIGURE 1
The Americas: Priority Routes

① Bering Strait
② Darien Gap

Main rail lines
— Existing
= Proposed

Source: *EIR*.

The New Silk Road Becomes the World Land-Bridge

FIGURE 2
The Great American Desert

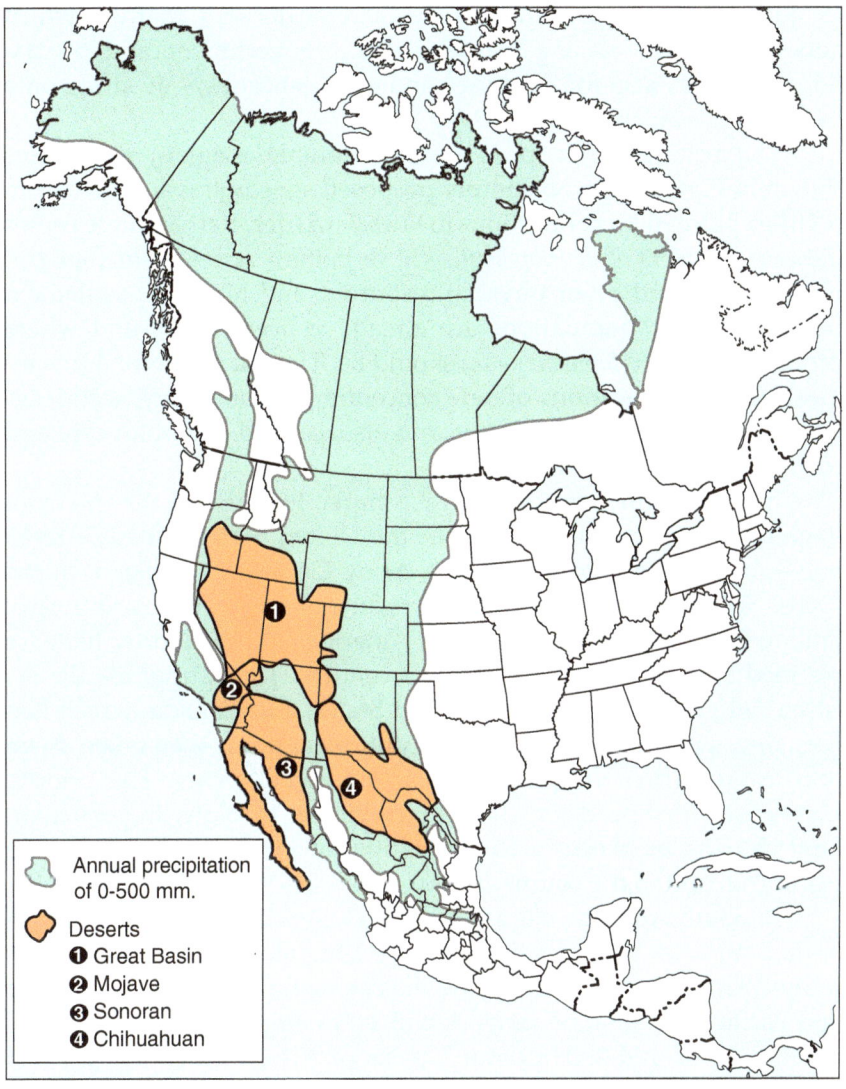

Source: *EIR*.

Darien Gap (see **Figure 1**) connecting Central and South America, where there is still no road—let alone a high-speed railroad—which must be built to cut through the jungle and link the Americas together, and provide a continuous rail connection from South America into all of Eurasia.

In South America proper, there is the world's greatest river, the Amazon, with the enormous Amazon jungle and its untold, untapped resources. And there are the Andes, a mountain range which runs the length of South America just east of its Pacific coast—a formidable barrier to the necessary construction of a trans-continental railroad across South America. But this project was also undertaken as a tri-national enterprise among Brazil, Peru, and China at the BRICS summit, a fitting answer to the way one Peruvian described her country: "Peru is God's challenge to engineers."

The outlook required is that of the Peruvian patriot Manuel Pardo, who as President during 1872-76, in alliance with Abraham Lincoln's networks in the United States, launched a major national railway project that included crossing the Andes. Their enemies sarcastically dubbed it "the Train to the Moon." But Pardo already knew, in 1860, the enormous role to be played by railroads:

"Join the three central lines by means of the fourth, and decide if, in ten years, a revolution will not have occurred in Peru, a revolution at once both physical and moral, because the locomotive—which, like magic, changes the face of the country through which it passes—also civilizes. And that is perhaps its main advantage: populations are put into contact. It does more than civilize; it educates. All the primary schools of Peru could not teach in a century, what the locomotive could teach them in ten years."

Rail Projects Link the Americas to the World Land-Bridge

The size of the landmass of North and South America combined (16,300 square miles, or 42,215 km²) ranks close to Asia, the largest continent (17,400 square miles, or 45,065 km²). The economic development potential of the Americas is vast, both in terms of the given natural resource base, and man-made "natural" resources—created through infrastructure projects. The maps shown here are a brief survey of selected key projects that *EIR* has promoted for decades, many of which have been on the drawing boards of various governments and international agencies also for decades—and even centuries!—awaiting only the policy go-ahead.

The priority high-speed (and preferably maglev) rail routes shown in **Figure 1** are not simply proposed speedy travel routes from point-to-point, with connections to Eurasia/Africa. Rather, these routes indicate *corridors of development*, whose pattern arises from topography, mineral and other physical resources, and historical settlement patterns (where populations are already concentrated), and where proposed new *development zones* should be. The rail route/corridors indicate intended locations of new concentrations of energy, water, agricultural and industrial activity, and also, centers for health care and cultural and educational activity.

In North America, the plan is simple. First, build the intercontinental lines planned for decades: the United States/Canada/Alaska line—already mapped out by the Army Corps of Engineers in the 1940s. Second, build the Pan-American rail connection southward, linking Central America and South America with the North, likewise planned for decades—in fact, over a century. Third, upgrade the existing rail grid in Mexico, the United States, and Canada, which had been developed as of the mid-20th Century, and was then taken down drastically during the past 40 "post-industrial" years. The priority routes for high-speed are shown (see **Figure 1**). Note, in particular, how Mexico City in central Mexico is interlinked with the entire northward grid, and to the south.

In South America, **Figure 3** shows key priority rail routes to be built, both to ring the continent, proceeding along the Andean spine in the West, as well as across the mountains, connecting the Atlantic and Pacific. This kind of network will act to *integrate growing economic activity*. As of mid-20th Century, parts of Argentina and Brazil had relatively dense regional rail networks, which were undermined over the past 40 years. A continental grid was never built at all. The little that currently exists is indicated on the map, often reflecting the classic colonialist policy of building a railroad leading only from a mine to the port, so that raw materials could be exported for foreign exchange, which was then used to pay the ever-growing foreign debt.

This overall rail project took a significant step forward at the July BRICS summit, where the idea of fulfilling the centuries-old dream of building a transcontinental railroad to connect the Atlantic and Pacific coasts of South America was taken up among Brazil, Peru, and China, in discussions between Chinese President Xi Jinping and Peruvian President Ollanta Humala, and then with Brazil's President Dilma Rousseff. An agreement was reached to open bidding for foreign, including Chinese, companies, to participate in the construction of one critical segment of that project: the "T"-shaped Palmas-Campinorte-Anapolis/Campinorte-Lucas route in central Brazil.

FIGURE 3
South America: Transcontinental Railroad

The importance of that segment within the overall project presented by *EIR* is clear from **Figure 3**, a schematic map first published by *EIR* in 1988. The northern terminus of Palmas is a stone's throw from the famous Carajás project in the middle of the Amazon jungle, the world's largest (and purest) iron ore deposit, which is now connected by rail only to the Atlantic port of São Luis. Once built, the western rail terminus of Lucas would then be halfway to the Brazil-Peru border, where the projected rail line would link up, in one option presented by *EIR*, with a Peruvian branch that would cross the Andes at Saramirisa—the lowest pass in that giant mountain range—and from there, to one or more Peruvian ports for shipment across the Pacific Ocean. This would drastically cut shipping time and costs from Brazil

FIGURE 4

A small section of the 1898 survey map for the Intercontinental Railway. After President McKinley was murdered, no railway or road was ever built connecting North and South America.

(and other Southern Cone countries including Argentina) to Eurasian powerhouses such as China, India, and Russia.

Even greater efficiencies, growth, and productivity can be achieved as this South American Transcontinental Railroad is able to connect *directly by rail* with Asia, as super-high-speed maglev rail lines are constructed through the Darien Gap and the Bering Strait.

There are various possible routes for a South American Transcontinental Railroad. (The one under discussion among China, Brazil, and Peru centers on São Paulo-Santa Fé do Sul-Cuiabá-Porto Velho-Pucallpa-Saramirisa-Bogotá-Panamá, with Andean crossings at either Pucallpa or Saramirisa. Another viable option, which has long been studied, is São Paulo-Santa Fé do Sul-Santa Cruz-Desaguadero-Saramirisa-Bogotá-Panamá, with Andean crossings at Desaguadero, Pucallpa, or Saramirisa.) Another viable option, which has long been studied, is São Paulo-Santa Fé do Sul-Santa Cruz-Desaguadero-Saramirisa-Bogotá-Panamá, with Andean crossings at Desaguadero, Pucallpa or Saramirisa. In fact, earlier versions of precisely this project were drawn up by the Intercontinental Railway Commission, started by U.S. Secretary of State James Blaine, which employed U.S. Army engineers to survey and project lines tying the United States through to Argentina and Brazil, presenting a completed map of the intended route project to President William McKinley in 1898 (**Figure 4**). The strongly pro-American System McKinley commemorated Blaine's plans as the future of humanity, when he spoke in 1901 at the Pan-American exposition in Buffalo—where he was then shot dead in a British-run operation.

Three Centuries of Great Water Projects

Figure 5 shows South America's priority water-improvement projects for intra-continental navigation, as well as flood control, power, irrigation, and other uses. The continent is well-endowed with navigable rivers (solid lines). The proposed canals (dotted lines) form key link-ups to form a continuous inland water route. The idea dates back at least to Alexander von Humboldt in the 19th Century, who conceived of linking up South America's three major river systems—the Orinoco, the Amazon, and the Rio de la Plata—with North America as well. He visualized the route going from the mouth of the Orinoco north through the Caribbean Sea, and into North America via

FIGURE 5
South America: Great Water Projects

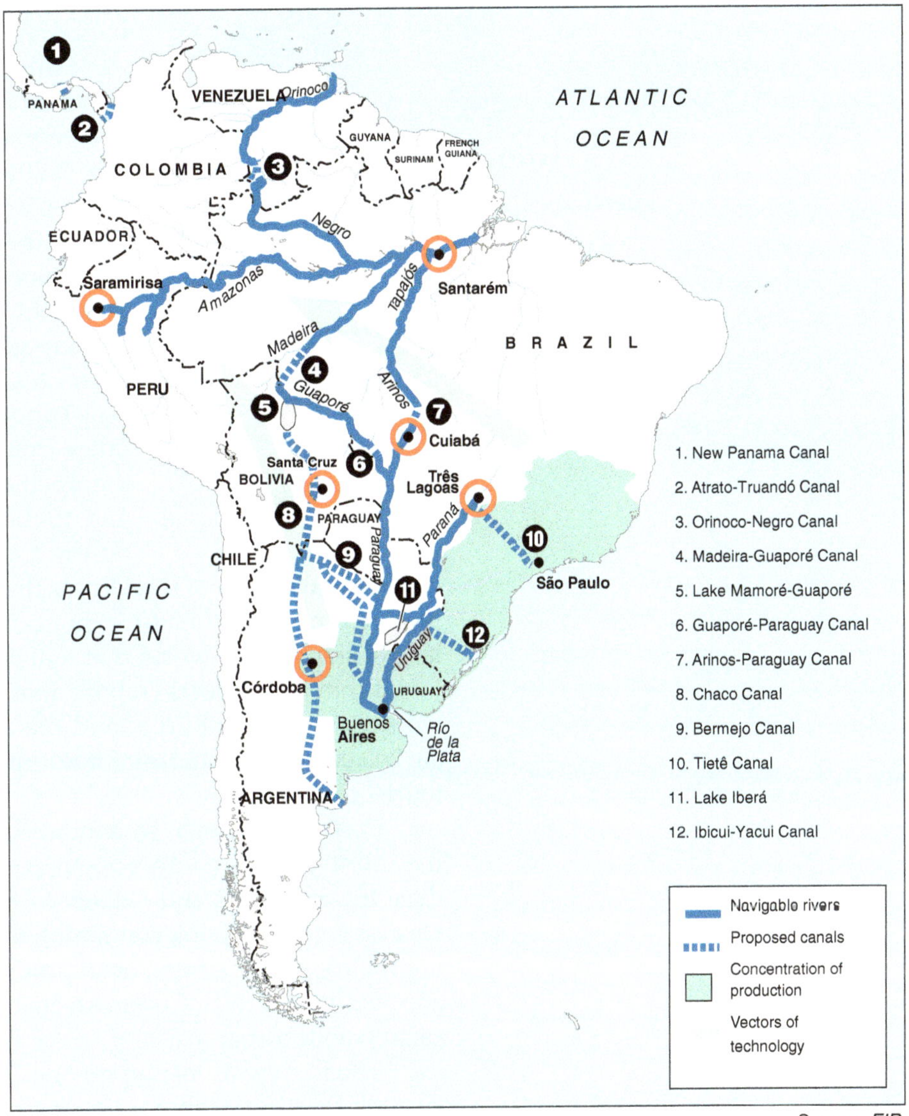
PANAMA
VENEZUELA
Orinoco
GUYANA
SURINAM
FRENCH GUIANA
COLOMBIA
ECUADOR
Negro
Saramirisa
Amazonas
Santarém
Tapajós
ATLANTIC
OCEAN
BRAZIL
PERU
Madeira
Guaporé
Arinos
Cuiabá
Santa Cruz
BOLIVIA
Três Lagoas
PARAGUAY
Paraná
Paraguay
São Paulo
CHILE
PACIFIC
OCEAN
Uruguay
URUGUAY
Córdoba
Buenos Aires
Río de la Plata
ARGENTINA
1. New Panama Canal
2. Atrato-Truandó Canal
3. Orinoco-Negro Canal
4. Madeira-Guaporé Canal
5. Lake Mamoré-Guaporé
6. Guaporé-Paraguay Canal
7. Arinos-Paraguay Canal
8. Chaco Canal
9. Bermejo Canal
10. Tietê Canal
11. Lake Iberá
12. Ibicui-Yacui Canal
Navigable rivers
Proposed canals
Concentration of production
Vectors of technology

Source: *EIR.*

the Mississippi and Tombigbee Basis, or via the East Coast—thus, an intercontinental "Great Waterway of the Americas."

More recently, in the late 20th Century, "The Great Waterway" was the name given by Brazilian expert Vasco Azevedo Neto, for the north-south link-up of the Orinoco to the Amazon system (No. 3 on the map), and the Amazon to the Rio de La Plata (No. 7). This inland waterway would also link up inter-modally with the railroad project described above. For example, the Amazon can be made navigable as far west as Saramirisa in Peru, where one of the possible trans-continental rail routes would cross the Andes to the Pacific coast.

The shaded "Production Concentration" area spanning parts of Brazil, Uruguay, and Argentina, refers to the concentration here of population, industry—in particular machine-tool capacity—science, and R&D; and output potential of all kinds (aviation, steel, automobile,

FIGURE 6
Nicaragua Interoceanic Canal

EIRNS

nuclear power, high-tech farming), which can provide needed technology transfer inland, throughout the continent—indicated by the shaded arrow-vectors.

Another great water project, the construction of an Inter-Oceanic Canal through Nicaragua (**Figure 6**), was announced on July 9, 2014 by Nicaraguan President Daniel Ortega, in the lead-up to the BRICS summit. The massive project will be carried out by the Chinese company HKND, but Russian President Vladimir Putin also made an unannounced stopover in Nicaragua on July 12, on his way to the Summit, to offer Russia's support as well. The canal will run 173 miles from the mouth of the Brito River on the Pacific Coast in southwestern Nicaragua, to the mouth of the Punta Gorda River on the Caribbean side. It will include two locks, and 65 miles of it will pass through Lake Nicaragua, and have a projected passage time of 30 hours, coast to coast, for the 5,100 of the largest ships in the world that will be able to use this canal.

Project engineers report that more than 50,000 construction workers will be required, and that, once in operation, it will generate 200,000 jobs, including for its sub-projects (an airport, two ports, a tourist center, etc.).

President Ortega, in announcing the selected route, stated that the country's entire educational system was being revamped to produce the engineers and skilled workers that the project will require. He also held up a book containing the feasibility studies for constructing such a canal that had been produced by the U.S. government, and adopted by the U.S. Congress, in 1896, detailing the benefits such a canal would bring.

The irony was lost on no one. China is actively involved in massive job-creating economic projects in Central America—the United States' proverbial "back yard"—while the United States under Obama has helped destroy that area with his policy of drug legalization, on top of decades of the British Empire's free-trade economic devastation.

That process has led, for example, to the fact that one-third of the population of El Salvador in Central America has fled the country to the United States to try to survive. In neighboring Honduras, official unemployment now stands at 60%, but knowledgeable Honduran sources say the reality is closer to 80%. In Mexico, about 18% of the total population has been driven to the United States over two generations by an economy incapable of providing elementary survival. Pope Francis, pointing to the staggering youth unemployment in many countries, including much of Ibero-America and southern Europe, described the situation as "unbearable." "We are excluding an entire generation of young people," he said.

Tripling Ibero-American Food Production

With adequate infrastructure, especially rail and water projects, Ibero-America is capable of nearly tripling its current levels of food production in a decade. **Figure 7** focuses on two areas with vast agricultural potential: the Colombian-Venezuelan Plains, and Brazil's Cerrado. The Amazon jungle lies between them.

The Colombian-Venezuelan plains are a continuous stretch of some 50 million hectares (212,000 square miles) in the Orinoco River basin. There is significant annual rainfall—in fact, too much in certain seasons—and there are major rivers which cross the region, including the Meta and the Guaviare. The land, once treated with lime (between 3 and 5 tons per hectare) to address the problem of acidity, is well-

FIGURE 7
South America: Great Rail and Agricultural Projects

suited for agriculture. Today, it is vastly underpopulated, underdeveloped, and largely controlled by London-promoted drug-trafficking armies. For example, the Colombian portion of the region (about 60% of the total for the two countries), constitutes 27% of Colombia's national territory, but has only 3% of its total population—some 1.5 million inhabitants. There are few roads in the region, and no railroads.

In size, the Colombian-Venezuelan Plains are equivalent to the combined area of the U.S. Great Plains states of Nebraska, Kansas, and Iowa.

Next, turn to the huge Cerrado region of Brazil, which is nearly four times larger than the Colombian portion. Its 205 million hectares (792,000 square miles) are equivalent to the three mentioned U.S. states, *plus* North Dakota, South Dakota, Missouri, Oklahoma, and Texas. Parts of the Cerrado are a bit more developed than the Colombian-Venezuelan Plains, but this is largely by the international grain cartels, which exploit huge tracts of soy beans, and process them almost entirely for export.

The Cerrado is a vast tropical savannah, a well-watered grassland, which constitutes 24% of Brazil's total land area of 846 million hectares—which, in turn, is 9% *larger* than the continental United States. Three main river systems drain the region: the Araguaia-Tocantins (into the Amazon basin); the Paraná (southward to the Río de la Plata basin); and the San Francisco (to the Atlantic Ocean). Like the Colombia-Venezuela Plains, with the right fertilizer and lime applications to the soils, the region's agro-climatic potential is vast. The temperature regime for much of the Cerrado will permit two, and sometimes three crops a year.

As we indicate in **Table 1**, some 50 million hectares out of the Cerrado's total of 205 million can be put under crop cultivation. This will produce about 210 million tons of grain per year. Similarly, in the Colombian-Venezuelan Plains, grain can be grown on some 15 million of its 50 million hectares, producing about 60 million tons.

If we add in the significant increase of irrigated land—and thereby food output—that can be achieved in Mexico with the implementation of the combined NAWAPA, PLHINO, and PLHIGON water projects, a total of 290 million tons of additional grain can be produced in Ibero-America. This will make for a *near tripling* of the current, inadequate output of 160 million tons of grain per year. Even if we factor in (1) replacing current imports (40 million tons) with regional production; (2) bringing food consumption levels up to the point where hunger is eliminated for the 40-50% of the population today suffering from it (another 60 million tons); and (3) providing for a 3% yearly population growth over the decade that it will take to bring these major projects online (90 million additional tons); the total required grain production of 350 million tons by 2018 will be more than matched by the 450 million tons

TABLE 1
Ibero-America's Great Agricultural Projects: Grain Production

	Total Land Area (Million Hectares)	Area Cultivated (Million Hectares)	Production (Million Tons)
Ibero-America today	2,058	51	
—Production			160
—Imports			40
—Consumption			200
—Consumption, no hunger			260
—Consumption, 2018			350
Colombia-Venezuela Plains	50	15	60
Cerrado	205	50	210
Mexico, NAWAPA-Plus	196	5	20
Sub-Total, 3 Projects		70	290
Total, Today + 3 Projects			450
—New Total as % of Today			281%

Source: FAO, *EIR*.

that will be produced. Regional food self-sufficiency is unquestionably an achievable goal.

Energy for Economic Development

Critical to "rediscovering" the Americas is the provision of plentiful, cheap energy, with a technological platform characterized by high and rising energy flux density. This means the appropriate combination of high-tech use of fossil fuel deposits, hydro-power potential where available, and most importantly, everywhere, the resumption of nuclear power development, leading directly into planet-wide cooperation on building a fusion-based world economy.

Soon after the 1953 announcement by President Eisenhower of the "Atoms for Peace" program, Argentina became the first nation to sign an agreement for cooperation on the peaceful uses of nuclear power. Its first reactor came online in 1974, the Atucha I; and its second, the Embalse, in 1983. As of 1979, four new plants were planned to go operational between 1987 and 1997, but the British Empire's "green" policy and IMF austerity dictates stopped all such programs—until recently. Atucha II finally came online in 2014.

In Brazil, the same British anti-nuclear policy was imposed, even though scientists were conducting experiments in nuclear fission there in the 1930s. Today, only two Brazilian plants are operational, Angra I (1982) and Angra II (2000).

In Mexico, President Jose Lopez Portillo (in office 1976-82) had plans for 20 nuclear plants. Today, there are two, both at Laguna Verde.

In all of the Americas, there were 126 nuclear generating plants operational in 2014: United States, 100; Canada, 19; Argentina, 3; and Brazil and Mexico with two each. Engineers had said 50 years ago, "2000 by 2000!"—the world needs 2,000 nuclear plants by year 2000. But as of 2014, there were only 437, with a mere 70 more under construction.

But a renewed drive for nuclear energy has emerged from the BRICS summit. Argentina is moving forward rapidly to build additional plants, as is Brazil, with Russian and Chinese collaboration.

But perhaps most moving—and indicative of the total paradigm shift that is now underway—is the vigor with which Bolivia has opted to go nuclear. As Vice President Alvaro Garcia Linea put it so eloquently at a late August energy conference in Santa Cruz:

"The use of and training in atomic energy is one of our obligations as a society and as a State. We have made that decision, and we are going to guide ourselves based on that decision. In the coming years, we will implement a program of nuclear energy, for peaceful purposes, with medical and agricultural goals, as the case may be, but we will have an elite, a core of brains, integrated with the world, to networks who work in the atomic field, which will allow Bolivia to learn about and use this fire of the XXI century: atomic energy.

"Nuclear energy is the fire of the XX and XXI centuries. It is the fire which our ancestors had 20,000 years ago, which allowed them to make philosophy, technical science, culture, agriculture. Knowledge of the atom, its regularities, its use, its functioning, is the touchstone of the XX and XXI centuries, the fundamental core of new knowledge and new technologies, new theories and new means of production....

"Bolivia cannot remain on the periphery, if this is the case, if knowledge of the atom ... is the sacred fire of the XX and XXI centuries, as fire was for the pre-agricultural civilizations of 20,000 years ago. Today a society which is respected—and we respect ourselves—cannot remain on the periphery, and we are not going to remain on the periphery....

"Fire in itself is not the destroyer," he said, and nuclear energy is not the destroyer. It can be "a creative productive force of life or a destructive force.... Nuclear energy exists independently of us. It functions in Nature, in the human body, in physical and chemical processes. The question, is if we have the ability, as society, to learn about it, to know it, to respect its force, and to know how to use it collectively and humanly for beneficial purposes....

"Let us break the mental and colonial chains; break them! Let us dare to leave the cave, as our ancestors did 20,000 years ago. Let us dare to assume our responsibility before the world, before our history and our society. Knowledge of nuclear energy is knowledge of the ABCs of nature....

"[Bolivia has] the technical, scientific, and moral obligation to take responsibility for the knowledge, use, understanding, and beneficial development of this fundamental force of Nature. It doesn't matter how long it takes us. We are going to do it, because we are convinced that that is how we cement the conditions for the technological development of Bolivians for the next 400 to 500 years."

FOR FURTHER READING

- Sept. 3, 1986: Ibero-American Integration: 100 Million New Jobs by the Year 2000!, Schiller Institute. Serialized in *EIR*. http://www.larouchepub.com/eiw/public/1986/eirv13n35-19860905/eirv13n35-19860905_018-ibero_americas_strategy_to_defea-lar.pdf
- May 9, 2003: Vernadsky and the Biogeochemical Development of North America's Desert, *EIR*. http://www.larouchepub.com/eiw/public/2003/eirv30n18-20030509/eirv30n18-20030509_004-vernadsky_and_the_biogeochemical.pdf
- Sept.26, 2003: Sovereign States of the Americas; Great Infrastructure Projects; *EIR*. http://www.larouchepub.com/eiw/public/2003/eirv30n37-20030926/eirv30n37-20030926_007-sovereign_states_of_the_americas.pdf
- Aug. 15, 2008: How to Triple Food Production by Developing High Speed Rail, *EIR*. http://www.larouchepub.com/other/2008/3532triple_food_rail.htm

North America: Restoring the American System

President Kennedy Would Be Building the World Land-Bridge

by Paul Gallagher and Marcia Merry Baker

October 2014

The United States economy is not productive—its productivity can only really be measured in comparison to its own past performance, by which measure it has fallen dramatically through an uninterrupted period of 50 years since the aftermath of President John F. Kennedy's assassination.

Therefore the United States urgently needs to join with the BRICS nations, particularly China, in the capitalization of new international banks for infrastructure; to draw credit from them; to issue national credit itself, for modern infrastructure-building "missions" of the kind which used to mark American history. President Kennedy, were he living and serving now, would be leading the mission to build the "world land-bridge" and its corridors of high-technology infrastructure development—as he would be pushing human space exploration through the Solar System.

The central banks of Europe and the United States, while flooding securities markets with vast floods of printed liquidity since 2008, proclaim the urgent need for giving their real economies a "total factor productivity shock." They associate this with "structural reform," or austerity programs, getting workers to produce more work in the same time and/or for less pay. Even by this degraded measure, productivity has not grown in the U.S. economy for the last 14 quarters. But that this measure itself is criminally incompetent, is shown by actual historical patterns of total factor productivity, which attempts to measure the growth of economies due to technological advance rather than simple application of more labor and more capital.

That this advance in general and real productivity must take place through the introduction of more modern and high-technology infrastructure platforms to the economy, is a necessity. This requires the powerful role of national government credit.

The highest annual rate of growth in America's history of productivity, thus measured, was the 3.30% of the 1930s under President Franklin Roosevelt's New Deal and his massive "Four Corners" infrastructure programs; followed by the 2.70% annual rate for the 1940s, and again during President Kennedy's 1960s. Today, U.S. total factor productivity

FIGURE 1
The 50-Year Disappearance of U.S. Infrastructure
Annual investment as % of GDP

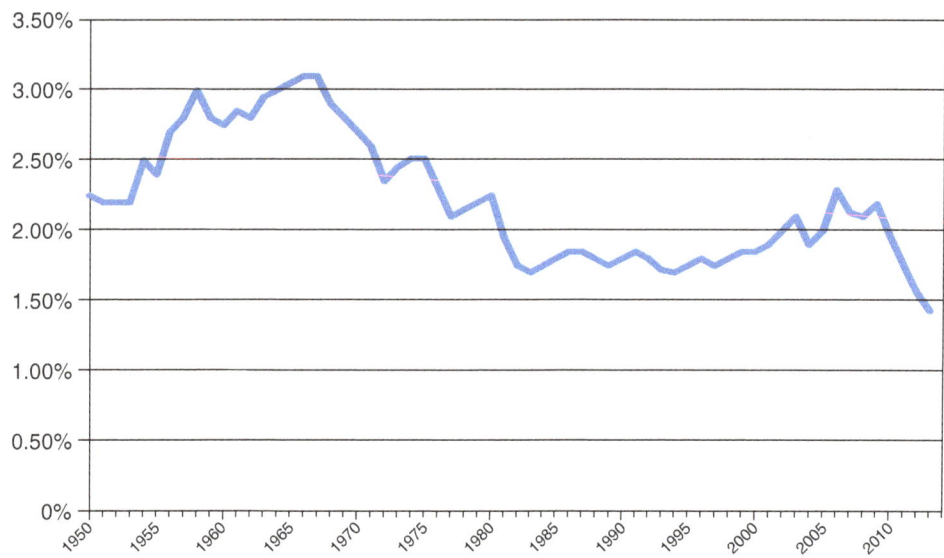

From its peaks in the 1930s under Franklin Roosevelt, and in the 1960s under John F. Kennedy, American infrastructure investment has all but disappeared—and the Obama "Stimulus Act" helped finish it off.

growth is estimated at "1% annually or below," where it has been for most of the period since 1972.[1]

And the major cause? United States investment in new infrastructure as a percentage of GDP, which again reached and exceeded 3% during Kennedy's 1960s, now scrapes the bottom among industrial countries at 1.4% (see **Figure 1**). Compare this to China's 8.8% average over the past 22 years 1992-2014.

The share of *productive employment* in the United States economy—taking the broadest definition of "productive," which includes work in engineering, transportation, and utilities—has dropped from 32.5% to 16.1% of the labor force since 1960, and from 19.5% to 10.2% of the work-eligible population.

The United States was the model for development, into the post-World War II years. In this period came the Atoms for Peace program, for advanced power and large-scale infrastructure projects internationally. American teams collaborated on building dams for hydro-power and irrigation, from Haiti to Afghanistan. Plans for nuclear power in Egypt, Iran, and throughout Southwest Asia were initiated by Detroit Edison, Westinghouse, and other private firms, working with the U.S. diplomatic corps. In North America, the Tennessee Valley Authority (TVA) model continued in the great California Water Project (1960 to 1973), and the upper Missouri River Basin project (Pick-Sloan Plan, 1944). In 1959, the St. Lawrence Seaway was completed, a transportation corridor to mutually serve Canada and the United States.

President Dwight Eisenhower's national interstate highway building program, with its dedicated capital source, was continued and expanded in the Kennedy years. The U.S. Apollo space program led the world to the Moon. The North American Water and Power Alliance (NAWAPA) was put forward as history's greatest water-management works, to benefit the entire continent. Kennedy's call was that "no drop of water in the West [of North America—ed.] should go to the ocean unmanaged." Nuclear isotope production for medicine and biology, and nuclear power production, took off, and nuclear desalina-

---

1. Congressional Budget Office, "Total Factor Productivity Growth in Historical Perspective," by Robert Shackleton, March 2013 (Paper 2013-01).

tion projects were launched. The U.S. public health system, centered on hospitals, was built up nationwide; TB, polio, and other diseases were conquered. Crop genetics advances in the Green Revolution foretold a future without famine.

Credit and Infrastructure

The United States also maintained essential control of its own issuance of currency and national credit, from the time President Franklin Roosevelt replaced the British gold standard with a gold-reserve system in 1933, through the strong capital and exchange controls of the post-War Bretton Woods System also initiated by Roosevelt.

The launching in British offshore centers of the so-called Eurodollar market, from approximately 1960, directly violated the rules of the Bretton Woods System. These accounts had the elevated interest rates and offshore speculative purposes of what has since been called a "carry trade." The Eurodollar market progressively drew the U.S. money supply offshore and robbed the Treasury of control of creation of its own currency. By 1980 approximately 80% of U.S. dollars were circulating, and effectively being created, outside the U.S. economy. The petrodollar, or "London dollar," had effectively replaced the U.S. dollar.

The last President who wanted to stop this massive speculative export of U.S. currency was Kennedy. Kennedy planned with aides to restore enforcement of the currency and capital controls of the Bretton Woods Agreement. After his death the destructive process was accepted.

President Richard Nixon made it an uncontrollable flood. The turning point into this devolution was 1971, immediately after Nixon was bullied by the British and by his Office of Management and Budget (OMB) chief/Treasury Secretary George P. Shultz into breaking up Roosevelt's Bretton Woods System and letting the dollar "float" speculatively against gold and other currencies. Nixon's and Shultz's actions triggered an explosion in the offshore markets for speculative U.S. dollar accounts—eurodollar/petrodollar markets.

When this happened, economist Lyndon LaRouche's first broadly published and widely noted economic forecast was proven extraordinarily prescient. In a series of published writings and pamphlets in the late 1960s, LaRouche had forecast that the 1960s successive crises of the British pound sterling, were being steered toward the forced breakup of the Bretton Woods fixed-exchange-rate system "at about the end of the 1960s decade." Under then-current policy trends, LaRouche wrote, Bretton Woods would be broken up and its destruction would be followed by a turn to fascist austerity against the United States economy.

A half-century later, LaRouche has promoted and hailed the emergence of the new development banks of China and the BRICS. Among other effects of this positive development dynamic, they give the United States the prospect, after more than 40 years, of regaining control of its currency and credit issuance, and using it for high-technology infrastructure development.

Today's Sad Reality

Since the 1960s forecast by LaRouche described above, the nearly 50-year slow-motion collapse of the U.S. productive economy and the standards of living of its once-productive citizens, has made that long-term forecast one of the most telling in economic history.

The United States has experienced no real economic growth since President Kennedy's assassination, and the living standards of two entire generations have fallen below those who preceded them. America has had no *national missions* for development of scientific, technological breakthroughs and the great projects of infrastructure which express them.

Nuclear power, NAWAPA, the space program, the drive to harness thermonuclear fusion power, have all been abandoned, or faded to economic insignificance, in the decades under the U.S. Presidents who have followed President Kennedy.

Fifty years later the United States economy is in a permanent low-productivity, cheap-labor, part-time/temporary morass, sometimes repugnantly called "the new normal." The median American household income in real-dollar terms was 13.7% lower in 2013 than its peak in 1972, even though cost-of-living measures have been changed since 1980 to soften the apparent effects of inflation. Under the old measures, median household income has fallen a full fifth, 20.7%, since 1972.

In fact, low and declining real wages and household incomes now dominate the economic *and the social reality* of the United States.

Entire, once-productive sections of the economic platform of the continent have been destroyed: for example, the steel centers of Monterrey, Mexico and Pittsburgh, Pennsylvania. The North American rail grid is dysfunctional—it cannot move out the High Plains Canadian and U.S. harvests. Detroit and other once great industrial and cultural cities are bankrupt ruins. Mexico, a grain surplus nation in the 1960s after the Green Revolution, now has an official hunger rate of 24%, and is grain-import dependent. The entire state of California has only 18 months of water left, without miracle rains.

Characteristics of the 565 million population of North America reflect the economic collapse. Canada's population is only 35 million, in a national land area which is the largest in the Americas, and second largest in the world. In Mexico, with a population of 119.7 millions—the largest Spanish-speaking nation in the world—an additional 12 million Mexicans have fled to the United States over recent decades, seeking a livelihood, because of the destruction of their homeland. In the U.S. population of 317 millions, the share of the working-age population now employed has plunged to 58.2%. Fully 15.9% of all Americans are in poverty, including one in five children. Life expectancy is now falling for citizens born in one of the many chronically impoverished areas in West Virginia, Oklahoma, the South, and elsewhere.

Every tenet of the Alexander Hamilton-based American System of economics (See Financing section) has been systematically violated for decades: (1) low technology, instead of advanced R&D; (2) cheap labor, instead of high-paid, productive jobs; (3) policy decisions based on manipulated opinion, not science; and (4) small-is-best superstition, instead of the implementation of large-scale infrastructure to create new capacity for the future.

Missions for North America

This is the starting point from which to understand what must be done to get the United States and North America on board with the World Land-Bridge impetus for development.

The necessary first actions are the reorganization under "Glass-Steagall principles" of the immense megabanks, which now absorb vast amounts of liquidity without lending; and the re-creation of U.S. institutions of national credit. These steps have already been described in detail in Part 2 of this Report.

This credit must be used to generate increases in economic productivity through the renewal of the continent's infrastructure. The key areas for investment, the parameters of which we outline below, are transport, water, power, and space.

I. Rail for Connectivity and Productivity

The North American continent lacks rail—let alone high-speed rail—connectivity, although it was the first continent with not one but five East-West transcontinental railroad corridors by 1890. The continental United States is unconnected with the great western plains provinces of Canada, and unconnected with Alaska. Rail connectivity to Mexico is only now being constructed—by China and Mexico! The entirety of North America is completely unlinked, even by highway, to South America.

The results for freight movement are very negative; trunk highways have become choked and structurally degraded by truck traffic as total road mileage per capita has declined 50% since 1965. Rail mileage (Class 1 plus Amtrak passenger) per capita has fallen from 90 in 1965 (and more than twice that 15 years earlier) to just 54 now. The North American rail system is essentially dysfunctional. In Spring 2014, fertilizer shipments were delayed past planting time in the northern High Plains of the United States and Canada. Then, come harvest, the trains could not move out the crops.

The proximate factor is that 48% of all cargo going by rail is not only coal, but oil from the fracking boom, per the Obama/London policy for the United States to be the new "Saudi Arabia of oil and gas." But total yearly shipments of rail cargo have declined from 2008 to 2013, from 14.960 million carloads down to 14.377 million, despite the near 1,000% spike in oil and fracking products haulage.

A. From the Bering Strait South

The Bering Strait project, currently a China-Russia collaboration which the United States should immediately join, would link, by hoops of steel, the entirety of the Americas to the entirety of Eurasia,

FIGURE 2
The Alaska Railroad

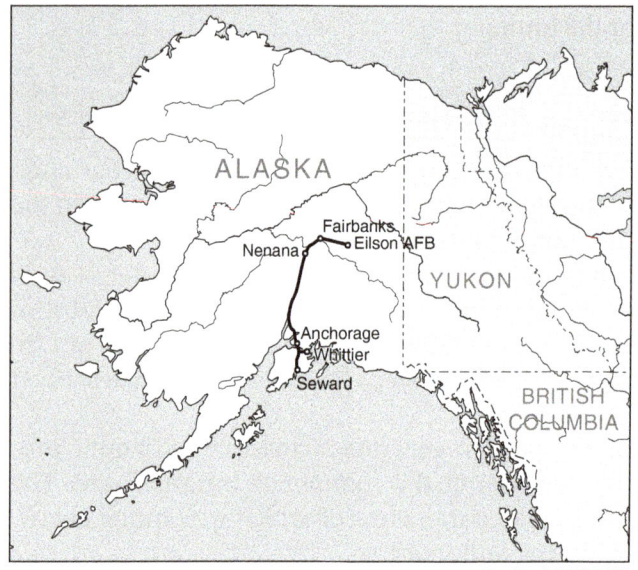

Alan Yue/EIRNS 2007

with the potential to connect to Africa. It would replace the Pacific's slow, outmoded, and vastly overburdened sea-rail routes with a geodesic high-speed-rail route. Goods produced in the American Midwest could be transported to China, or Russia, in 7-10 days, rather than the three weeks it currently takes by a combination of sea and rail.

A critical feature of the overall Bering Strait project, would be the development of a 4,876 km (3,030 mile) Alaska-Canada rail connector, which will contribute to moving the U.S. and Canadian physical economies from a deepening collapse process of several decades, onto an alternative path of growth. Building 4,876 km of track—and double that amount if the system is double-tracked—demands a huge quantity of goods, expressed as a *bill of materials*. This is an ordered array of goods—e.g., steel for tracks and railroad bridges, wood for ties and railroad structures, cement for culverts and other structures, and aggregates for cement manufacture and road-bed. The bill of materials for the Alaska-Canada rail connector will require the production of tens of millions of tons of goods. This will create 35,000 to 50,000 jobs in the building of the railroad, plus employment in the factories producing the steel, cement, copper and aluminum wire, power plants, locomotives, and other necessary components.

Figure 2 shows virtually the entirety of current railroad to the east and south of the Bering Strait crossing—the Alaska Railroad, built in 1914-23 by the U.S. government. The Alaska-Canada (Alcan) Highway, which was built under President Franklin Roosevelt's direction in 1942, extends from the Lower 48 states to Fairbanks, but goes no farther west.

The Alaska-Canada rail connector, with the construction of a development corridor extending 80 km (50 miles) on each side of the railroad, can transform the region in its entirety. Power lines, fiber-optic lines, and where necessary, freshwater pipes would be encased within the corridor. Cities, population, manufacturing, and scientific agriculture would be supported and advanced in this corridor as well. The Arctic North's nearby abundant, but largely untapped, mineral and raw material resources would be made accessible, by rail link, out of the frigid ground for rational use in the Arctic North and the world.

Figure 3 shows the plan for an Alaska-Canada rail connector system, as developed by Dr. Hal B.H. Cooper, Jr., a consulting engineer.

This rail system has two features to be noted. First, Prince George is a location where the North American rail grid nearly comes to an end. Starting in Prince George, the rail routes have been built out to Chipmunk and to Fort Nelson on the westerly and easterly branches, respectively. Both rail sections are owned by Canadian National Railway (CN). But some of the rail line to Chipmunk has already been torn up, and both lines would require substantial repair and upgrade as part of the Alaska-Canada rail connector plan.

FIGURE 3
Proposed Bering Strait/Alaska-Canada Rail Connector to Lower 48 States, Plus Existing Lines

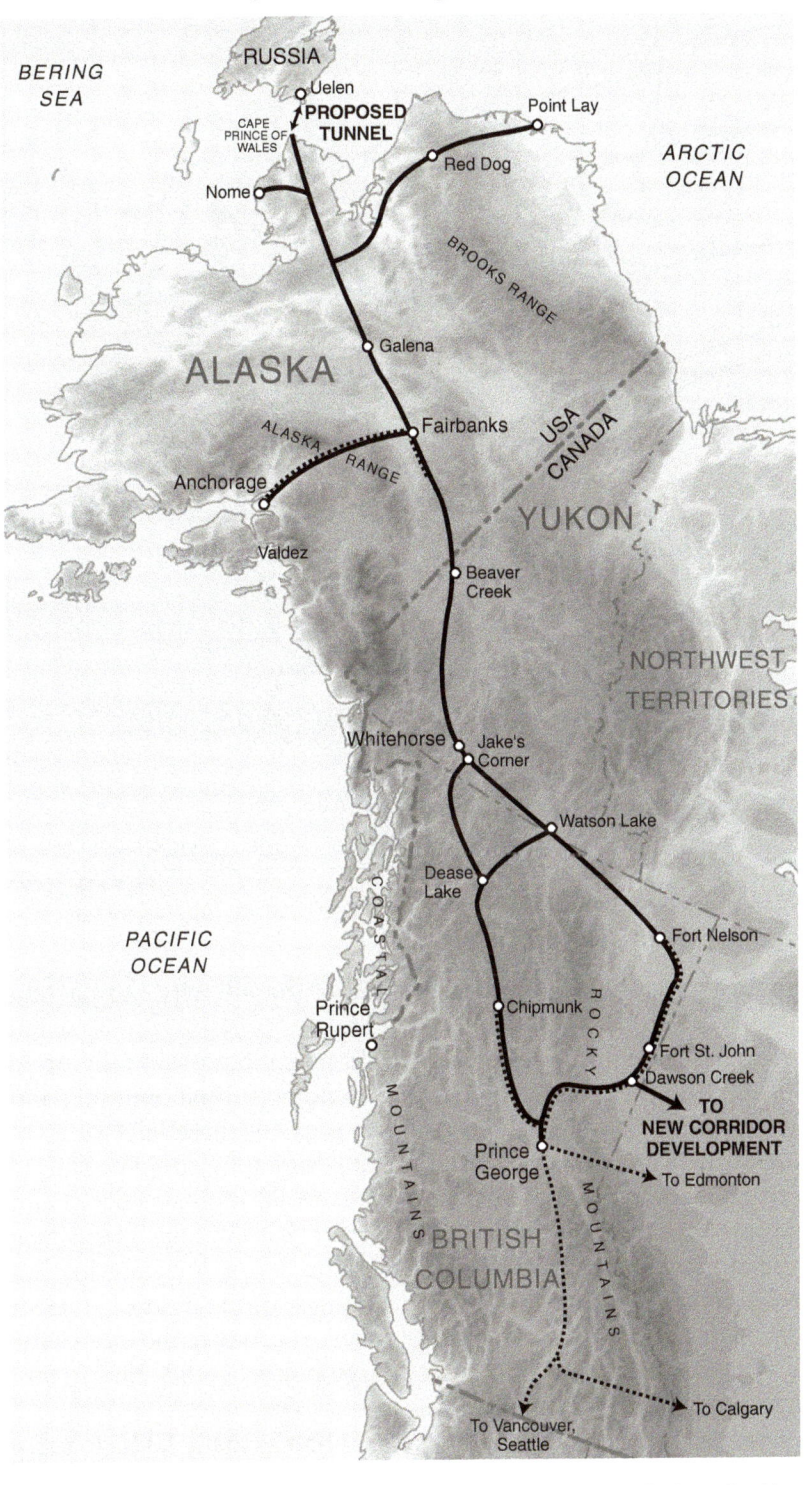

·····.· Existing Tracks   ⏵ Proposed Alaska Canada Railway Corridor

Alan Yue/EIRNS 2007

Second, by building the Alaska-Canada rail connector, we create the ability to move goods from Russia and China, as well as from Central Asia, Southwest Asia, and Europe, directly to the North American rail grid, and thus to the United States. The westerly branch would extend the system's reach due south to Vancouver, British Columbia; Seattle, Washington; and then to major cities in California. The easterly branch would enable goods to travel either from Fort Nelson to Chicago, or from Dawson Creek to North Dakota, and then to a projected rail corridor to Texas.

B. Electrifying the Network

Less than 1% of the rail route-miles in the United States is electrified. Thus the system, in order to qualify as efficient modern transport for both diversified freight and passenger traffic at high speeds, must not only be greatly expanded but also electrified.

True high-speed rail corridors—at travelling speeds for passengers of 250 kph (150 mph) or greater, and for freight at 145-175 kph (90-110 mph)—and, as quickly as possible, magnetically levitated train systems, will upgrade the whole U.S. economy. Both have a fundamental requirement: They run exclusively on electricity. For high-speed rail to operate, it must have electric-powered locomotives, and overhead catenary systems to transmit the electricity to the locomotives.

A national electrification program should concentrate on building and electrifying 42,000 critical rail-route miles (68,000 km) (**Figure 4**), in two phases: it would start with the electrification of 26,000 route-miles (42,000 km). These route-miles are selected because they are at the heart of America's rail system; they support, overwhelmingly, the greatest volume of freight and people. Although these 42,000 route-miles constitute only 29% of America's total rail route mileage, each year, they carry 65%

FIGURE 4
The Proposed 42,000-Mile-Long Network of National Electrified Rail

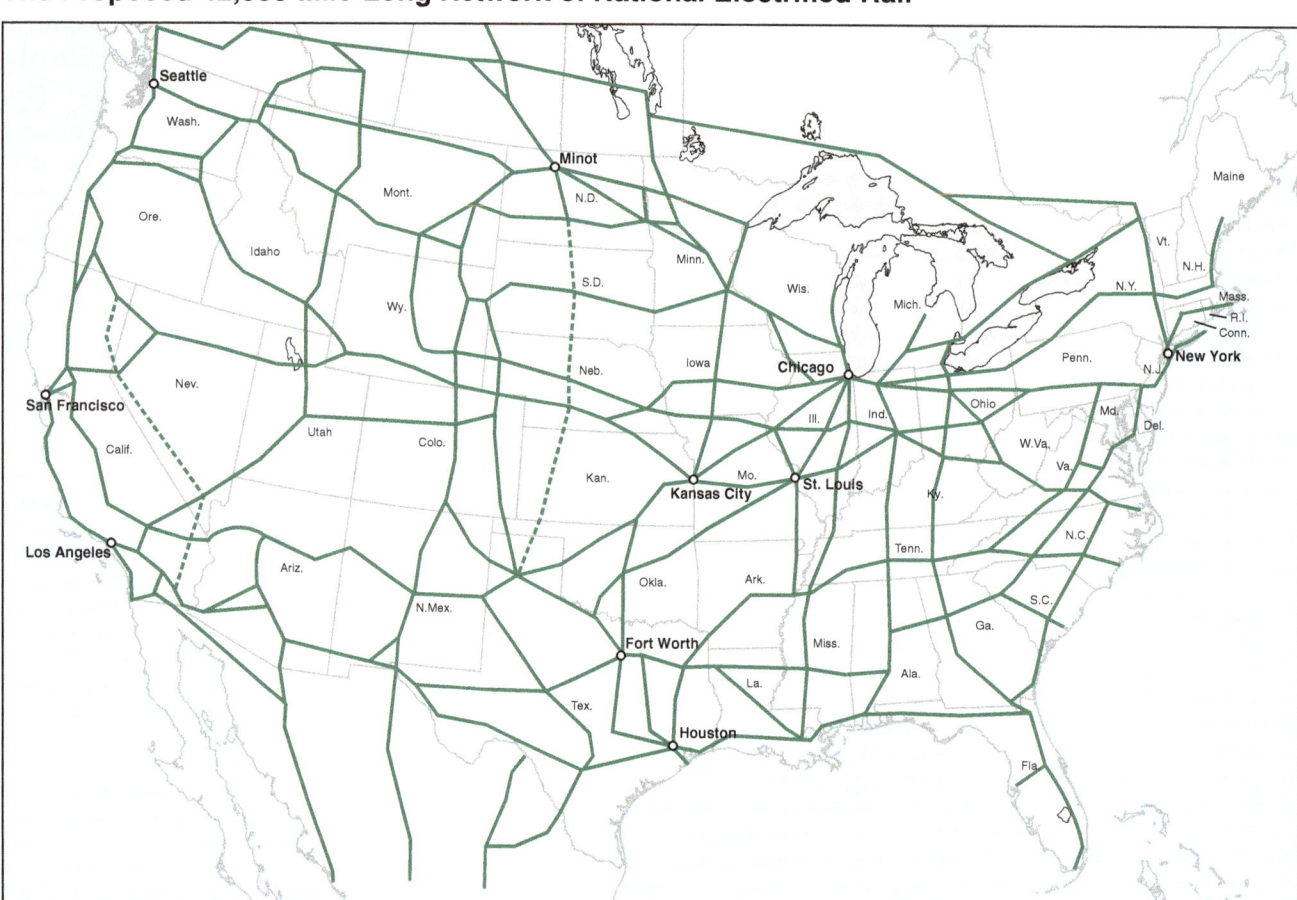

This route network of electrified intercity rail would transport freight and passengers, largely on existing (upgraded) rail lines.

of America's freight, and more than 70% of the intercity rail passengers.

The movement to electric rail would mean that eventually up to one-third of road-clogging truck traffic could be shifted onto rail, and that the current rail system, whose motive power is diesel-electric locomotives (which consume vast amounts of petroleum), could be shifted toward all-electric locomotives. Under advanced high-speed rail and maglev systems, *goods and people would move two to three times faster than they currently do*. Moreover, this will demand a huge increase in electricity generation. America will require mass production of nuclear power plants, and ultimately fusion power, to produce the electricity. Thus, America's transportation and energy policy would shift, in tandem, to higher efficiency and safety.

For example, the electrified train uses half as many BTUs to carry a ton-mile of cargo freight, as do steam-powered locomotives using oil as fuel, and also maintains very sizable energy efficiencies—on the order of five to 50 times relative to road and air—over other freight transport methods.

This entire project would employ several hundred thousand workers. It would require 15 years to construct, and cost more than half a trillion dollars, but its cost would represent but a fraction of the enhanced economic productivity it would impart back to the U.S. and North American economy.

II. A Water Management Emergency

A. Infrastructure Against Drought

In recent years the Western half of North America, including much of northern Mexico, has entered the grips of a drought that threatens its agriculture, food supply, industry, and even the ability to maintain civilized life in the worst-affected regions. The drought has still perhaps not reached the severity of the "Dust Bowl" conditions of the 1930s. But among institutional meteorologists there are many who forecast that solar patterns, and paleoclimatic history may make this condition last throughout the 21st Century; it is likely that the last, dry century was relatively "wet" for this region. Furthermore, it is not a new or unexpected development, so that plans for water-management infrastructure to attack drought on a grand scale are mature; they have been heard in Congressional hearings and from President Kennedy 50 years ago.

The greatest of them, the North American Water and Power Alliance (NAWAPA) or "Water from Alaska" scheme (**Figure 5**), has been put off for 50 years, and may not now be able to intersect this drastic and worsening drought situation in time. Among all its needs, North America needs to give the most urgent consideration, and investment, to infrastructure to provide multi-purpose water in the face of drought.

The extent and severity of today's drought in North America as of September 2014 is shown in **Figure 6**. In 11 Southwestern states, 98 million people are facing vanishing water, and all 22 states west of the Mississippi River are affected. California, with 37 million people, faces running out of water in 18 months. Many towns are now supplied by truck.

The food supply of the United States and internationally is jeopardized. California has accounted for 40% of all U.S. fresh fruits and vegetables, 20% of milk, and far higher percentages of specialty products—e.g., 90% of U.S. tree nuts come from the state, and 96% of tomatoes for processing. Texas alone had 13% of all cattle in the nation; now its herds are shrinking. The U.S. cattle herd has dropped from 111 million, down to 87 million, the lowest in more than 50 years.

All four of the Southwestern river basins, in what has been known as the Great American Desert, are in severe, and worsening, water deficit. On the Colorado River system, which serves 33 million people

FIGURE 5

NAWAPA XXI
NORTH AMERICAN WATER AND POWER ALLIANCE

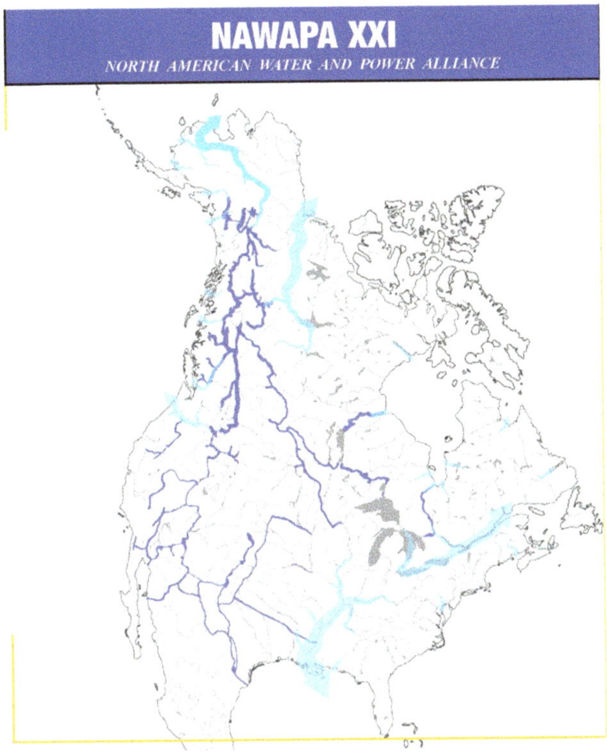

Source: Nuclear NAWAPA XXI; Gateway to the Fusion Economy, 21st Century Science & Technology, Special Report, 2013

Shown on the map of relative volume of river outflow, is the original 1960s North American Water and Power Alliance scheme, updated in 2013 with additional distribution systems.

FIGURE 6

North American Drought Monitor — September 30, 2014

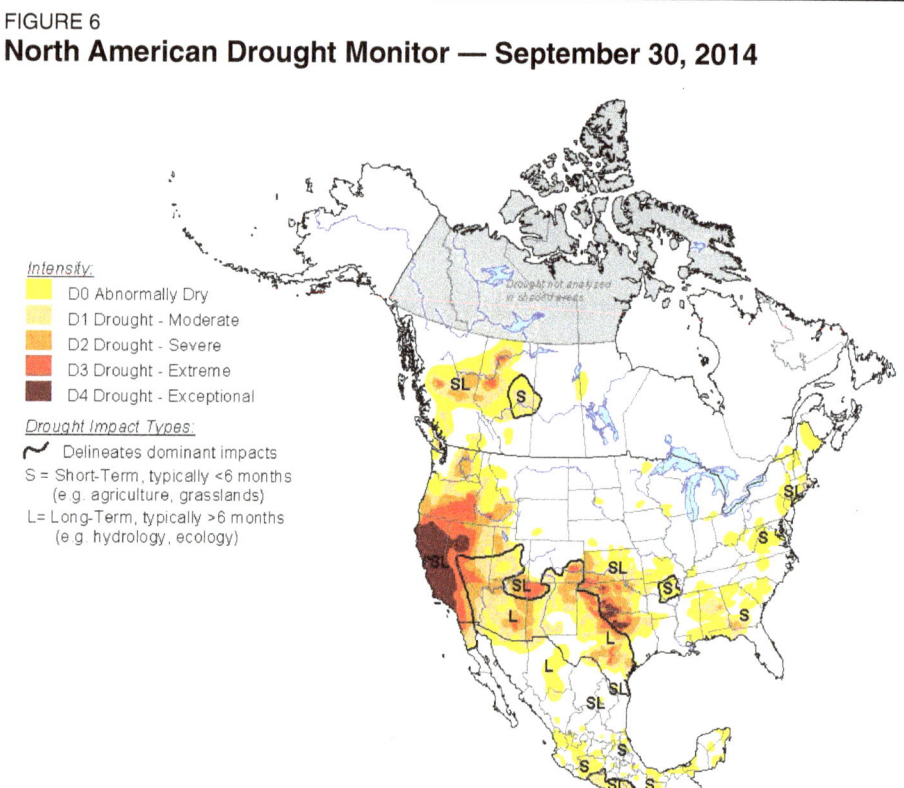

National Oceanic and Atmospheric Administration, October 2014

FIGURE 7

Trends in U.S. Total Water Withdrawals and Population, 1950-2005

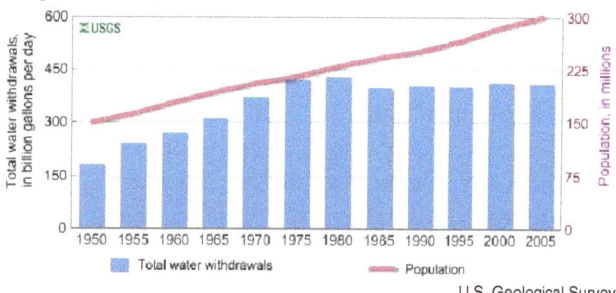

U.S. Geological Survey

in seven states, Lake Mead is at its lowest level since 1938, when it first started filling up behind the famous Hoover Dam. In California in 2014, for the first time ever, no water allocations are coming from its two management systems for the Sacramento-San Joaquin Basins—the California Water Project, and the Bureau of Reclamation's San Joaquin project. The same situation exists in the Rio Grande (Rio Bravo) Valley and the multi-state Great Basin.

Moreover, the drought follows a pre-existing decline in water throughput in the U.S. economy. **Figure 7** shows that the volume of water in use in the United States for all purposes (called "withdrawals," including for domestic, retail, industry, agriculture, and thermal cooling) is lower in absolute quantity than was used 30 years ago, when the population was smaller. Irrigation usage has declined; water for manufacturing, declined. Very little of this decrease is attributable to efficiency (higher output per unit of water). Instead, U.S. consumption has become dependent on imports of huge amounts of "virtual water" in foodstuffs and merchandise from abroad. This is untenable.

In terms of high-technology anti-drought infrastructure for North America, there is no "replacement" for not having built the NAWAPA scheme—the equivalent of ten TVA developments—to completion by the end of the last century, as promoted by Senator Frank Moss of Utah and, all too briefly, by President Kennedy. Throughout the past 50 years there has been sufficient precipitation and unused runoff in the most northwestern hydrologic districts of the continent, that a fraction of it would have slaked the drought deficits in the Southwest, Great Plains, and Mexico. But this may not continue to be the case, and the investment/building period for the vast NAWAPA water-management infrastructure is long.

Even given a prompt mandate to start building NAWAPA now, the hydraulic plan must at the same time be re-evaluated in terms of expected changes in solar cycles, which may alter Earth's water patterns, in a way that requires a re-designed continental hydraulic system.

B. Near-Term Desalination with Nuclear Power

The United States also has a long history of planning the use of desalination—in its most efficient, nuclear technological form—to deal with the Great American Desert; and this makes it possible to present **Figure 8**, a national, indicative plan of more than 40 sites.

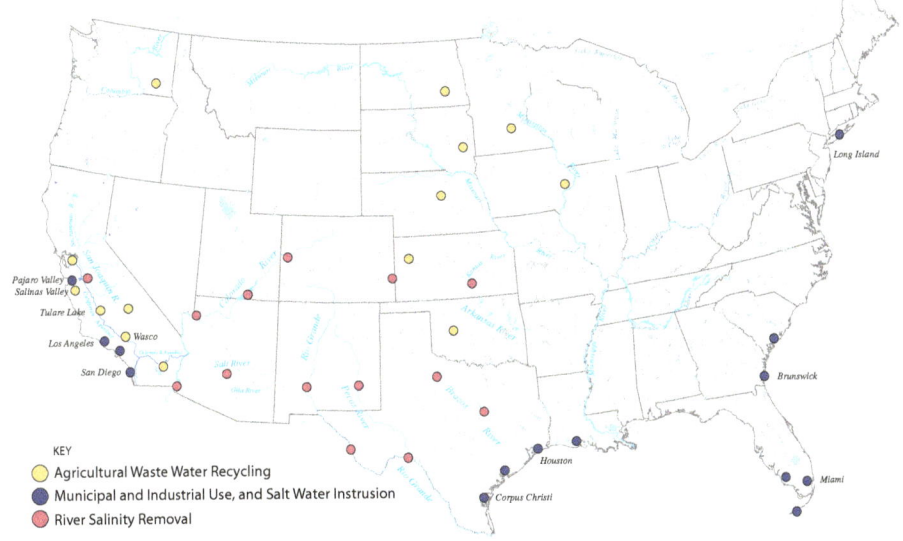

FIGURE 8
Proposed Locations for 42 Nuclear Desalination Plants

KEY
- Agricultural Waste Water Recycling
- Municipal and Industrial Use, and Salt Water Instrusion
- River Salinity Removal

21st Century Science & Technology, 2013

Nuclear-powered desalination was a goal of the 1950s Atoms for Peace program. The U.S. Office of Saline Water conducted pilot projects with Mexico, for the Pacific Southwest. President Kennedy commissioned a task force, which reported out in 1964, identifying which sites were priority for the go-ahead for full-scale operations.

But now, in response to the western water crisis, only non-nuclear desalination operations are in the works, dependent on local means and initiatives. Among California's 15 live projects, the Carlsbad facility on the Pacific Ocean, scheduled to open in 2016, will be the largest in the hemisphere, but still only provide 7% of San Diego's water needs (population 1.4 million). In Texas, the city of San Antonio (population 1.39 million) is building its first desalination plant, to try to gain a margin of additional water from an underlying aquifer.

It is now urgent to re-launch the Federal commitment for nuclear desalination, in collaboration with Mexico, and with China, where nuclear-desalination facilities on the North China Sea are in service, with more on the way.

By cutting out nuclear, water shortages were guaranteed. Since the 1960s, there has not been a water shortage, but rather a nuclear shortage. Today, in the short term, the interim-sized 567,800 m^3 per day (150 mgd) desalination plants proposed in the Kennedy era should immediately be built.

Coastal desalination for industrial and municipal use will provide for cities, offset demands on limited water for agriculture, and solve the problem of saltwater intrusion. Agricultural wastewater desalination combined with groundwater desalination will increase crop yields, and reclaim land abandoned due to high salinity levels and lack of water. Saline river water, a major problem in nearly every western river, can be treated.

Water quality in thousands of inland cities can be improved. As the larger size desalination plants become available, in addition to increasing the amount of all of the above, sufficient quantities of water could be made available for new agriculture.

In southern California, Pacific coast desalination plants can provide the equivalent of all or part of the region's demand from the Colorado River, thereby physically freeing its flow to meet use requirements in upriver states, and in Mexico.

Locations along the Atlantic and Gulf of Mexico littoral likewise require nuclear desalination facilities. At many coastal sites, pumping of fresh groundwater supplies over the decades and centuries, has resulted in saltwater intrusion when the extraction rate exceeds re-charge of wells. The brackish water requires expensive treatment. Already in 1979, this problem was a subject of Congressional study.

Ample supplies from nuclear desalination plants will offset the intrusion problem, as well as create new volumes of water for municipal and industrial use. Priority areas for these operations include San Diego and Los Angeles, California; Houston to Corpus Christi, Texas; Louisiana; Key West and southern Florida; Georgia; South Carolina; and Long Island, New York.

While these coastal applications of nuclear desalination would mostly indirectly supply agricultural use by offsetting demand and restoring groundwater supplies, larger desalination plants, such as the proposed 8,300 MWt reactors capable of producing 3.028 million m^3 per day (800 mgd), could produce water directly for agriculture. For example, the San Joaquin and Imperial Valleys, which are in terrible drought, could be supplied by coastal desalination, pumped to the California Aqueduct System, which could then be brought directly into the California Water Project system.

In climate dynamics, science and applied research must proceed on a crash basis to understand more of climatic patterns and causes, and identify potential modes of intervention to "improve" weather. For example, where should the placement be of scaled-up ionization systems to induce rainfall in the dryland west, already shown to be effective in North America, in the trials in central Mexico (see Part 2 "Solve the World Water Crisis")?

It is critical that the U.S. collaborate with researchers in China, where the national space program involves the highest level mission to conceptualize the solar system and beyond. Satellite and related investigations must be fully undertaken, in collaboration internationally, for purposes of advancing planetary science, whose applied research can include how to intervene to induce more rain, where and when required.

III. Power: Nuclear Expansion Urgent

The United States' need for nuclear power investments goes beyond the most urgent need—nuclear's unique suitability for combining power with desalination on a large scale to combat the advance of drought. In the period since nuclear power additions to the American power grid ceased in the mid-1990s, electricity production per capita broke a nearly century-long rise, and fell from 13.3 MWh/capita/year to 12.4 MWh/capita/year through 2013. The price index for electricity (1985 = 100 MWh) has nearly doubled in that period, from 109 in 1995 to 201 in 2013. Partly due to deregulation and inflation, this sharp price rise is associated with the reduction in production.

Figure 9 shows the broader picture (see also Part 2 of this report, "Energy Flux Density"). Whereas each previous introduction of a new fuel for power production has led to a leap in power available per capita, the environmentalist *blocking* of fission power's expected rate of expansion stopped the process of productive evolution of the U.S. economy.

Like Germany's "nuclear exit," but with less drastic results thus far, the United States has opted to increase carbon dioxide emissions, degrade local environments, and reduce the efficiency of power production, by seeking the goal of "becoming a new Saudi Arabia" with hydraulic fracturing of oil and gas. In addition, the United States has brought its railroad grid to the point of overt breakdown (as detailed above) by overwhelming it with shipments of coal, oil, and sand and chemicals for hydraulic fracking

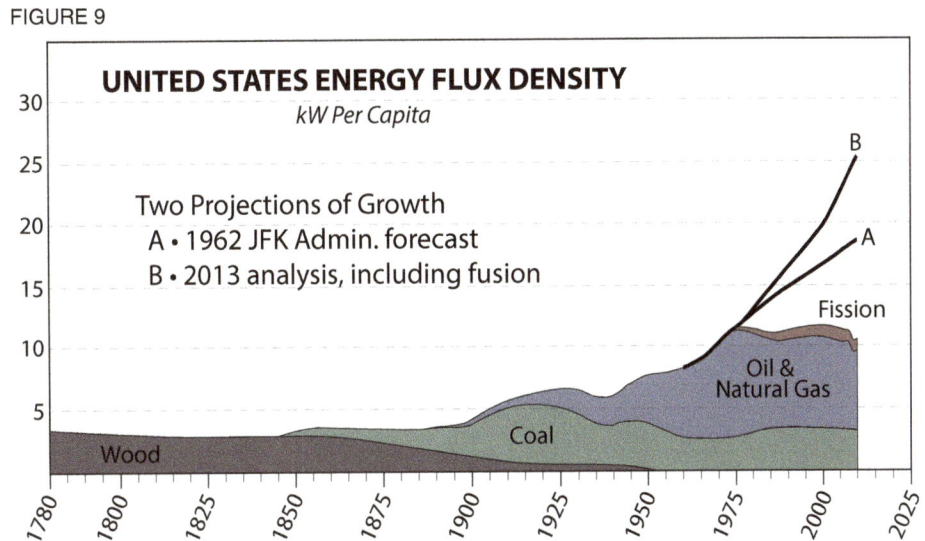

FIGURE 9

Benjamin Deniston, 2014, from US EIA and "Civilian Nuclear Power, a Report to the President" submitted to JFK by Leland Haworth.

Per capita power consumption for the United States from 1780 to 2010, divided by the major sources of power. The general growth trend is clear, until 1970, when the zero-growth insanity took over the United States. Two projections indicate what could and should have happened. Curve A is a 1962 projection made by the John F. Kennedy administration, which focused on the then-coming role of nuclear fission power. Curve B is an estimation of what was possible if the Kennedy vision had been pursued, followed by the development of controlled thermonuclear fusion (following the 1970s realization of the feasibility of fusion). These two curves, compared with the actual levels, show the 40-year growth gap which is a major source of the current economic collapse. (kW have been converted from BTUs/hour.)

Above all, U.S. national credit needs to be invested, with involvement of its national laboratories and their key contractors, in development of the "fourth generation" of nuclear plants, operating in compact fail-safe units at very high temperatures and cooled by gas flow.

With the operating temperatures of fourth-generation reactors projected at more than 1,000 degrees Celsius, nuclear-powered industrial complexes, with their array of reactors, can be designed to supply the heat needs of many different industries—industries that require steam at different temperatures and pressures, as well as direct heat at a higher range of temperatures.

While steam temperature requirements are typically in the range of 120-540°C (250-1,000°F), direct heat temperature requirements are often much higher, at 800-2,000°C (1,500-3,600°F). Steel production requires temperatures up to 1,370°C (2,500°F) with current technologies, and cement requires temperatures approximately up to 1,450°C (2,640°F).

The broad use of high-temperature fourth-generation nuclear reactors in these production units will change the entire industrial landscape. Not only will it serve to replace low-energy density coal, oil, and natural gas for many current industrial processes, but more importantly, it will create new types of industries and products, while reshaping and multiplying the productive output of existing industries. For example, among many other uses, the centralized mass production and delivery of hydrogen could replace two-thirds of the production cycle for ammonia, one of the most-produced inorganic chemicals in the world.

The same U.S. national laboratories continue to conduct work on thermonuclear fusion power, with funding levels that have never been consistent with real progress, and have been progressively cut down for decades, to virtually nothing today. So low is the American fusion power effort, relative to that of nations such as China, Russia, and Japan, that U.S. President Barack Obama has referred to fusion power as a "fancy, exotic" technology that America does not need. The opposite is true, as shown in Part 2, "Energy Flux Density."

The United States no longer has production capacity for nuclear power reactors, and its only reactor designer, Westinghouse Electric, is now a Japanese company. Thus there is another need for investment of national credit—and another opportunity for productive development—in the large number of former sites of auto/auto parts production, mothballed over the past decade. The machine-tool capacities of these plants, many of which are no more than 30 years old, has always been the highest in U.S. industry, making them universal producers and "the arsenal of democracy" (see **Figure 10**). Their retooling is essential for rebuilding the nuclear industry.

IV. Reviving Space Exploration

No priority would be higher for a new U.S. "President Kennedy," pursuing a new "Alliance for Progress" with the BRICS nations, than a revived U.S. mission in exploring the Solar System. The Apollo Project made the United States the world's space exploration leader then. And as noted at the outset, it was in those "Apollo" 1960s that total productivity in the U.S. economy reached heights second only to Franklin Roosevelt's reconstruction and war mobilizations.

The decline in the American space effort since 1969 is only too evident; it can be reversed in the context of "world land-bridge" investments, and by reversing "cold war" policy and embracing full cooperation with the now-leading Chinese, Indian, and Russian space programs, as well as the European ones.

As recently as 2005, NASA was still operating with a comprehensive roadmap toward developing the technology and capability to return a four-man crew to the surface of the Moon, targeting the icy south pole as an optimal destination. The Constellation program was to replace the Shuttle-centered program, slated to be mothballed in 2010. The shutdown of the Shuttle program was to create the "savings" to fund Constellation without increasing NASA's overall budget. Although it was known that this would create a time-gap where

FIGURE 10
Retooling Locations for the Mass Production of Nuclear Power Plants

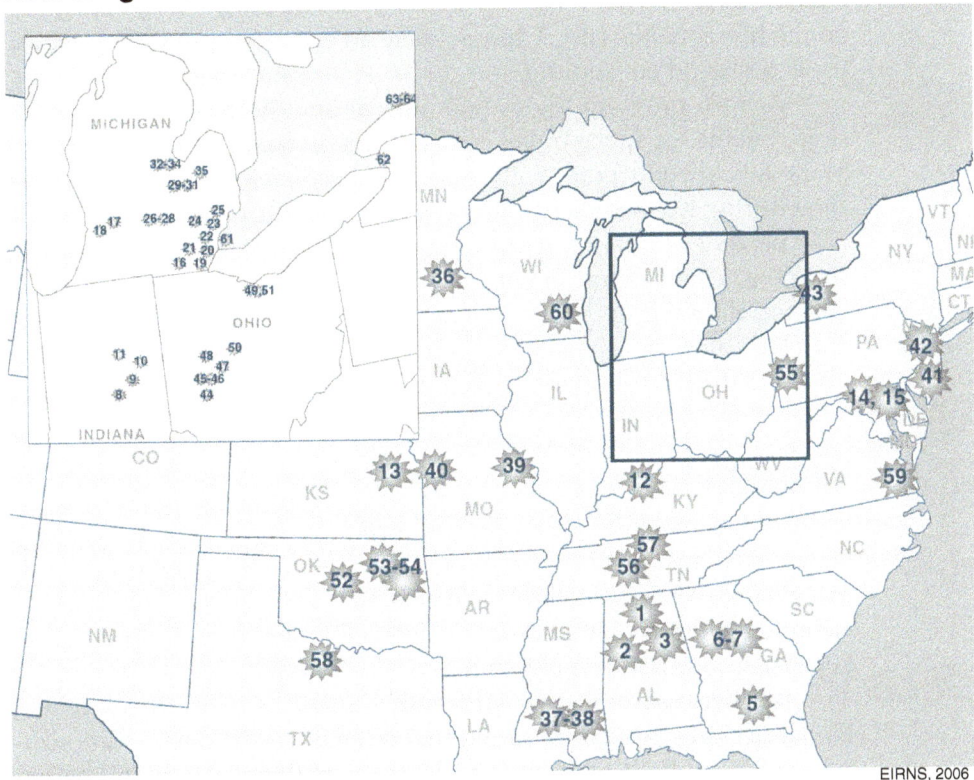

This map shows 64 specific locations for the mass production of nuclear plants, where idle auto capacity existed in 2005 that may still be potential locations for such conversion. Such plants could produce nuclear fuel rods, cranes, pumps, valves, pipes, and other components of nuclear power plants, electric locomotives, high-speed rail rolling stock, aluminum, plastic injection molding presses, mitre gates for locks and dams, parts for large earth-moving machines, pumping stations, and other infrastructure.

EIRNS, 2006

no American vehicle would be capable of getting astronauts and supplies to the International Space Station, it was accepted in the context of the U.S. embracing a greater mission.

In the 2005 period NASA officials still spoke definitively of where the U.S. space program was directed, if it got the investment; this is shown in statements from that period covered in the news.

"NASA briefed senior White House officials Wednesday on its plan to spend $100 billion and the next 12 years building the spacecraft and rockets it needs to put humans back on the Moon by 2018" (September 2005).

"NASA's plan envisions being able to land four-person human crews anywhere on the Moon's surface and to eventually use the system to transport crew members to and from a lunar outpost that it would consider building on the lunar south pole, according to the charts, because of the region's elevated quantities of hydrogen and possibly water ice."

"One of NASA's reasons for going back to the Moon is to demonstrate that astronauts can essentially 'live off the land' by using lunar resources to produce potable water, fuel and other valuable commodities. Such capabilities are considered extremely important to human expeditions to Mars which, because of the distances involved, would be much longer missions entailing a minimum of 500 days spent on the planet's surface."

The objectives can now become much broader.

As early as 2003-04, the goal of industrialization—focused on achieving Helium-3 production on the Moon for return to Earth as the breakthrough fuel for thermonuclear fusion power—was appearing on the horizon. Now the Chinese have an active program with which the U.S. should collaborate.

A crash program to develop nuclear and thermonuclear propulsion systems for space travel can accompany this, and make travel to Mars and beyond potentially feasible; these technologies were pursued during Kennedy's Apollo program, then dropped.

Leading Russian scientists have repeatedly proposed, especially since the "Chelyabinsk Event," an international program of "strategic defense of the Earth" (SDE) against space-object strikes. This requires both large networks of satellites surveilling near-Earth space in particular, and the new propulsion technologies capable of rapid response.

Postscript

With such infrastructure and productivity needs, the United States' response to the formation of the Asian Infrastructure Investment Bank (AIIB) is suicidal. The Obama Administration has tried to cripple or kill this new international development bank by pressuring key Asian countries not to join it.

But because both the AIIB and the BRICS New Development Bank are being formed as infrastructure lenders without conditionalities, and potentially on a large scale, they are the nuclei of an international development bank network that can provide the *many trillions of dollars-equivalent* in modern infrastructure needs of the world.

The United States' such needs are massive and glaring. They must, in fact, join the AIIB and the "BRICS Bank."

Major parts of this section were adapted from publications by Dr. Hal B.H. Cooper, Jr., Richard Freeman, and Michael Kirsch, cited below; research was also contributed by Marsha Freeman.

FOR FURTHER READING

"Build the Missing Link: Alaska-North America Rail," Hal B.H. Cooper, Jr., *EIR*, July 27, 2007.

"Infrastructure Corridors Will Transform Economy; Bering Strait Tunnel, Alaska-Canada Rail," Richard Freeman and Hal Cooper, *EIR*, Sept. 27, 2007

"A Plan to Revolutionize America's Transport," by Hal Cooper, *21st Century Science & Technology*, Summer 2005.

"Why Electrified Rail Is Superior," by Richard Freeman and Hal Cooper, *21st Century Science & Technology*, Summer 2005.

"The Nuclear NAWAPA XXI and the New Economy," by Michael Kirsch, in *Nuclear NAWAPA XXI; Gateway to the Fusion Economy*, 21st Century Science & Technology Special Report, 2013.

"The U.S. Economic Recovery Act of 2006," by Lyndon LaRouche, LaRouchePAC.

PART 12
Fighting for International Development

LaRouche's 40-Year Record Fighting for International Development

From 1970 to today, physical economist Lyndon LaRouche has an unparalleled record of proposals for reforming international financial institutions, and for launching the great development projects that can uniquely reverse the decline of the world economy. Leading instances of those initiatives are listed below. The impact of these ideas, and the political fight on their behalf, can be clearly seen in the current dramatic moves bythe BRICS nations to create a new financial architecture.

Financial Reform

1975: At a press conference April 24 in Bonn, Germany, LaRouche presents his plan for "the immediate establishment of an International Development Bank as an agreement among the three principal world sectors—the industrialized capitalist sector, the so-called developing sector, and socialist countries." He specifies that the immediate concentration of the investment thus made possible should be industrial development and expanded food production worldwide.

1976: The Group of 77 Developing Countries, meeting in Colombo, Sri Lanka in August, issues a call for a new world economic order, based on respect for sovereignty, technology transfer to the Third World, and mutually advantageous economic development proposals between the developed and developing world. This is followed in September at the United Nations, by a call for "new development banks" by Guyanese Foreign Minister Fred Wills.

1982: LaRouche addresses the exploding international debt crisis with his proposal for *Operation Juárez*, which outlines a specific Ibero-American plan for financial reorganization for development, based on principles that could be applied globally.

1988: Under the title "Development Is the Name for Peace," the Schiller Institute, founded by Helga Zepp-LaRouche, holds a conference in New Hampshire on Jan. 30-31, dedicated to elaborating the need to establish a "just new world economic order." Lyndon LaRouche addresses this conference on how the U.S. Presidency could establish such a new order, which he later dubbed a "New Bretton Woods."

1997: At a Jan. 4 webcast, LaRouche issues a rallying cry for a

FIGURE 1

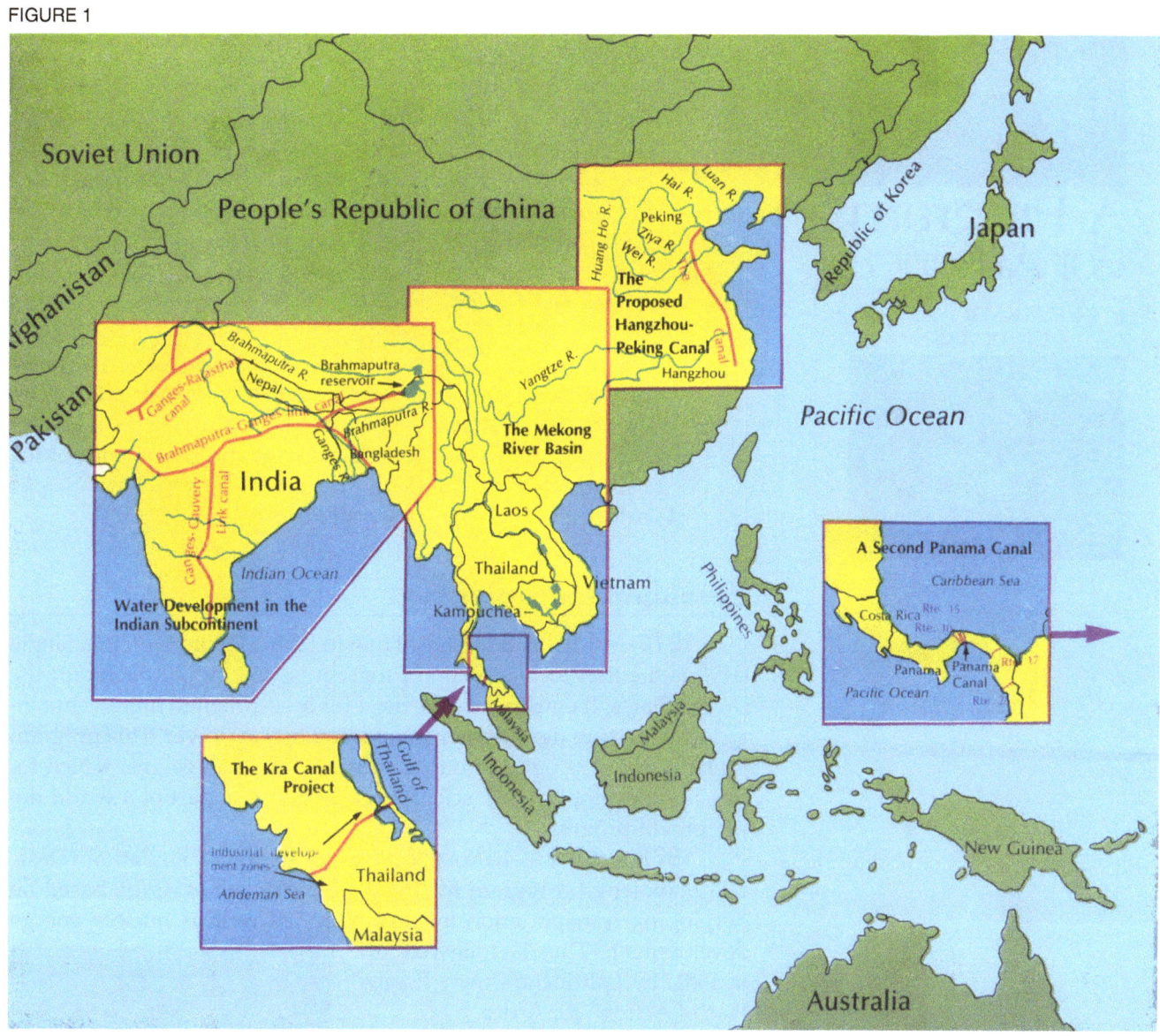

The principal projects outlined in Lyndon LaRouche's 1983 EIR *Special Report, "A Fifty-Year Development Policy for the Indian-Pacific Oceans Basin."*

New Bretton Woods, based on bankruptcy reorganization of the world economy, followed by establishment of an international credit system for global reindustrialization. In the months that follow, the LaRouche movement organizes a global movement of prominent political leaders and economists demanding this reorganization.

2008: In the face of the devastating global financial collapse, LaRouche, in the Fall, demands an urgent application of FDR's Glass-Steagall principle to banking systems throughout the world, but starting in the United States.

2014: On June 8, LaRouche issues his "Four New Laws to Save the U.S.A. Now!," which defines the urgent measures required to be taken by the U.S. Congress. These include:

"(1) *Immediate re-enactment* of *the Glass-Steagall law instituted by U.S. President Franklin D. Roosevelt, without modification, as to principle of action.*

"(2) *A return to a system of top-down, and thoroughly defined as National Banking.* "The tested, successful model to be authorized is that which was, under the direction of the policies of national banking which had been successfully installed under President Abraham Lincoln's superseding authority of a currency created by the Presidency of the United States (e.g., 'Greenbacks'), as conducted as *a national banking-and-credit-system placed under the supervision of the Office of the Treasury Secretary of the United States. . . .*

"(3) The use of a Federal Credit-system is to generate high-productivity trends in improvements of employment, with the intention to increase the physical-economic productivity and standard of living of the persons and households of the United States . . . ,"and

"(4) Adopt a Fusion-Driver `Crash Program.'"

Development Projects

1970—United States: "How to Lick a Depression in a Single Day" is the first reconstruction program issued by LaRouche for the United States. It emphasizes the need for investment in high-technology infrastructure development, including fusion power. This program is elaborated through LaRouche's Presidential campaigns, which focuses on developing U.S. scientific capabilities as part of a world development program.

1979—Africa: LaRouche's Fusion Energy Foundation releases a book-length program for the *Industrialization of Africa*, based on developing transportation infrastructure, as well as nuclear energy development. This is followed up in 1981, by LaRouche's own "Lagos Plan of Action" for Africa.

1979—India: *EIR* issues a study on "The Industrialization of India," commissioned by LaRouche, which defines how it can go "From Backwardness to Industrial Power in Forty Years."

1983—Asia/Pacific: LaRouche issues an *EIR* Special Report entitled "A Fifty-Year Development Policy for the Indian-Pacific Oceans Basin," which presents the conceptual basis for large-scale infrastructure projects, including water development in the Indian Subcontinent, the Mekong River Basin, the Kra Canal Project, the Hangzhou-Beijing Canal, and a second Panama Canal (**Figure 1**). These projects represent the "motor for development," LaRouche says.

1988—Ibero-America: The Schiller Institute issues a book-length study on "Ibero-American Integration, 100 Million New Jobs by the Year 2000!," which outlines the basis for an integrated agro-industrial modernization of the continent, including projects on water management, high-speed rail, increasing agricultural productivity, nuclear energy, and other investments in high-technology development.

1989—Europe: Lyndon and Helga LaRouche put forward the "Productive Triangle" development plan, in the face of the collapse of East Germany. It builds off LaRouche's October 1988 proposal for Western Europe to provide high-technology aid to deal with the food crisis in the East, and advancing to the development of high-technology development corridors between Moscow, Paris, and Vienna—an area which encompasses the most productive industrial centers in Europe.

1990—Southwest Asia: LaRouche releases his "Oasis Plan," as a basis for lasting peace between Israel and the Arab world, based on programs of nuclear desalination and industrial development for the entire region.

1990s—Eurasia: With the collapse of the Soviet Union in 1991, the LaRouches expand the concept of the Productive Triangle to become the Eurasian Land-Bridge, linking all of Eurasia through development corridors. One of the high points of this organizing occurs in 1996, during a conference sponsored by the Chinese government, which features plans for a "New Silk Road." This event is followed by many others dedicated to the Eurasian Land-Bridge over the following years.

2007—Russia/U.S.: LaRouche puts a special emphasis on the Bering Strait Tunnel during a May visit to Moscow, where he is a featured guest at the Russian Academy of Sciences in celebration of the 80th birthday of Prof. Stanislav Menshikov, a prominent Russian economist.

Development Model
The Tennessee Valley Authority: Great Projects Make Great Nations

By Marsha Freeman

August 2014

Many millions of the peoples in the nations along the Silk Road today face daily circumstances not unlike those faced by Americans in the southeast of the United States at the time of the Great Depression. When Franklin Roosevelt became President in March 1933, the people living in the seven-state region that defines the Tennessee Valley had no electricity, were subject to periodic floods, had seen much of their agricultural land destroyed, suffered from epidemic disease, and had little access to modern education, health care, or transportation. After decades of failed attempts, on May 18, 1933, President Roosevelt signed the legislation to create the Tennessee Valley Authority (TVA), in order to bring the people of the Tennessee Valley into the modern world, and to transform the Valley into a showcase for regional integrated economic development. The TVA still serves as a model for regional development today.

FDR's TVA—Integrated Watershed Planning and Development

The purpose of the TVA, as stated by the President and the law, was not only to halt the floods and tame the rivers, but to demonstrate the efficacy of "national planning for a complete river watershed involving many states and the future lives and welfare of millions." The TVA, the President stated, should be charged with the "duty of planning ... for the general social and economic welfare of the nation." This project would be "a great experiment for the benefit of generations yet to come," and "millions yet unborn." From its beginnings, the leadership of the TVA, under President Roosevelt's watchful protection, expected the TVA to be a model for the development of "a thousand valleys" in every corner of the globe. Throughout the 1930s and after World War II, the TVA was the best-known American great project, an approach to economic development which, with the help of the experts of the TVA itself, would be emulated around the world.

Six months following its creation, the TVA had engineers and workmen on site, building its first project—Norris Dam. Over the next

The Tennessee Valley Authority: Great Projects Make Great Nations

FDR signs the bill authorizing the Tennessee Valley Authority, in May 1933.

20 years, TVA built 20 dams, bringing electric power to homes and farms, and providing flood control, river navigation, and recreation to millions of citizens. But for the people of the Valley to progress, a dramatic up-shift in their physical well-being and cultural life was required. So the population could learn how to make use of mankind's "modern slave" of electric power, TVA sent college students and manufacturers to homes and farms to carry out demonstrations, and offered low-interest credit for the purchase of the electrical appliances that would free farmers and families from the drudgery of 19th-Century rural life.

TVA inoculated residents against the scourges of typhoid and smallpox, and tackled endemic malaria by eradicating mosquitoes. To create an educated workforce and populace, the TVA set up libraries in stores, post offices, and gas stations, and deployed bookmobiles near dam construction sites and around the countryside.

Reclaiming and reforesting damaged and deserted lands was a high priority in the once-fertile valley where farming was the primary economic activity. To create an up-shift in the energy and technology in agriculture, and taking advantage of now-available electricity, teams of chemists and chemical engineers began the operation of a TVA phosphate-based fertilizer production program. Thousands of demonstration farms were established on operating farms by 200 TVA experts, with TVA donating its new fertilizer. By 1935, TVA produced 24,000 tons of concentrated superphosphate fertilizer, which grew to 136,000 tons in 1953. The fertilizer was shipped all over the United States, and accounted for 24% of all fertilizer production between 1934 and 1955. But the most dramatic impact of the TVA's fertilizer research and production program would be in other nations.

It is estimated that 2-3 billion people, or more than one-third of the world's population, are alive today because of the development of synthetic fertilizer, more than 70% of which was developed by TVA's National Fertilizer Development Center in Alabama. Dr. Normal Borlaug, the father of the "Green Revolution," which has saved millions of people in developing nations from starvation, was on the Board of the Center, from 1994 to 2003.

What Visitors Want to See

From the first days it opened its doors, the TVA invited and received thousands of visitors from other lands, most from nations whose people faced deprivations similar to what had faced the people of the southeastern United States, before the TVA.

In 1944, TVA Chairman David Lilienthal wrote that the "more than 11 million people who have visited the TVA in recent years" included an agricultural commissioner from New Delhi, a Brazilian scientist, a Czech electrical expert, Israeli Prime Minister David Ben-Gurion, Indian Prime Minister Nehru, and President Gabriel Gonzales Videla of Chile. There were engineers and agriculturalists from China, and Russian engineers working with TVA technicians on Lend-Lease hydroelectric plants. Countries sent teams of their top engineers to learn how to transform their nations. After the conclusion of World War II, dozens of nations were ready to develop their own TVAs. TVA engineers fanned out around the globe.

The TVA outlined integrated economic development plans for the Valley of the Nile River, and for the Tigris and Euphrates Rivers, encompassing Turkey, Syria, Iraq, and Iran. There were plans to develop the Helmand River and its tributaries for the benefit of India and Pakistan. Former TVA project engineer James Hayes developed the plan for a project in Afghanistan. Extensive plans were developed for Mexico, Haiti, Chile, Costa Rica, Puerto Rico, Peru, Brazil, and Colombia by TVA engineers.

In the mid-1950s, former TVA General Manager Gordon Clapp headed a UN survey mission for the Middle East, to create an integrated development plan for the region. The result was the 1954 proposal for a "TVA on the Jordan," requiring the cooperation of Syria, Jordan, Israel, and Lebanon, to build a series of dams on the upper Jordan River and its tributaries. This network would provide electricity to the region, and hundreds of thousands of acres of land could be irrigated for cultivation. But the British-organized Suez crisis two years later ended the UN initiative. In fact, the great majority of the projects the TVA developed have yet to be brought to fulfillment.

Today, one of the greatest legacies of the TVA is the "TVA on the Yangtze," in China. David Lilienthal, describing the challenge, remarked, "The terms *gigantic* or *colossal* are not inappropriate for this plan, which dwarfs the TVA by comparison." With the urging of President Roosevelt, Lilienthal held discussions before the end of World War II with Chinese representatives, to plan the post-war reconstruction of China. Although it took decades for the project to become reality, reflecting the approach of the TVA, China's Three Gorges Dam project is a regional economic development plan. It consists of a network of dams on the Yangtze River and its tributaries, to tame the flow of this third-longest river in the world. The primary goal of the dam is to avoid tragedies such as the 1931 flood, which took the lives of 145,000 people.

But what the Three Gorges Dam is providing is not only protection from floods, but electricity for new homes and factories that are being built in entirely new cities. With the modern cities come schools, healthcare facilities, and modern communications and transportation. Farming will progress from animal-drawn plows to modern scientific methods, and, over time, a region that encompasses a population larger than that of the United States, will be transformed.

Although critics of the TVA, of which there have been many, proposed that when the dams had been built, and the rivers tamed, the work of the Authority had been completed, there were those with a

The Tennessee Valley Authority: Great Projects Make Great Nations

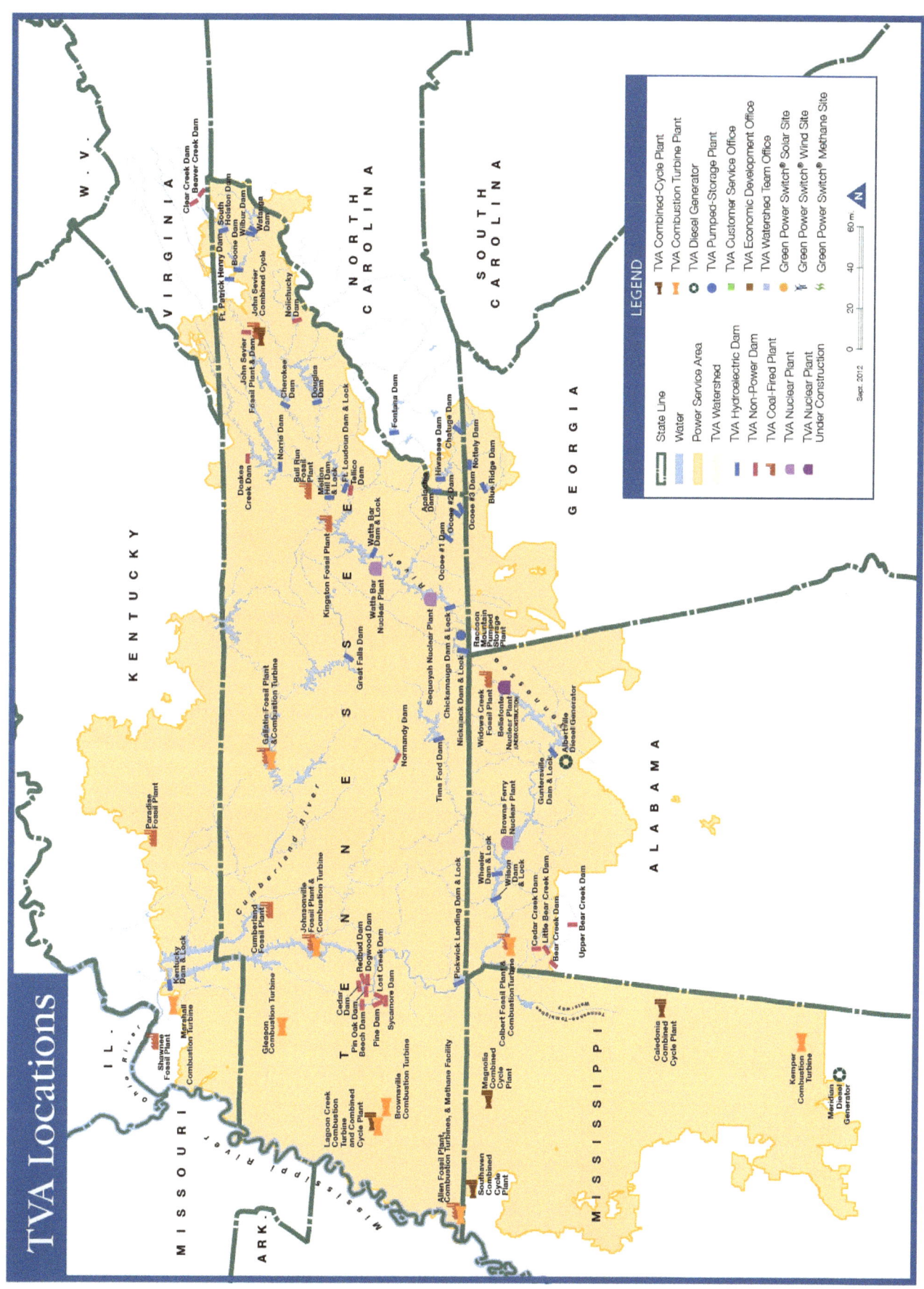

The New Silk Road Becomes the World Land-Bridge

broader vision, reflecting President Roosevelt's plans for the post-war transformation of formerly colonial nations.

Speaking at Muscle Shoals, Alabama on the 30th anniversary of President Roosevelt's signing of the TVA bill, on May 18, 1963, President John F. Kennedy located the broad national impact of the TVA's accomplishments, by noting that "a rising tide lifts all boats." Quoting Senator George Norris from Nebraska, who fought for decades for the TVA to be created, President Kennedy located the dedication of great projects to be "to generations yet unborn."

A Center for Frontier Technologies

Although the original concept of the TVA was centered on the control and direction of water resources, by the mid-1960s the scientists and engineers at the Authority, and the nearby Oak Ridge National Laboratory, were developing plans for the introduction of one of the two frontier technologies to come out of World War II—nuclear power. Applying the approach of integrated resource planning that was at the heart of the TVA, scientists developed the concept of "nuplexes," or nuclear-centered agro-industrial complexes. These would use the electrical energy and heat from fission plants to create new cities, and new centers of economic activity. A particular focus of the nuplex concept was the creation of fresh water through desalination, to tackle one of the most immediate global needs. In 1964, experts from the Valley traveled to India, Israel, Puerto Rico, Pakistan, Mexico, and the Soviet Union, to help plan desalination projects using the heat from nuclear power plants. At the same time, the Tennessee Valley region itself was being prepared to become the largest nuclear power plant construction project in the world, with 17 plants planned at seven sites in Tennessee, Alabama, and Mississippi.

But President Roosevelt's vision of how the world would change following the war, through the decolonization of much of the developing world, in large part was sabotaged either directly or indirectly by the British Empire, which had no intention of relinquishing its power.

Today, virtually all of the TVAs in the "thousands of valleys," which had been envisioned in the 1930s, still remain undone. The approach of integrated regional resource and economic development, based on the up-lifting of the standard of living of the population, and the creation and application of frontier technologies, is, 70 years later, still the model for transforming the nations along the Silk Road.

FOR FURTHER READING

"Roosevelt's TVA: The Development Program That Transformed a Region and Inspired the World," by Marsha Freeman, *21st Century Science and Technology*, Summer 2011.

"The Tennessee Valley Authority: a model for world development," by Marsha Freeman, *EIR*, December 21, 1990

Development Model
Deng Xiaoping's China Miracle

By Jeffrey Steinberg

September 2014

Deng Xiaoping

With the death of Mao Zedong on September 6, 1976 and the arrests the following month of the "Gang of Four," Deng Xiaoping rapidly established himself as China's new "supreme leader." Despite the fact that the only formal top position he would hold was Chairman of the Central Military Commission, it was Deng's vision of a modern China that was the single most significant factor in the revolutionary changes that China underwent in the next 35 years.

According to Deng biographer Ezra Vogel, from the moment he returned to power, Deng prioritized the development of a scientific and technological cadre over all other responsibilities and initiatives. He emphasized to colleagues that if China were able to train a generation of world-class scientists and engineers, within 30 years China would be able to emerge as a leading nation.

In pursuit of this enormous goal, Deng completed the normalization of relations with the United States that had begun with the 1972 Kissinger-Nixon diplomatic opening. On January 1, 1979, the United States officially recognized the People's Republic of China as the one China.

Deng knew the magnitude of the challenge that China was facing. Following the start of the Cultural Revolution in 1966, nearly a generation of young people was deprived of any higher education. The campaign of the Gang of Four against "bourgeois intellectuals" led to restrictions on enrollment of students not from a worker or peasant background, and the loss of a good portion of the teaching cadres, many of whom, designated as of "bourgeois background," were sent to the countryside to perform manual work and be "re-educated." Chairman Mao called for shorter study times, and demanded that students who finished their course work return to work in the factories or on the collective farms. Although there were still some universities in operation during that period, the level was far below what it had been, or would later become. This was the situation for almost 10 years.

Then in 1977, with the first vestiges of the Deng reforms taking root, many universities were reopened, and for the first time since 1965, students were allowed to take college entrance examinations. This first group of college entrants in more than a decade, dubbed the "Class of 1977" (of which China's current Premier, Li Keqiang was one), became the basis for China's remarkable development since that time.

Already in 1975, Deng had begun to revive the Chinese Academy of Sciences, bringing back many of the teaching cadres that had been sent to the countryside. However, many of those had died during the dark days of the Cultural Revolution.

With China's "reform and opening up [gaige kaifang]" under Deng, and the establishment of relations with the United States in 1979, China agreed to become a source of cheap labor production for the Western economies. Western investors, eager to slash wages and increase profits, thought they would do what they had been doing south of the border in Mexico and in the rest of Ibero-America—bringing down their costs by paying slave-labor wages to workers.

The Chinese Government was prepared to temporarily accede to these conditions, but Deng also had a long-term strategy. China would agree to what were disadvantageous terms, but would forge the means by which it could work its way out of those conditions as quickly as possible, targeting key areas of science and technology in which it intended to "leapfrog to a higher stage of development."

Taking a page from the U.S. SDI program, China in 1986 developed its "863 program for research and development." The government chose seven key scientific areas in which it would put the nation's resources, with the intent of making major scientific and technological breakthroughs. These areas were space, lasers, energy, biotechnology, new materials, automation, and information technology (IT).

By 2009 the "863 program" was funding 110 new programs, including in IT, manufacturing, materials, resources and environment, earth observation satellites, transportation, biology, energy, and agriculture. All told, $200 billion was invested under this one program into scientific research, resulting in 2,000 domestic and international patents, and the emergence of an indigenous Chinese IT sector that is today world-class.

The Chinese Government also took another page from the U.S. model and established a National Science Foundation, similar to the one in the United States. In 1997 China upgraded its science research program with the "973 Basic Research Program." This program had the following objectives: (1) Support multidisciplinary and fundamental research of relevance to national development; (2) Promote frontline basic research; (3) Support the cultivation of scientific talent capable of original research; and (4) Build high-quality interdisciplinary research centers.

These programs helped to launch China into new fields of development by leaps and bounds. With their help, China has developed the world's first light quantum telephone network, has grown the first living mice from induced pluripotent stem cells, and has built the fastest computer in the world, the Tianhe 1-A.

And although half of China's population is still living on the land in impoverished, but improving conditions, the ordinary Chinese citizen has not been forgotten. The notion of the "people's livelihood," or the "general welfare," as an American would call it, is an integral part of the Chinese program, an issue which has been continually stressed by President Xi Jinping. Currently 9% of the Chinese labor force is college-educated. By 2020 the Chinese intend to bring that percentage up to 20%.

Deng Xiaoping's Science "Long March"

Immediately upon being reinstated to all of his former party and government posts at the Third Plenum of the Tenth Party Congress on July 17, 1977, Deng made it clear that he intended to focus his attention on science, technology, and education. He identified specific scientific programs—nuclear energy, computers, polymers, semi-conductors, astronautics, and lasers—as the first priorities. "Science and technology are a force for production," he told an audience at one of the first party events he addressed on his return.

At the time, China had 200,000 scientific and technological workers, where the United States had 1.2 million. Deng sought every opportunity to meet with visiting Chinese-American scientists, including Lee Tsung-Dao, Yang Zhenning, and Samuel Ting, to discuss detailed plans. He insisted on placing top priority on building a nuclear accelerator to start training a generation of nuclear physicists and engineers. Although Deng had not attended a university, his wife and three of his five children had obtained degrees in physics from Beijing University.

Within a month of his return, on August 3, 1977, Deng convened a Forum on Science and Education, to begin the reorganization and expansion of scientific institutions. He insisted that professional scientists be among the directors of all centers. In addition to reviving the Chinese Academy of Sciences (CAS), he founded a new Chinese Academy of Social Sciences (CASS). He reinstated the State Science and Technology Commission, and ordered the drafting of a new Seven Year Science Plan. During March 18-31, 1978, Deng held a conference on scientific and technological policy that inaugurated 108 new projects.

Deng insisted that Chinese scientists be provided with the necessary laboratory facilities, salaries, and resources to rapidly revive the core work in the hard sciences. He said at the time: "We should select several thousand of our most qualified personnel within the scientific and technological establishment and create conditions that will allow them to devote their undivided attention to research. Those who have financial difficulties should be given allowances and subsidies.... We must create within the party an atmosphere of respect for knowledge and respect for trained personnel. The erroneous attitude of not respecting intellectuals must be opposed. All work, be it mental or manual, is labor."

To accelerate the advancement of Chinese science, Deng sent many of the brightest Chinese students abroad to study in the best universities. He explicitly set out to reconstitute an elite based on merit. Officials were sent around the world to gather up university texts. Leading Chinese scientists living and working abroad were encouraged to return to China, or to visit to lecture. He insisted that Chinese scientists who remained abroad, should be still welcome as patriotic Chinese who would ultimately contribute to China's scientific advancement.

A crucial aspect of Deng's shift was his rehabilitation of the great

The revival of the philosophical tradition of Confucius (shown here) has greatly contributed to China's educational and scientific progress.

Chinese scholar/philosopher Confucius. Under Mao, Confucianism had been attacked and banned, under the charge that it represented the "tyranny of the landlord class." In 1982, Deng started a Confucian revival, including state-sponsored conferences on his thought. Deng considered Confucianism the core of Classical Chinese culture, and crucial to the new scientific education system that he was in the process of setting up.

China's "Iwakura Mission"

During 1871-1873, at the start of the Japanese Meiji Revolution, a group of top Japanese business leaders, scientists, economists, and government officials traveled around the world. The 51 participants in the mission visited 15 countries under the leadership of Tomomi Iwakura. The Iwakura Mission, as it came to be known, took a decade to produce a 12-volume report summarizing the lessons they learned on the advancement of science, technology, management, and governance. The enterprise significantly contributed to the Japanese industrial revolution.

Deng Xiaoping took a similar approach.

In 1975, Deng made a five-day visit to France, where he got a first-hand opportunity to see the tremendous advances that had been made by Western European states. Between 1977 and 1980, as he was relaunching China's economy, Deng sent many delegations abroad to study the methods of economic growth, scientific advancement, and education. After hearing back from some of the first of the delegations, Deng noted, "Recently our comrades had a look abroad. The more we see, the more we realize how backward we are."

In spring 1978, Deng dispatched four "study tours" to Eastern Europe, Hong Kong, Japan, and Western Europe (normalization with the United State had not yet been finalized). The most important of the study tours was led by Gu Mu, a respected economist, who brought a 20-person ministerial delegation to Western Europe. They visited 15 cities in five countries on continental Western Europe, focusing almost exclusively on industrial production, scientific research facilities, and government agencies in charge of the economy and investment. The delegates were stunned at the openness they encountered. They came back with initial offers for more than $20 billion in foreign investment in China. By June 30, 1978, Gu Mu had completed a written report to the Politburo.

On July 6, 1978, the State Council convened a Forum on the Principles to Guide the Four Modernizations (science and technology, agriculture, industry, and national defense). It was led off by Gu Mu's report on the findings of his and other travel missions. The Forum ran through Sept. 9, allowing for the findings to be disseminated widely throughout the government and party structures. The Forum was chaired by Li Xiannian, the leading State Councilor for economic policy.

The foundations for China's spectacular growth were thus set by these initial actions by Deng Xiaoping, bringing the Chinese onto a path that today establishes them at the apex of scientific accomplishment globally.

Development Model

The South Korea Model: How to Transform an Impoverished Nation into a Modern Economy

by Michael Billington

August 2014

Park Chung-hee (1917-79), President of the Republic of Korea (1963-79).

South Korea was transformed during the 1960s and 1970s from one of the poorest nations on Earth, suffering from 35 years of Japanese colonial domination before 1945, and three years of destruction during the Korean War of 1950-53, into one of the world's pre-eminent industrial powers, with a standard of living among the highest in the world for urban and rural citizens. This was accomplished largely through the efforts of one man, Park Chung-hee, who served as President from 1961 until his death by assassination in 1979. It should be noted that although he has also been castigated by many in his own country, and by many followers of the British free-trade model abroad, as a tyrant, he won every one of the five elections in which he contested for the Presidency following his assuming power in a coup in 1961. In 1998, Park was voted in a national poll to be the best President ever by more than 75% of the South Korean population.

The importance of examining the "Korean Miracle," is to demonstrate that the methods used by Park can be usefully understood as a form of the American System of Political Economy, even if Park had to fight every step of the way against many of the policies demanded by Washington. The Korean Model can be an invaluable aid in the necessary transformation of poor and underdeveloped nations around the world.

The Korean Model and the American System

Park Chung-hee's system, developed over time after his relatively bloodless coup against a weak South Korean government in 1961, was based on principles that are strikingly similar to those of the American System, developed by Alexander Hamilton and implemented by such Presidents as John Quincy Adams, Abraham Lincoln, and Franklin Roosevelt. These principles were described by Joong-kyung Kim[1] at

---

1. Kim Joong-Kyung, an official at the Korean Development Institute (KDI), is the son of Kim Chong-nyon, the Chief of Staff to Park Chung-hee during his Presidency, and one of the "triumvirate" described below.

a conference in Washington, D.C. in June 2010, titled "Recasting the Korean Model," as follows:

- Directed credit, selective industrial promotion, and export-push trade policies;
- A carrot and stick approach in linking the government's support with performance-based standards of success, both in industry and in rural development;
- Selective support to the firms with the potential to become industrial champions in the heavy and chemical industries;
- Emphasis on technical and vocational high schools and training centers;
- Material support for the rural sector based on the Green Revolution in agricultural science, linked to government-provided construction supplies to villages that helped themselves—the so-called Saemaul Undong, or New Village Movement.

The results speak for themselves. In 1961, the per-capita income of South Korea was 101st out of 125 countries. Per-capita income in North Korea (where most of the industry was developed under Japanese colonial rule) was three times higher at that time. Per-capita income in South Korea is now 13th in the world. Between 1961 and 1980, South Korean gross domestic product exploded from $12 billion to $57 billion, with an average 8.5% growth rate—the fastest in the world. Electricity generation expanded ten-fold, while life-expectancy increased from 55 to 66 years. While there were only 4,500 engineers in the country in 1960, there were 45,000 by 1980. Other parameters are equally impressive.

The concept of {directed credit,} to the purpose of increasing the technological productivity of the population, is the core of the American System of Political Economy, as opposed to the British system of unregulated free trade under monetary policies determined by the private banking system. Also central to both the American System and Park Chung-hee's Korean Model was the concept of shared growth—ensuring that all members of the society, rural and urban, entrepreneurs and workers, participated in the nation's progress, through uplifting the productivity of the nation as a whole.

Park and the Meiji Restoration

Major General Park organized a coup and took power in 1961. The newly installed Kennedy Administration in the United States extracted a pledge from Park that he would hold free elections within two years, but otherwise offered America's support to Park's plans for development, inviting him to Washington in November of that year.

Park declared an "administrative democracy," to meet Korea's social and political reality, rather than introducing what he considered unworkable West European democracy (Kim Hyung-A, 2004). He immediately established an Economic Reconstruction Committee, identifying six key industries to be promoted: cement, synthetic fibers, electricity, fertilizer, iron, and oil refineries. He strongly sup-

President John F. Kennedy meets with Gen. Park Chung-hee in the White House, Nov. 14, 1961.

ported nuclear power development, which had been launched by Syngman Rhee in the 1950s under President Eisenhower's "Atoms for Peace" program. Park entertained the development of a nuclear weapons program, but later dropped the idea.

Park's first Five Year Plan generally ignored the advice coming from the International Monetary Fund (IMF) and many of the Americans who encouraged South Korea to emphasize handicrafts, labor-intensive agriculture, and small export industries, in favor of the rapid development of heavy industry and mechanized agriculture.

By 1963 Park had established himself as an effective political leader and chose to resign from the military and run for President as a civilian. While maintaining close relations with President Kennedy, welcoming U.S. economic aid and investment, Park declared that he was campaigning against the "pre-modern, feudalistic, flunky-ist opposition," and for a "nationalist democracy" as opposed to a European-style populist democracy. He declared a Korea-first policy to protect the nascent industries (like the protectionism at the heart of the American System of Alexander Hamilton). Park won the election by a narrow margin, but his party took 110 of the 175 legislative seats.

With the electoral victory, Park began his Korean Model policies in earnest. One of the first steps was to establish relations with Japan, although this evoked huge opposition and mass riots across the country due to the historical anger against the often brutal Japanese colonial occupation. Park negotiated a treaty with Japan in 1965, and Japan quickly became South Korea's largest trading partner and a major source of foreign investment.

Park's reforms included directing credit and providing tax breaks selectively to successful companies. As historian Gregg Brazinsky wrote: "By mobilizing domestic capital and promoting exports, the Republic of Korea dramatically raised growth rates, after two decades of futile efforts to do so by U.S. and Korean leaders" (Brazinsky, 2007). The large family-owned conglomerates (the chaebol, such as Hyundai and Samsung) played a leading role, but the accusations against Park of favoritism and cronyism are generally bogus. These large firms had to prove themselves capable of competing with the best of the foreign corporations, or face a cut-off of preferential credit. Seven of the ten such conglomerates active in 1965 had vanished by 1975 (Kim Hyung-A, 2004).

Shared Growth: The New Village Movement

In the early 1970s, Park launched the programs that would come to characterize the Korean Model, based on the concept of Shared Growth. Under Japanese occupation, North Korea (which borders Manchuria) was developed as an industrial center, while the South was predominantly agricultural. Under Park, protective tariffs and directed credit fostered the development of heavy industry, and created huge numbers of industrial jobs, while at the same time a remarkable program was launched to transform the stagnant rural economy, where 63% of the population lived, and to integrate the agricultural sector with the industrial economy. This was known as the New Village Movement, or Saemaul Undong, a program that nearly doubled farmer income and increased productivity by 50% within a decade.

This approach is a major reason that undeveloped nations of Africa are looking to the Korean Model as a means of escaping their entrenched poverty. It contrasts with the current Western policies toward Africa, which maintain the stench of the colonial era, whether from Western government agencies, the United Nations, and IMF, or the private funds of billionaires such as Bill Gates and George Soros. Their approach is the same: Aid the small-holder farms through mini-grants and marketing schemes to sell their cash crops abroad, and similar aid which keeps the farmers small and poor, based on the premise that major infrastructure programs, such as dams, transportation systems, and power supplies are not "appropriate" for underdeveloped nations.

Government of the Republic of Korea

President Park Chung-hee at a construction site in a rural village. The New Village Movement, or Saemaul Undong, modernized villages and developed their infrastructure.

Park had a different vision. While the national government built the required infrastructure nationwide, and funded a Green Revolution program in agricultural science, Park began a self-help approach in the rural villages that promised the peasants a means to participate in the national reconstruction effort. Saemaul Undong teams went to each of more than 34,000 villages in 1970 with 300 bags of cement, some basic machinery, and some advisors, telling the village leaders that they should use the supplied materials to build roads, irrigation systems, and other needed local infrastructure. In a year's time, the Saemaul Undong team would evaluate the progress in each village. Those that used the government-issued supplies well, received 500 more bags of cement, some iron-reinforcing rods, and new equipment. Those that did not meet the grade, were given nothing. Access to scarce but increasing supplies of electricity were also apportioned according to proven success.

The lesson was soon learned. Although only half the villages qualified for the continued government support in the first year, by the end of the decade most villages were fully participating. By 1980, 97% of the villages were electrified, compared to only 12% in 1964. The income of the poorest farmers increased by 76% over the 1970s, while that of the larger farmers doubled. Reforestation projects saved vast areas that had been denuded of vegetation by the war. Rice yields increased by 50% between 1970 and 1977, bringing South Korea's rice yield per hectare up to that of Japan. The growth in total agricultural output leaped from an average of 3.4% per year in the decades after World War II, to 6.8% annually during the 1970s.

Heavy and Chemical Industries

Another hallmark of the Korean Model was the Heavy and Chemical Industries (HCI) policy, which not only directed credit to these industries, but also extended some of the principles of the New Village Movement to industry. The chaebol that are today recognized internationally as leading industrial innovators, such as Hyundai and Samsung, rose rapidly under the HCI program, while those that were not competitive were allowed to fail. The primary focus was on five industries: machinery; shipbuilding and transport; iron and steel; chemicals and fertilizers; and electronics.

Park drew from the Japanese experience in heavy industry development during 1957-67, and received help from the United States, especially in the defense industries that were connected to the HCI. However, Park had to do battle with many of the U.S. advisors (and some of his own U.S.-trained technocrats) in order to achieve the industrial transformation of the 1970s. Economic advisors at the U.S. Operations Mission in Seoul had advised Park to promote small and medium-sized enterprises (SME) rather than heavy industry, as "more appropriate" for the scale of the Korean economy, and to lessen the dominance of the central government in economic planning, while also advising the government to privatize the state-sector industries (Kim Hyung-A, 2004).

Park rejected the advice. He created the Korean Development Institute (KDI) in 1971, to bring together a team of economists dedicated to rapid industrialization of the country, and also to lure back to Korea some of those foreign-educated economists who otherwise might have chosen to stay abroad to further their careers. Within the administration, Park established a triumvirate of himself, Kim Chong-nyon as the economic manager, and O Wonchol, who oversaw the defense aspects of the industrial buildup.

The HCI also integrated scientists from the universities into the drive for industrial excellence. Forty-six leading academics in physics and chemistry were hired as advisors to the Ministry of Commerce and Industry, which interfaced with the HCI. Missions were deployed to the United States and Japan in 1973 to seek investments for the heavy industries project, but domestic industrial firms that wished to participate in HCI were required to raise at least 30% of the required investment on their own, with no more than half to come from foreign sources.

Hyundai, one of the companies supported by the Heavy and Chemical Industries (HCI) program under President Park Chung-hee, is now one of the largest industrial and construction conglomerates in the world.

Beginning in 1973, five major industrial complexes were established, focused respectively on machinery, petrochemicals, shipbuilding, electronics, and steel. The process was not limited to Korea. Alon Levkowitz, an Israeli scholar, wrote that the chaebol were able to build huge infrastructure projects in the Middle East at low cost, backed by financial support from the Korean government. Levkowitz noted that this development aid came "with an absence of any perceived political agenda or ideological aspirations to influence the governments of the Mideast" (Levkowitz, 2011). Similar projects were launched across Southeast Asia.

Park arranged with France in 1975 to build a nuclear fuel reprocessing plant in South Korea, as well as two nuclear power plants. The U.S. government, having abandoned the Atoms for Peace policies of the Eisenhower and Kennedy administrations, complained loudly that reprocessing capacities, although essential for any nuclear nation and fully legal under all international nuclear energy agreements, would move South Korea closer to the capacity to produce nuclear weapons. Park responded, publicly, that South Korea could in fact produce a nuclear weapon, but had not yet chosen to do so. If the United States were to remove its nuclear umbrella, Park said, Korea would build a weapon itself.

The anti-science mafia in the United States went ballistic. U.S. Secretary of Defense Donald Rumsfeld threatened Park with a cut-off of support for Korea's ambitious nuclear power program, and Washington coerced France to renege on the reprocessing deal. Korea turned to

Canada for its nuclear reactors, and built its own heavy water fuel rod plant. In 1976, Park established the Korean Nuclear Engineers, which took over from the American company Burns and Roe as the primary nuclear advisors, and set up a Nuclear Fuel Development Corporation.

Today, South Korea has become a major exporter of nuclear power reactors of its own design. But agreements that were forced on Seoul by the United States in the past, limiting its freedom to produce a full-cycle nuclear fuel capacity, continue to deprive South Korea of its lawful rights as a modern scientific and industrial nation. Ongoing negotiations to rectify this injustice are being dragged out by the U.S. side.

Assassination

After Park's assassination at the hands of his own Korean CIA chief in 1979, a study was discovered in his home titled: "A Plan for Remodeling Korea for the 2000s." Some of his plans were carried out under subsequent governments, but, as historian Kim Hyung-A has reported, in the years immediately following Park, Korean economics was increasingly dominated by American-trained "neo-liberal technocrats."

But today, the spirit of the Korean Model is alive and well. Korea is a leading player in the development of infrastructure and heavy industry across Southeast Asia, in the Mideast, and increasingly in Africa, with few, if any, strings attached. China has also studied and learned from the Korean Model in regard to its own development and its policies toward trade and investment in the developing sector nations, while also becoming a major trading partner of South Korea itself.

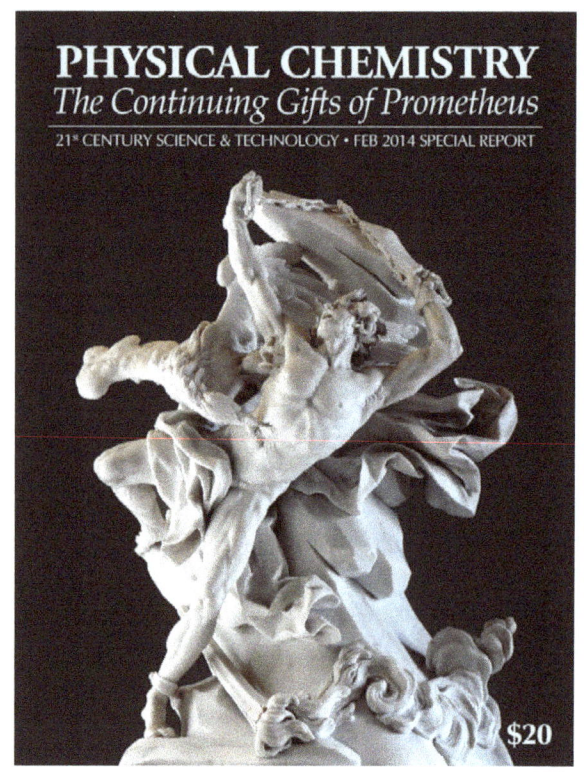

21st Century Science & Technology

The Continuing Gifts of Prometheus brings to life the stunning progress made in physical chemistry over the course of mankind's history, in the context of the ongoing conflict between Prometheus, who gave fire and "all the arts" to man, and Zeus who was determined to destroy humanity.

Physical Chemistry is the application of higher forms of "fire" (such as nuclear "fire" today) to transforming the phyical world. A Promethean culture today will fully develop a nuclear economy, including mining the Moon for the ideal fusion fuel, helium-3.

Get your copy today from Amazon.com $20

The latest in the series of

NEW PARADIGM

conferences by the Schiller Institute

SEE
newparadigm.schillerinstitute.com
for videos and texts

PART 13
Epilogue

Dump Geopolitics, and Create a Future for Humanity

October 2014

In the preceding pages, you have gotten the picture: The time is ripe for action to realize a new international economic order that will create a future worthy of mankind's nature as the only creative species.

There is a growing awareness, even among policy layers in Europe and the United States, that the old paradigm of monetarist economics and geopolitical confrontation has failed, and that a new course must be taken if disaster is to be avoided. That new course is represented by the nations of the BRICS, which have launched a process of cooperation around the very great projects necessary to save mankind from the present scourges of poverty, famine, and war. These nations are already pulling others into their orbit, and spreading optimism throughout the planet. Indeed, the BRICS and associated nations now represent more than half the world's population.

A reflection of the motion within the trans-Atlantic region to seize the time, was the Oct. 18-19 30th anniversary conference of the Schiller Institute, held in Frankfurt, Germany, which brought together more than 350 people from Europe to deliberate on "The New Silk Road and China's Lunar Program: Mankind Is the Only Creative Species!" At the conclusion of the conference, the participants adopted the following resolution:

> Mankind experiences presently a deep civilizational crisis, where the foundations of society in many parts of the world have eroded, and established codes of international relations have broken down. On top of this, we are faced with mortal dangers, each of which could lead to the potential extinction of the human species:
> There is first, the Ebola pandemic, which is already out of control in Africa, for which there is no cure, and which is threatening to become more threatening than the Black Death of the 14th Century.
> There is second, the terrorist threat for the whole world, and genocide spreading from the so-called IS Caliphate, not only demonstrating a subhuman barbarism, but explicitly

The leaders of the BRICS, joined by the South American heads of state, in Fortaleza, Brazil, July 16, 2014.

threatening Russia and China, and therefore, becoming the potential trigger to blow up all of Southwest Asia and even leading to a new world war.

And there is third, the absolute certainty that the completely bankrupt trans-Atlantic financial system is about to blow up, threatening also to throw much of the world into a Dark Age.

In light of these three mortal dangers, it is a question of life and death of humanity to stop the immoral and imbecilic policies of geopolitics and confrontation against Russia and China. Instead, we have to shift the agenda to the common aims of mankind, and work together with Russia, China, India, and other nations, to defeat these mortal threats. We call on all forces of reason in Europe and the United States, to join the emerging new economic order of the BRICS and the New Silk Road. Let's work to establish an inclusive peace order, with the participation of every nation on the planet, a peace order for the 21st Century, worthy of mankind as the only known creative species in the universe. Let us grow up into the adult age of humanity, where love, creativity, and beauty define the values of our common human family.

With such a commitment, the New Silk Road will lead to a new paradigm for the survival of civilization in the immediate weeks and months ahead.

SUBSCRIBE TO

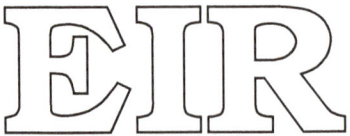

Executive Intelligence Review
EIR Online

EIROnline gives subscribers one of the most valuable publications for policymakers—the weekly journal that has established Lyndon LaRouche as the most authoritative economic forecaster in the world today. Through this publication and the sharp interventions of the LaRouche Movement, we are changing politics worldwide, day by day.

EIR Online includes the entire magazine in PDF form, plus up-to-the-minute world news.

EIR DAILY ALERT SERVICE

EIR's new Daily Alert Service provides critical news updates and analysis, based on EIR's 40-year unparalleled track record in covering global developments.

SUBSCRIBE (e-mail address must be provided.)

EIR Online

- ☐ **$360** for one year
- ☐ **$180** for six months
- ☐ **$120** for four months
- ☐ **$90** for three months
- ☐ **$60** for two months

EIR DAILY ALERT SERVICE

- **$500** one month (introductory)
- **$3,000** six months
- **$5,000** one year (includes EIR Online)

I enclose $ _____ check or money order

Make checks payable to
EIR News Service Inc.
P.O. Box 17390, Washington, D.C. 20041-0390

Please charge my ☐ MasterCard ☐ Visa
                  ☐ Discover   ☐ Am Ex

Card Number _____

Signature _____

Expiration Date _____

Name _____
Company _____
Address _____
City _____ State ____ Zip _____ Country _____
Phone (____) _____
E-mail _____

EIR can be reached at: **www.larouchepub.com/eiw**
e-mail: **fulfillment@larouchepub.com**   Call **1-800-278-3135** (toll-free)

Executive Intelligence Review

EIR News Service Inc.
P.O. Box 17390, Washington, DC 20041-0390

ORDER MORE
The New Silk Road Becomes The World Land-Bridge

Single Copy
Hardback $300 + $8 s/h
Paperback $250 + $8 s/h
PDF $200
E-book available soon

Multiple Copies & Other Discounts
Contact us for discount rates for textbook, library, legislative and other special uses

Reach the authors: authors@worldlandbridge.com

PHONE
800-278-3135
M-F, 10am to 10pm, or message.

ON-LINE
http://store.larouchepub.com

EMAIL
fulfillment@larouchepub.com

MAIL
EIR
P.O. Box 17390
Washington, DC 20041-0390

Name _____
Company _____
Address _____
City _____ State _____ Zip _____
Country _____
Phone _____
E-mail _____

Payment Options
☐ I enclose $ _____ check or money order
Make checks payable to
EIR News Service, Inc.
Please charge my ☐ MasterCard   ☐ Visa
                  ☐ Discover   ☐ AmEx
Card Number _____
Exp. Date _____
Signature _____

Mail to: EIRNS, P.O. Box 17390, Washington, DC 20041-0390